Comparison of *Ab Initio* Quantum Chemistry with Experiment for Small Molecules

The State of the Art

Comparison of *Ab Initio* Quantum Chemistry with Experiment for Small Molecules

The State of the Art

Proceedings of a Symposium held at Philadelphia, Pennsylvania, 27-29 August, 1984

edited by

Rodney J. Bartlett

Quantum Theory Project, University of Florida, U.S.A.

D. Reidel Publishing Company

A MEMBER OF THE KLUWER ACADEMIC PUBLISHERS GROUP

Dordrecht / Boston / Lancaster / Tokyo

Library of Congress Cataloging in Publication Data

Main entry under title:

Comparison of ab initio quantum chemistry with experiment for small molecules.

Based on a symposium held at the American Chemical Society meeting in Philadelphia, Pa.
Includes index.
1. Quantum chemistry–Congresses. 2. Molecules–Congresses. I. Bartlett, Rodney J. II. American Chemical Society.
QD462.A1C66 1985 541.2′8 85-19631

ISBN-13: 978-94-010-8917-3 e-ISBN-13: 978-94-009-5474-8
DOI: 10.1007/978-94-009-5474-8

Published by D. Reidel Publishing Company,
P.O. Box 17, 3300 AA Dordrecht, Holland

Sold and distributed in the U.S.A. and Canada
by Kluwer Academic Publishers,
190 Old Derby Street, Hingham, MA 02043, U.S.A.

In all other countries, sold and distributed
by Kluwer Academic Publishers Group,
P.O. Box 322, 3300 AH Dordrecht, Holland

Softcover reprint of the hardcover 1st edition 1985

CONTENTS

FOREWORD

At the American Chemical Society meeting in Philadelphia, Pennsylvania, U.S.A., a symposium was organized entitled, "Comparison of *Ab Initio* Quantum Chemistry with Experiment: State-of-the-Art." The intent of the symposium was to bring together forefront experimentalists, who perform the types of clean, penetrating experiments that are amenable to thorough theoretical analysis, with inventive theoreticians who have developed high accuracy *ab initio* methods that are capable of competing favorably with experiment, to assess the current applicability of theoretical methods in chemistry. Contributions from many of those speakers (see Appendix A) plus others selected for their expertise in the subject are contained in this volume.

Such a book is especially timely, since with the recent development of new, more accurate and powerful *ab initio* methods coupled with the exceptional progress achieved in computational equipment, *ab initio* quantum chemistry is now often able to offer a third voice to resolve experimental discrepancies, assist essentially in the interpretation of experiments, and frequently, provide quantitatively accurate results for molecular properties that are not available from experiment.

Recent successes of *ab initio* theory include the singlet-triplet energy separation in methylene, the prediction of a non Van der Waal's bound ground state for Be_2, and the experimental identification via theoretical predictions of numerous cations like H_3O^+ and NH_4^+, to name just a few well-known examples. But there are also apparent failures, as in many studies of Be_2 that missed the inner minimum, the equilibrium geometry and binding energy in the Cr_2 dimer, and questions surrounding the activation barrier in the $F+H_2$ chemical laser reaction, among others. The successes and failures of *ab initio* theory are a constant subject of contention among experimentalists and theoreticians. This book attempts to provide a partial answer to what can be accomplished at the current state-of-the-art and identifies limitations where the theory still requires development.

Many research areas, sparked by the synergism among theorists and experimentalists, are addressed in this volume. These include such exciting topics as: the nature and bonding in small metal clusters and their catalytic activity; the spectroscopic identification and characterization of weakly bonded molecules like CO_2•NCH; many detailed studies of reaction paths and activation barriers in chemical reactions, including the characterization of reaction intermediates; the availability of very complete spectroscopic data for classes of small molecules like alkali hydrides, alkali dimers, and alkaline earth hydroxides,most of which would not be available from experiment alone; and careful predictions of energy transfer processes for molecular processes through the cooperative efforts of quantum chemists, who predict potential energy surfaces, and dynamicists, who use these surfaces to calculate the probabilities ("cross-sections") for different events to occur.

Although it is impossible for any one book to cover the wealth of contributions ab initio theory can now make to chemistry, or to assess its limitations, this volume provides a selection of topics by leaders in the field, both experimentalists and theoreticians, where theory has been instrumental in explaining, quantifying, and predicting properties of molecules. The book also identifies classes of problems for which detailed experimental data are forthcoming, and which will challenge future theoretical efforts. We hope the papers herein will serve as a reference point from which future progress may be measured.

Rodney J. Bartlett
Professor of Chemistry and Physics
University of Florida
and
Chairman of the Subdivision of
Theoretical Chemistry,
American Chemical Society

ACKNOWLEDGMENTS

The American Chemical Society symposium and this Proceedings on "Comparison of Ab Initio Quantum Chemistry: State-of-the-Art," was sponsored by Floating Point Systems of Portland, Oregon, U.S.A. Their help made it possible to prepare this volume and to present a symposium of very high quality. The participants, the authors, and I greatly appreciate their assistance.

This entire volume has been typed, supervised and organized by Ms. Jody-Kate Fisher. Considering the corrections and changes forced upon her in three or more levels of proofing, the innate difficulty with equations and scientific terminology, not to mention an often recalcitrant word processor, her effort was stupendous. Without Jody, we could not have undertaken this project. Thanks!

Also, I imposed upon members of my research group to help me in proofing this volume to attempt to ensure as much accuracy as possible. Hence, let me thank Dr. Sam Cole, Dr. George Fitzgerald, Dr. Robert Harrison, Dr. Bill Laidig, Mr. David Magers, Mr. Alan Salter and Mr. Gary Trucks for their help in preparing this volume.

Rodney J. Bartlett

MOLECULAR INFORMATION FROM ESR SPECTRA: TRANSITION METAL MOLECULES

W. Weltner, Jr. and R.J. Van Zee
Department of Chemistry
University of Florida
Gainesville, Florida 32611

ABSTRACT. Electron spin resonance (ESR) spectroscopy at low temperatures can provide detailed electronic and magnetic properties of reactive molecules, such as their multiplicities, zero-field splittings, and hyperfine interaction constants. This has proved to be particularly valuable in the study of small transition metal molecules and clusters where *ab initio* theory is also being applied. A review of the experimental and theoretical knowledge of those molecules is brought up to date.

1. INTRODUCTION

Electron spin resonance (ESR) is a form of high resolution spectroscopy, and when it is applicable it can provide, directly, information that is usually obtained only by detailed analysis of optical spectra. ESR provides the most information when applied to gas-phase molecules, as done so successfully by Radford [1] and Carrington [2] and their coworkers, but the problems associated with maintaining a high enough concentration of reactive molecules in an ESR cavity have severely limited its application to gases. One can, however, trap most such recalcitrant species in rare gas matrices at low temperatures and measure their spectra by the conventional methods used for frozen glasses [3]. The spectra are then those of rigidly-held, randomly-oriented (not necessarily! [4]) molecules isolated from each other by a relatively nonperturbing medium. The loss in

R. J. Bartlett (ed.), Comparison of Ab Initio Quantum Chemistry with Experiment for Small Molecules, 1–16.

rotational information in the gas phase must then be considered to be compensated by the wealth of magnetic and electronic information that can still be obtained from the solid state spectra. These matrix ESR spectra can establish the multiplicity and symmetry of the ground electronic state, g tensor components, zero-field splittings, hyperfine interaction constants and approximate spin densities, nuclear quadrupole coupling constants, and spin-rotation constants [3].

Figure 1 illustrates, in a very approximate way, the transititions observed and the significance of the Zeeman splittings occurring when a $^5\Sigma$ molecule (such as Sc_2 [5]) is placed in a magnetic field. Observation and analysis of fine structure transitions will yield g tensor components and zero-field splitting parameters (here b_2^o and b_4^o) and further splittings within these levels due to magnetic nuclei produce hyperfine structure (hfs). Then all of the molecular parameters can be derived in the spin Hamiltonian, which is, for this S = 2 molecule:

$$H = g_{\parallel}\beta H_z S_z + g_{\perp}\beta(H_x S_x + H_y S_y) + b_2^o[S_z^2 - \frac{1}{3} S(S + 1)] +$$

$$+ \frac{1}{60} b_4^o(35S_z^4 - 155S_z^2 + 72) + \sum_i [A_{\parallel}^i S_z^i I_z^i + A_{\perp}^i(S_x^i I_x^i +$$

$$+ S_y^i I_y^i)].$$

Furthermore, the identity of the molecule being observed is established if hyperfine structure data from all nuclei can be obtained. For example, in the study of the $^2\Sigma$ TiCo molecule [23], the hfs due to each nucleus can be detected if the molecule is prepared from isotopically-enriched ^{47}Ti (see Fig. 2), thus identifying the molecule being observed as diatomic TiCo.

2. TRANSITION METAL MOLECULES

A review of the status of our knowledge of these molecules up to January, 1984, has recently appeared [7] in which all experimental and theoretical work on well characterized transition-metal molecules has been cited and discussed. As pointed out there, although many diato-

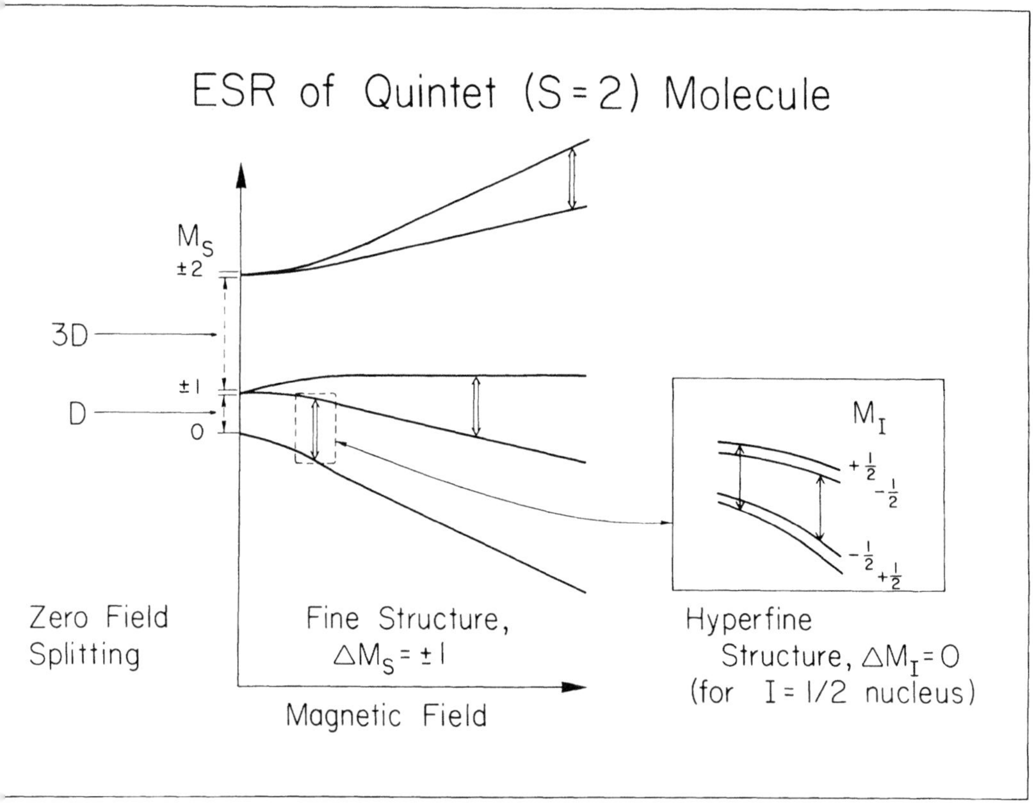

Figure 1. Depiction of Zeeman effect and the approximate transitions observed in the ESR spectrum of a $^5\Sigma$ molecule.

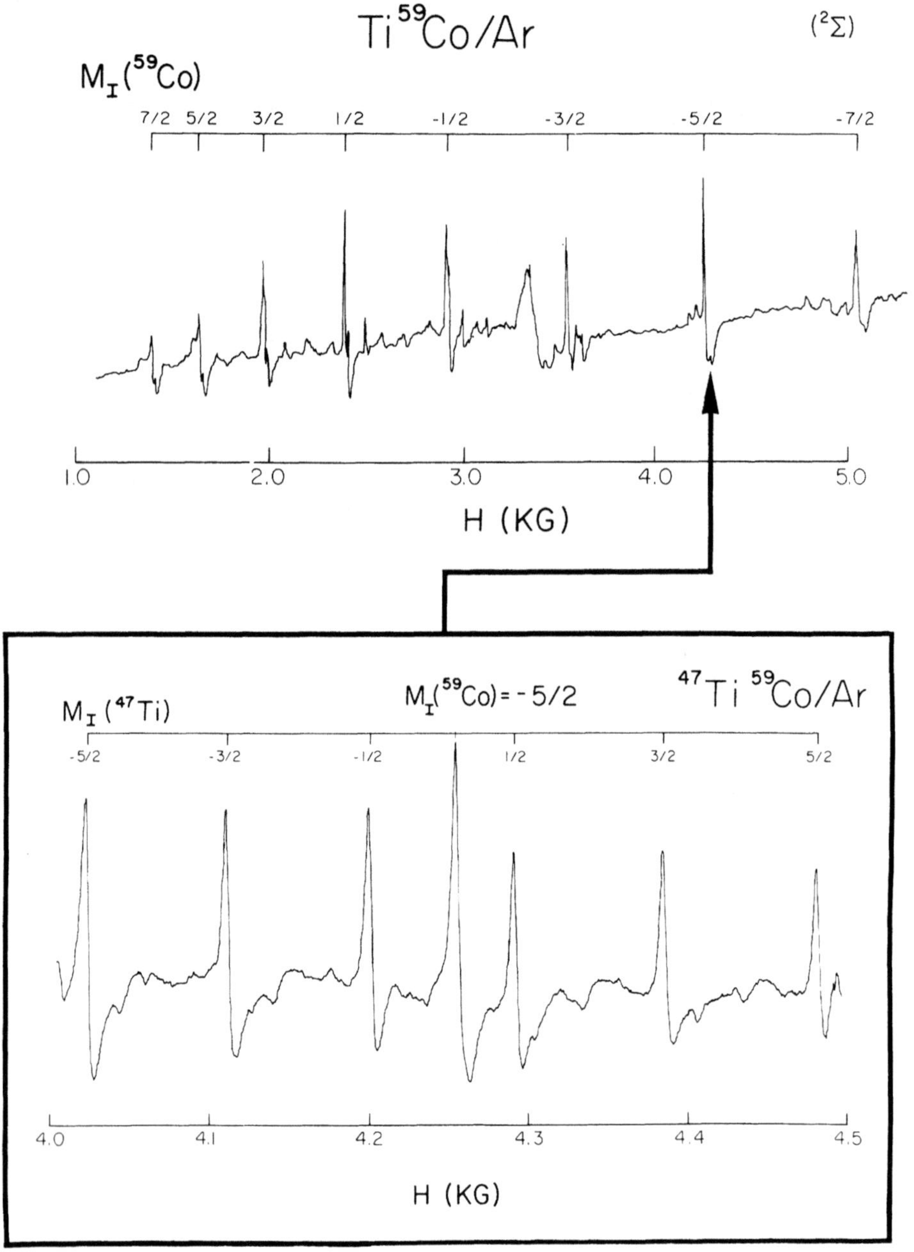

Figure 2. Top: ESR spectrum of the $Ti^{59}Co$ ($^2\Sigma$) molecule where $I(^{59}Co) = 7/2$. Bottom: additional splitting of each line when titanium metal enriched to 80% in ^{47}Ti ($I=5/2$) is vaporized along with cobalt metal [23].

mics fall in this category, detailed information about larger molecules is scarce. This is in spite of the detection of clusters of transition metals containing up to more than 20 atoms in supersonically-expanded [8] and thermally quenched [9] beams. All-electron theory has been applied, usually to diatomics, but even to clusters of 13 atoms.

Since that review many relevant publications have appeared. The Proceedings of a Bunsen-Kolloquium on "Experiments on Clusters" in October, 1983, has been published as a volume of the Berichte der Bunsen-Gesellschaft für Physicalische Chemie [10]. Among those papers is a review of their supersonic beam work on transition metal molecules by Morse and Smalley [11], and a preliminary report, later published in full [12], by Flad, Igel, Preuss and Stoll on Cu_n, Cu_n^+, Ag_n, Ag_n^+ ($n \leq 4$) calculations. Here Table 1 cites that work and other recent experimental and theoretical papers published since January 1984 and/or omitted from the earlier review.

The role of ESR in the characterization of transition metal molecules is made more apparent by considering the diatomics formed from first-row elements. Figure 3 shows an array of all possible such diatomics, similar to one published earlier [7], but here also indicating the sources of information. Those molecules in bold frames are well characterized in that at least their ground electronic states have been established; one notes that seven have been determined via matrix ESR work. V_2 [25] and Cr_2 [26] have been observed in the gas-phase and their optical spectra analyzed. Ni_2 [15] has also been produced in a supersonically-cooled beam, but the complexity of its spectrum has made analysis difficult. A lowest $^1\Gamma_g$ or $^3\Gamma_u$ state, as suggested by theory [27], is consistent with the observations. The ground states given for diatomics containing Fe have been inferred from their Mössbauer spectra. In general, deduction of the ground state from such data is difficult, as shown by the calculations of Guenzburger and Saitovich [28] and Montano, et al [29]. TiFe and the three other molecules ScCo, TiNi, and VCo have been indicated as $^1\Sigma$ in this array. This state has been deduced from their adjacent $^2\Sigma$ molecules ScNi and

Table 1. Recent experimental and theoretical research on transition metal molecules[a]

Cr_2	Exptl:	optical spectra in matrices, absorption and emission	Pellin and Gruen [13]
	Theor:	comparison of HF-LCAO, GVB, Xα-LCAO, and LDF-LCAO	Messmer [14]
		LCAO-LSD and Xα	Baykara et al. [14a]
Ni_2	Exptl:	spectroscopy in supersonic beam	Morse et al. [15]
Cu_2	Exptl:	optical spectra in solid neon, absorption and emission	Kolb et al. [16]
	Theor:	CAS SCF/SDCI or POLCI	Bauschlicher [17]
		ECP/UHF-SCF/LSD	Flad et al. [12]
		LCAO-MO-SCF	Cingi et al. [18]
Pd_2	Theor:	all electron HF-CI	Shim and Gingerich [19]
Ag_2	Theor:	ECP/UHF-SCF/LSD	Flad et al. [12]
		relativistic SCF-Xα-Dirac-SW	Rabii and Yang [20]
		relativistic ECP/LSD	Martins and Andreoni [21]
		LCAO-MO-SCF	Cingi et al. [18]
Au_2	Theor:	relativistic ECP	Pitzer [22]
		relativistic SCF-Xα-Dirac-SW	Rabii and Yang [20]
Mo_2	Theor:	comparison of HF-LCAO, GVB Xα-LCAO, and LDF-LCAO	Messmer [14]

Table 1 (cont.)

	LCAO-LSD and Xα	Baykara et al. [14a]
ScNi, TiCo		
	Exptl: ESR in matrices	Van Zee and Weltner [23]
$Cu_{3,4}$,$Ag_{3,4}$		
	Theor: ECP/UHF-SCF/LSD	Flad et al. [12]
$CuAg_4$ and Cu_2Ag_3		
	Exptl: ESR in cyclohexane matrices	Howard et al. [24]
Cu_{13}		
	Theor: Xα-SW and HF	Messmer [14]

[a] This is an extension of the review in reference 7. See footnote 3 there for the meanings of the initialisms used here. Additional ones are LDF, local density functional and SDCI, single double configuration interaction.

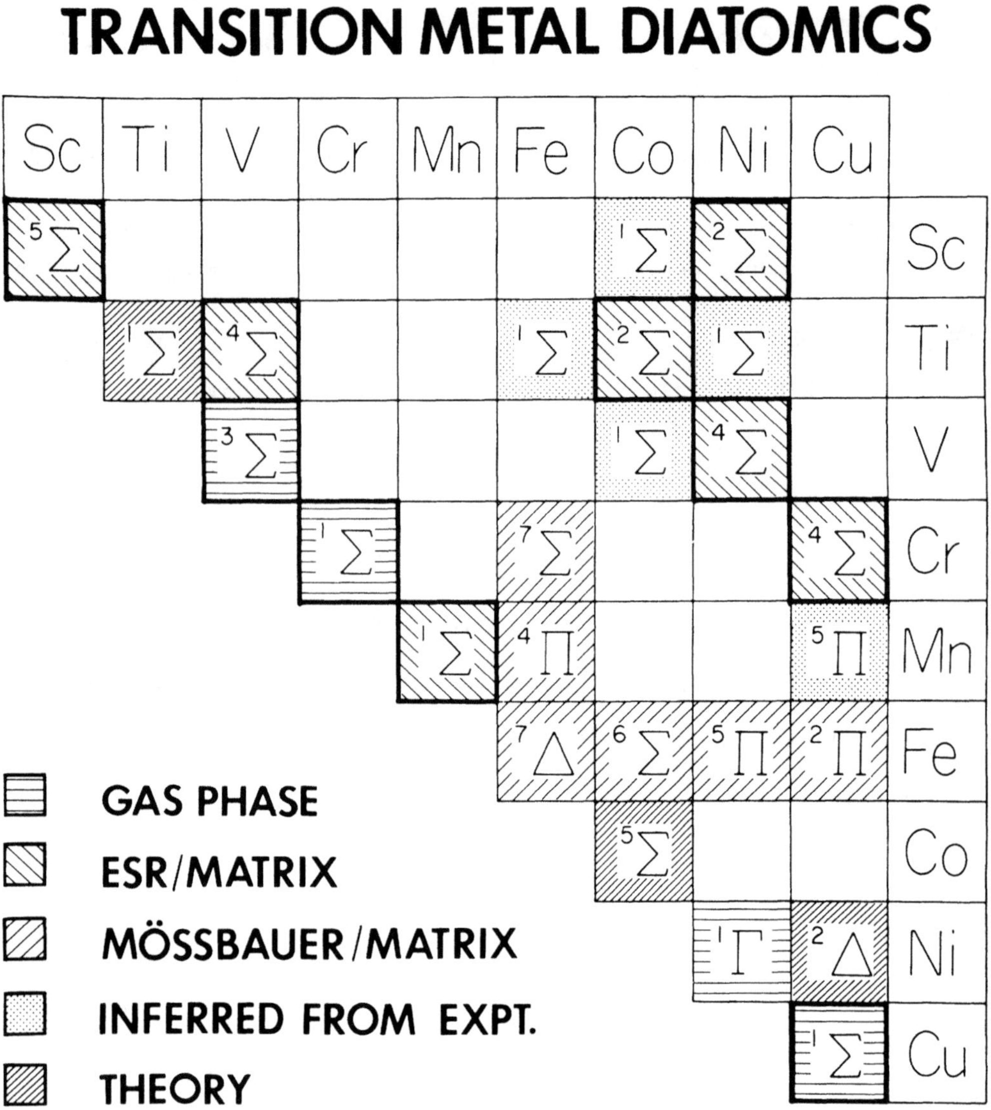

Figure 3. An array of all possible diatomic molecules formed from the first-row transition metals. The ground states of those in bold frames have been established experimentally. The source of information about each molecule is indicated.

TiCo by addition or subtraction of one electron in a half-filled σ orbital. ScCu would also fit into this category, but the stability of the d^{10} shell of Cu makes the validity of such an extrapolation less likely, so that it was not assigned a ground state in the array. Ab initio theory has been applied to all of the homonuclear molecules in this array and particularly to Cr_2, Ni_2 and Cu_2. We refer the reader to the earlier review [7] for discussion and references. Recently Walch, Bauschlicher et al. [30] have extended their calculations on Sc_2, Ti_2, V_2 and Cr_2 to TiV [6].

Sc_2 and Mn_2 exhibit quite different bonding and also illustrate a good interaction between theory and experiment. Sc_2 is the simplest transition-metal diatomic, but theory has predicted several possible ground states. The ESR spectrum was unambiguous, since natural Sc is 100% ^{45}Sc (I = 7/2), and it definitely established a $^5\Sigma$ ground state [5]. Following that, Walch and Bauschlicher showed that it was necessary to consider molecular states formed from the ground state ^2D atom and an excited ^4F atom in the calculation and then one obtained the strongly bound $X^5\Sigma$ state [31].

Mn_2 is an entirely different case because of the large promotion energy of the $3d^54s^2 \rightarrow 3d^64s$ excitation in the Mn atom [32]. This dictates against strong bond formation. The calculations made by Nesbet in 1964 suggested a very weak bond with r = 2.88Å and antiferromagnetic coupling of the ten d electrons to yield a $^1\Sigma$ lowest state but with an exchange coupling constant J of only 4 cm^{-1} [33]. This picture was essentially corroborated by the variations in the ESR spectra taken between 4 and 70°K [34]. Again, because of the hfs, the identity of the molecule being observed was assured. As the temperature was raised, more and more fine-structure transitions were observed corresponding to successive population of the S = 1, 2, and 3 states. From the variation in intensity with temperature J = −9(3) cm^{-1} was derived, in excellent agreement with Nesbet's value. Furthermore, analysis of the spectral lines assigned to each spin state yielded anisotropic exchange contributions expected to vary as $\langle 1/r^3 \rangle$, where r is the interatomic distance in Mn_2. r was found to 3.4 Å, indicating van der Waals bonding, even weaker than the bonding predicted by Nesbet at r = 2.88 Å.

One can note another characteristic of the array in Fig. 3, i.e., that isoelectronic molecules lie along diagonals perpendicular to the homonuclear diagonal. On this simple basis one could predict the ground states of all of the molecules from the homonuclears and the first off-diagonals to them (i.e., ScTi, TiV, etc.). From the ground states listed in Fig. 3, one sees that there is some evidence that the isoelectronic principle holds, although there are also several discrepancies. Accurate electronic data on more molecules are needed to see whether it is valid here.

As mentioned earlier [7], extension of the Brewer-Engel theory of transition-metal alloys to these molecules [35] also appears to be verified. Thus, low-spin, multiply-bonded molecules appear to be formed between elements at opposite ends of the Periodic Table.

Although there are no ESR data on the heavier homonuclear diatomics, there are interesting results on the heteronuclears formed from Cr and Mn bonding with Cu, Ag, and Au. Cr(d^5s) and Ag($d^{10}s$) would be expected to form an s-s bond and yield a $^6\Sigma$ ground state. This was found to be the case for CrAg and also for CrAu [36]. However, the interpretation of the ESR spectrum of CrCu has not been so straightforward, and Baumann, et al. [36], have concluded that it does not have a multiplicity of six but that it is $^4\Sigma$. This implies multiple bonding involving the d orbitals on Cu, which is unusual because of the stability of the closed d shell on that atom. The bonding in this molecule should then be of interest to theorists.

The Mn series is similar but different. Among the three molecules MnCu, MnAg, and MnAu, only the ESR of MnAg was detected in a matrix at 4°K, although the spectra of the two metal atoms could be clearly observed in the other cases [37]. $^{55}Mn^{107,109}Ag$ was definitely identified as a $^7\Sigma$ molecule, as might be expected from the bonding of a d^5s^2 and $d^{10}s$ atom. Its zero-field splitting parameter D = (+)0.20 cm^{-1}, and one expects D to be a strong function of the spin-orbit coupling constants of the bonded atoms. Then the fact that MnAu was not detected, although probably more strongly bound, can be rationalized by assuming that it has a large value of $D > 8\ cm^{-1}$. For this odd-multiplicity molecule the X-band quantum (~9 GHz) would then not be large enough to allow a transition to occur. This kind of rationale cannot be applied to MnCu since one would guess that $D < 0.20\ cm^{-1}$ for it. However, another explanation is possible, namely that it has an orbitally-degenerate ground state, such as $^5\Pi$ or $^5\Delta$. In such cases, the matrix ESR spectra will be broadened beyond detection. This view is somewhat encouraged by the apparent anomalous behavior of Cr with Cu, as mentioned above. It remains to be seen whether _ab initio_ theory will clarify the situation for CrCu and MnCu.

The triatomic Au_3 recently studied by Howard, et al. [38], was inadvertently omitted in the earlier review, but not unexpectedly its properties are very similar to those of Cu_3, Ag_3 and also Cu_2Ag [39], which are in turn similar to the alkali-metal triatomics Na_3 and K_3 researched so well by Lindsay, et al [40]. Indeed, from the hfs in the ESR spectrum of $^{197}Au_3$ trapped in solid benzene at 77°K it is established that the molecule contains two equivalent atoms and a third unique atom. The one odd electron largely occupies an antibonding s orbital placing about 40% of the s character on each of the two outer equivalent atoms and very little on the central Au atom. It is suggested that it has an obtuse-angled C_{2v} symmetry and a $^2B_2$ ground state, since all of its properties are similar to those derived from the ESR spectra of Cu_3 and Ag_3 by these authors. In the latter molecules there was evidence of a g tensor anisotropy indicating a non-axial ground state. Also, the extraordinary spectroscopic study of Cu_3 in a supersonically-cooled beam by Smalley, et al. [41] indicates that it is distorted from D_{3h} symmetry, as found via the ESR spectrum. Some theory has derived an acute-angled ground state for these molecules [42], but Basch's pseudopotential study of Ag_3 is in agreement with the $^2B_2$ state [43].

Knight, et al. [44], have also characterized the triatomic Sc_3 and Y_3 molecules from their ESR spectra in matrices. The former has an equilateral triangle structure with a $^2A_1'$ ground state. The spectrum of Y_3 differs markedly from that of Sc_3, but can be fit again assuming a bent (C_{2v}) structure with X^2B_2; however, the hyperfine structure does not establish whether it is obtuse or acute angled. La_3 was not observed but it was probably trapped so that it is tempting to extrapolate the trend and suggest that La_3 is a linear $^2\Pi$ or $^2\Delta$ molecule. The orbital angular momentum in the ground state would then lead to large g-tensor anisotropy and broaden its spectrum beyond recognition.

A recent, rather thorough ESR investigation has been made of rare-gas matrices containing chromium metal. Besides the signals of Cr atoms in two matrix sites, a puzzling spectrum of an axial molecule with $S = 3$ and $|D| = 0.11$ cm^{-1} was observed. This, of itself, was not surprising, but the hyperfine structure produced when $^{53}Cr(I=3/2)$ was trapped indicated that in this molecule most of the spin, or more precisely most of the s character among the six unpaired electrons, was on one unique Cr atom. The molecule being observed is reasoned to be either a trigonally-distorted tetrahedral Cr_4, square-based pyramidal Cr_5, or a severely perturbed Cr atom bound axially to symmetrical, planar Cr_3 or Cr_4. There are features of all three of these models which are objectionable. A detailed account of the research has just appeared [45].

Pentatomic transition-metal molecules have only been characterized by ESR spectroscopy, among them are Mn_5 [34] and Cu_5, Ag_5 [46]. Recently, Howard, et al. [24] have extended their ESR studies of Cu_5 and Ag_5 to mixed copper-silver pentamers and have observed under the same conditions (cyclohexane matrix, 77°K) the spectra of two molecules which are most probably $CuAg_4$ and Cu_2Ag_3. These authors had previously concluded that Cu_5 and Ag_5 were Jahn-Teller distorted from trigonal-bipyramid (D_{3h}) structures to C_{2v} symmetry with most of the s spin density shifted to two equivalent atoms in the distorted triangle base. This resulted in a $^2B_2$ ground state (although a $^2B_1$ state with large s spin density on the two axial atoms is not completely eliminated). The mixed molecules are also judged to have trigonal bipyramid structures but undistorted. Then $CuAg_4$ is either of

Table 2. ESR of polyatomic transition metal molecules.

Molecule	Structure	Symmetry	Ground State	References
Sc_3		D_{3h} at 4 K	$^2A_1$	Knight et al. [44]
Cu_3, CuAgCu Ag_3, Au_3		C_{2v} at 77 K	$^2B_2$	Howard et al. [38,39]
Y_3		C_{2v} at 4 K	$^2B_2$	Knight et al. [44]
Cr_4		C_{3v} or C_{4v} at 4 K	^7A	Van Zee et al. [45]
Cu_5, Ag_5 ($CuAg_4$,Cu_2Ag_3)		C_{2v} at 77 K	$^2B_2$	Howard et al. [24,46]
Mn_5		D_{5h} at 4 K	^{26}A	Baumann et al. [34]
Sc_{13}		I_h at 4 K	^2A_g	Knight, et al. [44]

symmetry C_{2v} ($^2B_2$) or C_{3v} ($^2A_1$) and Cu_2Ag_3 either C_{2v} ($^2B_2$) or D_{3h} ($^2A_2''$). Thus, although not unambiguously assigned since hyperfine splittings from all nuclei were not resolved, these experiments do provide further needed information about the larger mixed-metal clusters.

The only other pentatomic transition-metal molecule, Mn_5, has also been observed via ESR in matrices and found to have a multiplicity of 26 [34]. Its structure could not be determined because hyperfine splittings were not resolved. Since it showed strong orientation in the matrix a pentagonal ring conformation seems most likely.

Finally, ESR has detected a cluster of Sc metal containing at least nine atoms. Knight, et al. [44], have observed a broad Lorentzian line at g = 2.0 with extensive hfs gradually decreasing in intensity on the wings of the line. It is proposed that this may be a stable cluster of 13 atoms in the form of an icosahedron, cube octahedron, or hexagonal close-packed polyhedron. Some support for this comes from an Xα calculation of Salahub indicating that hcp Sc_{13} would have an S = 1/2 ground state [47]. However it is possible that the three polyhedra structures have nearly the same energy, and the molecule could even be fluxional at 4°K.

A summary of the present status of our knowledge of triatomic and larger clusters of transition-metal atoms is given in Table 2.

Acknowledgments

The authors gratefully acknowledge the support of the National Science Foundation. We also wish to thank many colleagues working in this field of research, particularly L.B. Knight, Jr., and C.A. Baumann, for contributions, communications and preprints.

References:

1. H.E. Radford, Phys. Rev. 122, 114 (1961); ibid 136A, 15 (1964); J. Chem. Phys. 40, 2732 (1964).
2. A. Carrington, D.H. Levy and T.A. Miller, Adv. Chem. Phys. 18, 149 (1970).
3. W. Weltner, Jr., Magnetic Atoms and Molecules (Van Nostrand Reinhold, 1983).

4. P.H. Kasai, E.B. Whipple and W. Weltner, Jr., J. Chem. Phys. 44, 2581 (1966) and reference 3, pages 95-102.
5. L.B. Knight, Jr., R.J. Van Zee and W. Weltner, Jr., Chem. Phys. Lett. 94, 296 (1983).
6. R.J. Van Zee and W. Weltner, Jr., Chem. Phys. Lett. 107, 173 (1984).
7. W. Weltner, Jr. and R.J. Van Zee, Ann. Rev. Phys. Chem. 35, 291 (1984).
8. J.B. Hopkins, P.R.R. Langridge-Smith, M.D. Morse and R.E. Smalley, J. Chem. Phys. 78, 1627 (1983).
9. S.J. Riley, E.K. Parks, C.-R. Mao, L.G. Pobo and S. Wexler, J. Phys. Chem. 86, 3911 (1982).
10. Proceedings of a Bunsen-Kolloqium "Experiments on Clusters" (October, 1983) Ber. Bunsenges. Phys. Chem. 88, 188 (1984).
11. M.D. Morse and R.E. Smalley, Ber. Bunsenges. Phys. Chem. 88, 228 (1984).
12. J. Flad, G. Igel-Mann, H. Preuss and H. Stoll, Ber. Bunsenges. Phys. Chem. 88, 241 (1984); Chem. Phys. 90, 257 (1984).
13. M.J. Pellin and D.M. Gruen, J. Chem. Phys. 79, 5887 (1983).
14. R.P. Messmer, J. Vac. Sci. Technol. A2, 899 (1984).
14a. N.A. Baykara, B.N. McMaster and D.R. Salahub, Mol. Phys. 52, 891 (1984).
15. M.D. Morse, G.P. Hansen, P.R.R. Langridge-Smith, Lan-Sun Zheng, M.E. Geusic, D.L. Michalopoulos and R.E. Smalley, J. Chem. Phys. 80, 5400 (1984).
16. D.M. Kolb, H.H. Rotermund, W. Schrittenlacher and W. Schroeder, J. Chem. Phys. 80, 695 (1984).
17. C.W. Bauschlicher, Jr., Chem. Phys. Lett. 97, 204 (1983).
18. M.B. Cingi, D.A. Clemente and C. Foglia, Mol. Phys. 53, 301 (1984).
19. I. Shim and K.A. Gingerich, J. Chem. Phys. 80, 5107 (1984).
20. S. Rabii and C.Y. Yang, Chem. Phys. Lett. 105, 480 (1984).
21. J.L. Martins and W. Andreoni, Phys. Rev. A 28, 3637 (1983).
22. K.S. Pitzer, Int. J. Quant. Chem. 25, 131 (1984).
23. R.J. Van Zee and W. Weltner, Jr., High Temp. Sci. 17, 181 (1984).
24. J.A. Howard, R. Sutcliffe and B. Mile, J. Phys. Chem. 88, 2183 (1984).
25. P.R.R. Langridge-Smith, M.D. Morse, G.P. Hansen, R.E. Smalley and A.J. Merer, J. Chem. Phys. 80, 593 (1984).
26. D.L. Michalopoulos M.E. Geusic, S.G. Hansen, D.E. Powers and R.E. Smalley, J. Phys. Chem. 86, 3914 (1982); S.J. Riley, E.K. Parks, L.G. Pobo and S. Wexler, J. Chem. Phys. 79, 2577 (1983); V.E. Bondybey and J.H. English, Chem. Phys. Lett. 94, 443 (1983).
27. See H. Basch, M.D. Newton and J.W. Moskowitz, J. Chem. Phys. 73, 4492 (1980) and references given there.
28. D. Guenzburger and E.M.B. Saitovich, Phys. Rev. B24, 2368 (1981).
29. H.M. Nagarathna, P.A. Montano and V.M. Naik, J. Am. Chem. Soc. 105, 2938 (1983).
30. See the paper by S.P. Walch and C.W. Bauschlicher, Jr. in this Symposium.
31. S.P. Walch and C.W. Bauschlicher, Jr., J. Chem. Phys. 79, 3590 (1983).

32. L. Brewer and J.S. Winn, Symp. Faraday Soc. No. 14, 126 (1980).
33. R.K. Nesbet, Phys. Rev. A 135, 460 (1964).
34. C.A. Baumann, R.J. Van Zee, S.V. Bhat and W. Weltner, Jr., J. Chem. Phys. 74, 6977 (1981); 78, 190 (1983).
35. See L. Brewer, Science 161, 115 (1968).
36. C.A. Baumann, R.J. Van Zee and W. Weltner, Jr., J. Chem. Phys. 79, 5272 (1983).
37. C.A. Baumann, R.J. Van Zee and W. Weltner, Jr., J. Phys. Chem. 88, 1815 (1984).
38. J.A. Howard, R. Sutcliffe and B. Mile, J. Chem. Soc. Chem. Commun., 1449 (1983).
39. J.A. Howard, K.F. Preston and B. Mile, J. Am. Chem. Soc. 103, 6226 (1981); J.A. Howard, K.F. Preston, R. Sutcliffe and B. Mile, J. Phys. Chem. 87, 536 (1983); J.A. Howard, R. Suttcliffe and B. Mile, J. Am. Chem. Soc. 105, 1394 (1983).
40. D.M. Lindsay, D.R. Herschbach and A.L. Kwiram, Mol. Phys. 32, 1199 (1976); G.A. Thompson and D.M. Lindsay, J. Chem. Phys. 74, 959 (1981).
41. M.D. Morse, J.B. Hopkins, P.R.R. Langridge-Smith and R.E. Smalley, J. Chem. Phys. 79, 5316 (1983).
42. S.C. Richtsmeier, J.L. Gole and D.A. Dixon, Proc. Natl. Acad. Sci. USA 77, 5611 (1980); S.C. Richtsmeier, R.A. Eades, D.A. Dixon and J.L. Gole, "Metal Bonding and Interactions in High Temperature Systems," eds. J.L. Gole and W.C. Stwalley, ACS Symp. Series No. 179, 1982, pg. 177.
43. H. Basch, J. Am. Chem. Soc. 103, 4657 (1981).
44. L.B. Knight, Jr., R.W. Woodward, R.J. Van Zee and W. Weltner, Jr., J. Chem. Phys. 79, 5820 (1983).
45. R.J. Van Zee, C.A. Baumann and W. Weltner, Jr., J. Chem. Phys. 82, 3912 (1985).
46. J.A. Howard, R. Sutcliffe and J.S. Tse, Chem. Phys. Lett. 94, 561 (1983); J.A. Howard, R. Sutcliffe and B. Mile, J. Phys. Chem. 87, 2268 (1983).
47. D.R. Salahub, Impact of Cluster Physics in Material Science and Technology, ed. J. Davenas, The Hague, Nijhoff, to be published.

THEORETICAL STUDIES OF TRANSITION METAL DIMERS

Stephen P. Walch*, Eloret Institute,
Sunnyvale, CA 94087
and
Charles W. Bauschlicher, Jr., NASA Ames Research Center,
Moffet Field, CA 94035

ABSTRACT. The results of CASSCF calculations are presented for the Sc_2, Ti_2, V_2, Cr_2, Cu_2, TiV, Y_2, Nb_2, and Mo_2 molecules. CASSCF/CI calculations were also carried out for Sc_2, Ti_2, Cu_2, and Y_2. The CASSCF procedure is found to generally provide reliable R_e and ω_e values. However, D_e values are systematically underestimated at the CASSCF level; the worst case is Cr_2 where the CASSCF curve is not bound although a shoulder is observed in the region near the experimental R_e. The CASSCF procedure is shown to provide a consistent set of calculations for these molecules from which trends and a simple qualitative picture of the electronic structure may be derived. These calculations confirm the $^5\Sigma_u^-$ ground state of Sc_2 and the $^4\Sigma^-$ ground state of TiV and lead to predictions for other molecules in this series. So far only the $^3\Sigma_g^-$ ground state symmetry and the bond length of V_2 have been confirmed by experiment.

1. INTRODUCTION

Transition metals and transition metal (TM) compounds are currently of considerable interest because of their relevance to catalysis and to material science problems such as hydrogen embrittlement and crack propagation in metals. In spite of such interest, progress in understanding the chemistry of TM molecules has been slow due to

* Mailing address: NASA Ames Research Center, Moffett Field, CA 94035

R. J. Bartlett (ed.), Comparison of Ab Initio Quantum Chemistry with Experiment for Small Molecules, 17–51.

both experimental and theoretical difficulties. For a review of the current state of both theory and experiment for TM molecules the reader is referred to the review article by Weltner and Van Zee [1]. Recent progress in the experimental characterization of TM molecules has come from both matrix isolation and gas phase spectroscopic approaches, while progress in the theoretical treatment of TM molecules has occurred as a result of improved methods of treating electron correlation. These improvements in part involve more efficient computational methods such as improved MCSCF [2], CASSCF [3] and direct CI [4] techniques but also qualitative ideas such as the GVB [5] method have provided a conceptual framework for understanding electron correlation effects in these molecules.

A dramatic example of the importance of electron correlation for transition metal compounds is given by the NiCO $X^1\Sigma^+$ state. This state is unbound by 2.6 eV at the SCF level, while the triplet states are only slightly repulsive [6]. The Mulliken populations for the triplet states show a 3d population of 9, while the $^1\Sigma^+$ state has a 3d population near 10. If an MCSCF approach is used, the 3d population for the triplet states is only slightly changed, but the 3d population for the $^1\Sigma^+$ state is qualitatively different, 9.5. This arises because of the mixing of the Ni $4s^1 3d^9$ (^1D) and $3d^{10}$ (^1S) states, and the $^1\Sigma^+$ state is now the ground state of NiCO. It would take a very large CI to overcome the $3d^{10}$ bias of the SCF reference. In this case, the CASSCF wavefunction yields results in remarkable agreement with a multi-reference singles and doubles CI calculation [6] and a D_e of 1.3 eV which is quite consistent with experiment (1.3±0.6 eV [7]).

MCSCF treatments do not always yield results as good as those obtained for NiCO but they usually do represent a major improvement over SCF results. For an especially difficult case, the Cr_2 molecule, SCF leads to a potential curve which is unbound with respect to the SCF atoms by 10 eV in the region near the experimental R_e. An MCSCF treatment shows the outer well arising from the 4s-4s bonding but is still unbound by 1.4 eV in the region near R_e where one expects multiple 3d-3d bonds [8,9]. While this might appear to be a major failing of the MCSCF approach, if Cr_2 is compared to Mo_2 [8] and V_2 [9], both of which are bound at the MCSCF level, useful information about the nature of the bonding in Cr_2 can be inferred. Thus, unlike the chemistry of the first two rows of the periodic table, where computational chemistry is achieving accuracy comparable with experiment, for some transition metal compounds only qualitative accuracy is possible. However, so little is known about many of these systems that even qualitative information can contribute to a basic understanding of the chemistry. This is somewhat pessimistic since the MCSCF procedure does lead to good R_e and ω_e values for many of the TM dimers even if the binding energies are systematically underestimated and can treat states arising from the same occupation to very similar accuracy. Thus in these cases T_e values should be good. When comparing very different states, sometimes a combination of theory and experiment can combine to resolve a question. For example for Ti_2 theory can calculate the R_e and ω_e of the low-lying $^1\Sigma_g^+$ and $^7\Sigma_u^+$ states but cannot clearly establish which is the ground state. However, combining the calculated results with the experimental ω_e leads to a clear assignment of a $^1\Sigma_g^+$ ground state and provides new information about the chemistry.

Many people have investigated individual transition metal dimers (see review article of Weltner and Van Zee [1]). However, as noted for Cr_2, a consistent set of calculations can reveal trends which greatly contribute to the understanding. Goodgame and Goddard have considered Cr_2 and Mo_2 [8], Shim and co-workers [10-11] have considered several transition metal dimers for those metal atoms with more than half filled nd shells, and these authors have considered many of the dimers for those metal atoms with less than half filled nd shells where strong nd-nd bonding is common. In this article we summarize our previous work, report on several unpublished systems and describe the trends for the bonding in the transition metal dimers on the left side of the first two transition metal rows. Here we concentrate on the CASSCF results since this level of calculation is possible for all the systems and provides a useful qualitative understanding. However, for some systems we also are able to carry out multi-reference singles and doubles CI(MRSDCI) calculations using the CASSCF orbitals and these calculations serve to calibrate the CASSCF approach. For other systems, such as Cr_2, the CASSCF + MRSDCI approach is not possible. This has led to several attempts to understand the bonding using even more qualitative approaches [12-14]. These include the extended MCSCF approaches [12-13] and approaches based upon physical arguments [14].

Section 2 describes some trends for the transition metal atoms which are important for understanding the bonding of the TM dimers and also discusses qualitative features of the bonding in the TM dimers. Section 3 discusses the technical details of the calculations. Section 4 discusses the potential curves and spectroscopic constants obtained for the first row TM dimers while Section 5 describes the same information for the second row TM dimers and discusses some of the differences between the first and second row TM dimers. Section 6 discusses the electron correlation problem for Cr_2 and discusses some possible future approaches that might be used. These future approaches are discussed in relation to some of the current more approximate techniques. Finally Section 7 presents the conclusions from this work.

2. QUALITATIVE FEATURES OF THE BONDING IN THE TM DIMERS

We first consider some features of the TM atoms, in particular the relative (n+1)s and nd orbital sizes, the nd-nd exchange terms, and the ordering and separations of the low-lying atomic states [15]. These properties of the atoms are important because they control the atomic states available for bonding, the strength of the nd-nd bonds and the nature of the bonding, i.e. whether the bonding orbitals involve predominantly (n+1)s or nd character. Table 1 shows the following properties i) $\langle r_{(n+1)s}\rangle/\langle r_{nd}\rangle$, obtained from numerical Hartree-Fock calculations for the $(n+1)s^1nd^{m+1}$ state ii) the $(n+1)s^2nd^m \rightarrow (n+1)s^1nd^{m+1}$ excitation energy and iii) the $(n+1)s^2nd^m \rightarrow nd^{m+2}$ excitation energy. Looking first at the relative orbital sizes

Table 1a. Relative orbital sizes and excitation energies for first row transition metal atoms. The orbital expectation values are taken from Numerical Hartree-Fock calculation, while the excitation energies are taken from experiment.

Atom	$\langle r_{4s}\rangle/\langle r_{3d}\rangle$	$4s^2 3d^m \rightarrow 4s^1 3d^{m+1}$	$4s^2 3d^m \rightarrow 3d^{m+2}$
Sc	2.03	1.43	4.19
Ti	2.32	0.81	3.35
V	2.51	0.25	2.47
Cr	2.69	-1.00	3.40
Mn	2.81	2.14	5.59
Fe	2.95	0.87	4.07
Co	3.08	0.42	3.36
Ni	3.22	-0.03	1.71
Cu	3.36	-1.49	--

Table 1b. Relative orbital sizes and excitation energies for second row transition metal atoms. The orbital expectation values are taken from Numerical Hartree-Fock calculation, while the excitation energies are taken from experiment.

Atom	$\langle r_{5s}\rangle/\langle r_{4d}\rangle$	$5s^2 4d^m \rightarrow 5s^1 4d^{m+1}$	$5s^2 4d^m \rightarrow 4d^{m+2}$
Y	1.61	1.36	3.63
Zr	1.79	0.59	2.66
Nb	1.92	-0.18	1.14
Mo	2.05	-1.47	1.71
Tc	2.16	0.41	--
Ru	2.29	-0.87	0.22
Rh	2.42	-1.63	-1.29
Pd	2.54	-2.43	-3.38
Ag	2.67	-3.97	--

we see that for the first transition row the ratio $\langle r_{4s}\rangle/\langle r_{3d}\rangle$ increases monotonically from 2.03 for Sc to 3.36 for Cu. The same trend is seen for the second transition row but here the 5s and 4d orbital sizes are more comparable. The increase in $\langle r_{(n+1)s}\rangle/\langle r_{nd}\rangle$ as one moves across a row is due to increased shielding of the nuclear charge by the nd electrons.

Looking next at the excitation energies, the ground state for the first few elements of the first and second transition rows is the $(n+1)s^2nd^m$ state. The trends in the $(n+1)s^2nd^m \rightarrow (n+1)s^1nd^{m+1}$ excitation energies depend on two competing factors i) the stabilization of the nd level with respect to the (n+1)s level as the nuclear charge increases and ii) the number of nd-nd exchange terms. Up to the point that the nd shell is half filled both of these factors favor the $(n+1)s^1nd^{m+1}$ state for larger nuclear charge since movement of an s electron into the d shell increases the number of nd-nd exchange terms. This leads to a monotonic decrease in the excitation energies from Sc to Cr and from Y to Mo. Starting with Mn or Tc, the excited state has one more doubly occupied nd orbital in each case than the ground state. Thus, s to d promotion leads to loss of nd-nd exchange terms and a strong preference for the $(n+1)s^2nd^5$ state resulting in a discontinuity between Cr-Mn and Mo-Tc. However, from Mn to Cu and Tc to Ag, there is once again a monotonic decrease in the excitation energy due to the stabilization of nd with respect to (n+1)s for the increasing nuclear charge. From Table 1 one also sees that the nd^{m+2} state is an excited state for the first transition row but becomes a low-lying state for the right half of the second transition row. In fact $4d^{10}$ is the ground state configuration for Pd.

For the molecules considered here the atomic states which are sufficiently low-lying to be available for bonding are the $(n+1)s^2nd^m$ and $(n+1)s^1nd^{m+1}$ states. These two states lead to three different atomic asymptotes: i) $(n+1)s^2nd^m + (n+1)s^2nd^m$, ii) $(n+1)s^2nd^m + (n+1)s^1nd^{m+1}$ and iii) $(n+1)s^1nd^{m+1} + (n+1)s^1nd^{m+1}$. The accessibility of these asymptotes of course depends on the $(n+1)s^2nd^m \rightarrow (n+1)s^1nd^{m+1}$ excitation energy. Thus for the first transition row we find that for Sc_2 the low-lying states arise from the first two asymptotes with $4s^13d^{m+1}$ + $4s^13d^{m+1}$ too high in energy to lead to the ground state, while for Ti_2, V_2, Cr_2 and Cu_2 the ground states arise from the $4s^13d^{m+1}$ + $4s^13d^{m+1}$ atomic asymptote.

For the first transition row, because the 4s orbital is significantly larger than the 3d orbital the predominant interaction in the TM dimers at large internuclear separation (R) is between the 4s orbitals with very little 3d interaction. For states arising from the $4s^23d^m$ + $4s^23d^m$ atomic asymptote this interaction is basically repulsive and only a shallow well at large R_e (approximately 8.0 a_o) arising from the 4s→4p near degeneracy effect is observed. For states arising from the $4s^23d^m$ + $4s^13d^{m+1}$ atomic asymptote the 4s interaction is weakly bonding at intermediate R but is repulsive at small R leading to intermediate R_e values (approximately 5.0 a_o). At these R_e values the 3d-3d overlaps (S) are small which favors one-electron 3d bonds, (i.e., one electron per bonding orbital as in H_2^+, a bond order of one half) whose bonding terms vary with distance like S, over two-electron 3d bonds (i.e., two electrons per bonding orbital as in H_2, a bond order of one), whose bonding terms vary with distance like S^2 [17]. Finally for states arising from the $4s^13d^{m+1}$ + $4s^13d^{m+1}$ atomic asymptote the 4s-4s portion of the interaction is attractive and also appears to not become repulsive until well inside the optimal 4s-4s bonding radius. Thus, states arising from this atomic asymptote favor short R_e regions where the 3d-3d overlaps are large enough to permit two-electron 3d bonding.

The strength of multiple one-electron nd bonds is determined mainly by overlap considerations, since the resulting singly occupied orbitals are orthogonal and high spin coupled, leading to no loss of atomic exchange interactions. The relative overlaps of nd orbitals are nd$\sigma \approx$ nd$\pi >$ ndδ. Thus, the ndσ and ndπ orbitals are filled first. An example of such bonding is the $^5\Sigma_u^-$ state of Sc_2, which has three one-electron 3d bonds of σ and π symmetry.

The strength of multiple two-electron nd bonds is determined both by the overlap of the orbitals and by the size of the atomic exchange terms which are lost due to bond formation. Overlap considerations primarily determine the occupation of the bonding orbitals. Since two electron bonds vary with R as S^2, and one-electron bonds as S, the ndδ bonding is even less favorable relative to ndσ and ndπ than for one-electron bonds. Thus, for Ti_2 the $^1\Sigma_g^+$ state which arises from $4s^13d^3$ + $4s^13d^3$ has three two-electron 3d bonds with bonding orbitals of σ and π symmetry. It is interesting that in the region where the ndσ and ndπ orbitals have large overlaps leading to two-electron bonds, the overlaps of the ndδ orbitals are so small that one-electron bonds are favored. For example, in going from Ti_2 to V_2 the two electrons which are added to the 3dδ preferentially couple high spin leading to a $^3\Sigma_g^-$ ground state. When two additional electrons are added into the 3dδ orbital as in the ground $^1\Sigma_g^+$ state of Cr_2, the overall bonding becomes much weaker. This shows that doubly occupying the 3dδ orbitals is unfavorable; although part of the difference between V_2 and Cr_2 involves 3d-3d exchange interactions.

One measure of the atomic exchange interaction is the relative energies of the atomic states as the coupling of the nd electrons is changed (we use Moore [16] to estimate the importance of this effect). For Sc, $4s^13d^2$, the energy needed to change the 3d coupling from ^3F in the ^4F state to ^1D in the ^2D state is 0.8 eV. For Ti $4s^13d^3$, $4s^13d^3(^2G)$ ^3G is 1.0 eV above $4s^13d^3(^4F)$ ^5F. By Cr, with the maximum open 3d orbitals, this value has reached 2.5 eV. This means that the 3d-3d bonds are expected to be weaker for Cr_2 unless the bond length shortens to increase the 3d-3d overlap.

For the TM elements with more than half filled 3d shells, the formation of 3d bonds becomes much less favorable for two reasons. First, the ratio of 4s to 3d orbital sizes is larger for the right half of the first transition row. Secondly, the presence of doubly occupied 3d orbitals leads to repulsive interactions which effectively cancel any bonding interactions from the 3d shell. Thus the bonding here is dominated by the 4s electrons.

For the second transition row the qualitative features of the bonding from the corresponding three atomic asymptotes are similar.

However, the more comparable 5s and 4d orbital sizes as well as the smaller 4d exchange terms lead to stronger 4d bonding. Hence the 4dδ overlaps appear to be more comparable to the 4dσ and 4dπ overlaps as evidenced by a low-lying state of Nb_2 with occupation $4d\sigma_g^1 4d\pi_u^4 4d\delta_g^3$. Also, for the right hand side of the second transition row, the $4d^{m+2}$ atomic states are available for bonding. An additional important difference is that relativistic effects become quite important for the right half of the second transition row (see Ref. [18], for example). It is for this reason that the current all-electron studies are restricted to the lighter elements of the second transition row.

3. CALCULATIONAL DETAILS

As noted in the previous section, both the $(n+1)s^2nd^m$ and $(n+1)s^1nd^{m+1}$ states of the atoms are important in the bonding in the dimers. Hay [19] was one of the first to realize that the basis sets optimized for the $(n+1)s^2nd^m$ states yield a very poor description of the $(n+1)s^1nd^{m+1}$ states, but by optimizing one additional d function for the $(n+1)s^1nd^{m+1}$ occupation, the separation between the $(n+1)s^2nd^m$ and $(n+1)s^1nd^{m+1}$ occupations was very close to that obtained in numerical Hartree-Fock calculations. When (n+1)s bonding is present, due to the near degeneracy of the (n+1)s and (n+1)p orbitals, the basis set requires p functions in the same region. Since the (n+1)p orbital is not occupied, such functions are missing in the atomic basis sets. One common approach is to optimize two additional p functions for the $(n+1)s^1(n+1)p^1nd^m$ occupation and then multiply the exponents of the additional functions by from 1.2 to 1.5 to bring them more into the region of the (n+1)s function [20]. The final concern in the basis set is f functions to help describe any d-d bonds. For Cu_2 with a 4s-4s bond or Sc_2 with one-electron d bonds, the addition of f polarization functions has very little effect. However, McLean and Liu [21] have shown that for Cr_2, with two-electron d bonding, the addition of f functions lowers the energy by more than 1 eV at the SCF level. Thus it is important to have f functions, especially when comparing one state which has one-electron nd bonds with another that has two-electron nd bonds, such as in Ti_2.

The basis sets used in this work meet these requirements. For the first row transition elements the basis set is a (14s11p6d3f)/[8s6p4d2f] segmented contraction based on Wachter's basis set [20] plus his 4p (multipled by 1.5) and the Hay diffuse 3d [19]. The 4f functions are optimized at the CI level for the $4s^1 3d^{m+1}$ state of the atom (note that for the Cu_2 calculations the three f functions were contracted to one). For the second transition row elements the basis set is a (17s13p9d2f)/[6s5p5d1f] general contraction based on Huzinaga's basis set [22] plus 5p (multiplied by 1.3), diffuse 4d, and 4f functions as optimized by Walch, Bauschlicher, and Nelin [23]. The 3s function arising from the d, as well as the 4p functions arising from the f functions were removed.

In this work we use the CASSCF approach to perform the MCSCF calculations. Ideally one would like to include the nd, (n+1)s and (n+1)p orbitals in the active space to allow the formation of both the s-s bond and nd-nd bonds and to account for the (n+1)s to (n+1)p near degeneracy. More extensive correlation could then be included with a CI calculation. However, even the CASSCF treatment is not always possible. The first approximation we make is to remove the (n+1)p orbital from the active space, except for Sc_2 and Y_2 which arise from the $(n+1)s^2nd^m + (n+1)s^1nd^{m+1}$ atomic limit. Here the (n+1)p orbital should be included in the active space since the (n+1)s to (n+1)p near degeneracy is important for the $(n+1)s^2nd^m$ state of the atom. For the remaining systems which arise from the $(n+1)s^1nd^{m+1} + (n+1)s^1nd^{m+1}$ atomic limit the (n+1)p does not contribute for the atom and tests show that omitting the (n+1)p leads to less binding energy at the CASSCF level (due to to angular correlation effects which vanish at large R), but virtually the same D_e when correlation is added. Thus, this approximation affects only those calculations for which only CASSCF calculations were performed.

The second approximation is to restrict the number of active electrons in σ, π and δ symmetry. The constraints are similar in spirit to those in the GVB wavefunctions [5]. As an example of the

GVB wavefunction consider the state of Ti_2 which has the configuration $4s\sigma_g^2 3d\sigma_g^2 3d\pi_{xu}^2 3d\pi_{yu}^2$, i.e. a quadruple bond with 4sσ, 3dσ, and 3dπ bonds. The configurations in the GVB wavefunction consist of a geminal product of the three possible configurations constructed from the bonding and antibonding orbitals of each pair, i.e. 81 configurations in the case of four pairs. This wavefunction dissociates to neutral atoms and also allows spin recoupling. In the CASSCF calculations it is not possible to do precisely the GVB calculation since the configuration space must be a product of full CI calculations within each symmetry type. Consistent with these constraints we restrict four electrons to the four σ orbitals, two electrons to the two π_x orbitals, and two electrons to the two π_y orbitals. The δ orbitals would not be active in this calculation since they are forced to have no electrons by our constraints. For the other cases the constraints are not specifically stated but can be inferred from the principle configuration given.

The last approximation which was made only for the second transition row dimers is to freeze the deep core orbitals, the n=1,2,3 shells and 4s orbitals at their atomic forms, based on a high spin SCF calculation at R=50.0 a_o, while the 4p and valence (5s, 5p, and 4d) orbitals are optimized at each distance. Since the core orbitals are expected to be atomic like, this approximation will have negligible effect on the results, but results in some computational savings.

When it is possible, we follow the CASSCF calculations with multireference singles and doubles CI calculations. In these calculations only the electrons arising from the (n+1)s and nd shells of the atom are correlated. The reference configurations are selected as the most important configurations in the CASSCF wavefunction. In general the most important configurations are found to be the leading terms in the GVB geminal expansion. Different reference lists were used near R_e and at long range. The long distance points were computed using the high spin SCF wavefunction as the reference configuration, while the points near R_e use the CASSCF wavefunction.

The calculations for the first transition row dimers use the

MOLECULE- [24] SWEDEN [25] programs while the calculations for the second transition row dimers use BIGGMOLI- [26] SWEDEN.

4. FIRST ROW TRANSITION METAL DIMERS

Figure 1 shows calculated potential curves for low-lying states of Sc_2 while Table 2 shows calculated spectroscopic constants for all the first row TM dimers. The dominant configurations for the states considered here are:

$$^5\Sigma_u^- \; 4s\sigma_g^2 \; 3d\sigma_g^1 \; 4s\sigma_u^1 \; 3d\pi_{xu}^1 \; 3d\pi_{yu}^1$$

$$^3\Sigma_g^- \; 4s\sigma_g^2 \; 4s\sigma_u^2 \; 3d\pi_{xu}^1 \; 3d\pi_{yu}^1$$

$$^1\Sigma_g^+ \; 4s\sigma_g^2 \; 4s\sigma_u^2 \; 3d\sigma_g^2$$

$$^3\Sigma_u^+ \; 4s\sigma_g^2 \; 4s\sigma_u^2 \; 3d\sigma_g^1 \; 3d\sigma_u^1$$

The initial study of Sc_2 by Walch and Bauschlicher [17] found only weakly bound states arising out of the $4s^23d^1 + 4s^23d^1$ asymptote in contrast to mass spectrometric experiments [27] which indicated strong bonding (De = 1.1±0.2 eV). A $^5\Sigma_u^-$ state was found which was bound by about 0.8 eV with respect to the $4s^23d^1 + 4s^14p^13d^1$ atomic limit, but unbound with respect to $4s^23d^1 + 4s^23d^1$. However at about the same time that this work was published, matrix isolation studies by Knight, VanZee and Weltner [28] indicated a bound $^5\Sigma$ state of Sc_2. From ESR studies it appeared that this state arose from the $4s^23d^1$ + $4s^13d^2$ atomic asymptote which had not been studied in detail in the previous theoretical studies. A reexamination of this system [30] revealed a new $^5\Sigma_u^-$ state which had been missed in the previous study because its R_e (about 5.0 a_o) is much shorter than the R_e values for the states studied previously (about 7.0 a_o). The $^5\Sigma_u^-$ state turned out to be of considerable theoretical interest because it exhibited multiple 3d bonding (three one-electron bonds) and constituted the

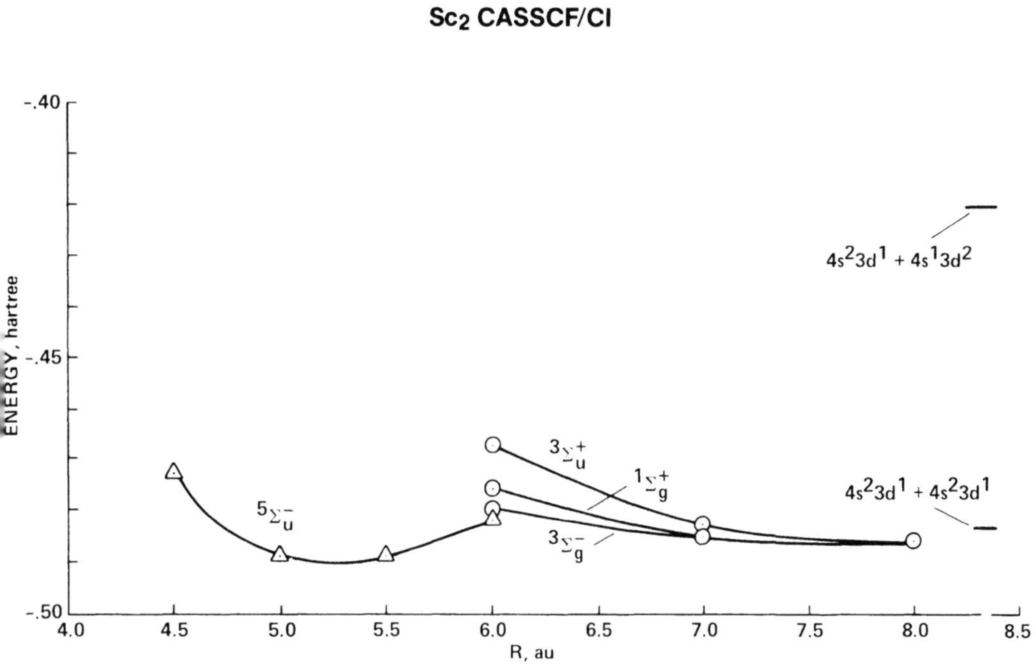

Figure 1. Calculated CASSCF/CI potential curves for selected states of Sc_2.

Table 2. Calculated spectroscopic constants for first row transition metal dimers

Molecule	R_e(Å)	ω_e(cm^{-1})	d_e(eV)	T_e(eV)
Sc_2 $^5\Sigma_u^-$				
CI	2.79	184.	0.44	
Exp.		238.9[a]	1.13[b]	
Ti_2 $^1\Sigma_g^+$				
CASSCF	1.97	436.	0.37	
CI	1.97	438.	1.94	0.40
Exp.		407.9[a]	2.91[c]	
Ti_2 $^7\Sigma_u^+$				
CI	2.63	205.	1.53	0.00
V_2 $^3\Sigma_g^-$				
CASSCF	1.76	564.	0.60	0.00
Exp.	1.76[d]	537.5[a]	2.29[d], 2.97[c]	
V_2 $^1\Gamma_g$				
CASSCF	1.80	486.	0.58	0.02
V_2 $^3\Delta_g$				
CASSCF	1.77	413.	-0.32	0.92
Cr_2 $^1\Sigma_g^+$				
CASSCF			-1.4	
Exp.	1.68[e]	480.[a]	1.56[f]	

[a] Ref. [29]
[b] Ref. [27]
[c] Ref. [31]
[d] Ref. [32]
[e] Ref. [49-51]
[f] Ref. [34]

first theoretical evidence of multiple 3d bonding in a first row transition metal dimer. The D_e given in Table 2 is obtained from the CI calculations by computing the binding energy of the $^5\Sigma_u^-$ state with respect to the $4s^2 3d^1 + 4s^1 3d^2$ asymptote and then subtracting the experimental asymptotic separation to give the D_e with respect to $4s^2 3d^1 + 4s^2 3d^1$. The resulting D_e of 0.44 eV and ω_e of 184 cm^{-1} are in reasonable accord with experimental values of D_e = 1.1 eV [27] and ω_e = 238.9 cm^{-1} [29].

Adding two electrons to the 3dδ orbitals of the $^5\Sigma_u^-$ state of Sc_2 and high spin coupling the 3d electrons leads to the $^7\Sigma_u^+$ state of Ti_2. The Ti_2 $^7\Sigma_u^+$ state, like the Sc_2 $^5\Sigma_u^-$ state comes from the $4s^2 3d^m + 4s^1 3d^{m+1}$ asymptote. These states have long bond lengths (R_e greater than 5.0 a_o) and small vibrational frequencies (ω_e about 200 cm^{-1}). The states of Ti_2 which are considered here have the configurations:

$$^7\Sigma_u^+ \; 4s\sigma_g^2 \; 3d\sigma_g^1 \; 4s\sigma_u^1 \; 3d\pi_{xu}^1 \; 3d\pi_{yu}^1 \; 3d\delta_{xyg}^1 \; 3d\delta_{(x^2-y^2)g}^1$$

$$^1\Sigma_g^+ \; 4s\sigma_g^2 \; 3d\sigma_g^2 \; 3d\pi_{xu}^2 \; 3d\pi_{yu}^2$$

For Ti_2 in Table 1, one sees that the excitation energy to the $4s^1 3d^{m+1} + 4s^1 3d^{m+1}$ atomic asymptote is about half as large as for Sc_2. Thus the $^1\Sigma_g^+$ state which arises from the $4s^1 3d^3 + 4s^1 3d^3$ atomic symptote becomes a competitor for the ground state of Ti_2. The bonding here is a triple two electron 3d bond (3dσ, $3d\pi_x$ and $3d\pi_y$). This leads to a short R state, R_e = 3.72 a_o and ω_e = 438 cm^{-1}. The bond length here is not known experimentally but the experimental vibrational frequency is 407.9 cm^{-1} [29] which is consistent with the $^1\Sigma_g^+$ state of Ti_2, but inconsistent with the $^7\Sigma_u^+$ state. Based on this we tentatively assign the ground state of Ti_2 as $^1\Sigma_g^+$ although the calculations place $^1\Sigma_g^+$ 0.40 eV above $^7\Sigma_u^+$.

Table 2 also shows the calculated D_e of Ti_2. Here we see that CASSCF obtains only a small percentage of the binding energy. The CI wavefunction consists of a MRSDCI from nine references (more than 180,000 CSFs). The calculated binding energy here is 1.94 eV compared to an experimental D_e of 2.91 eV [31] (all referenced to $4s^1 3d^3$ + $4s^1 3d^3$). Thus even when more extensive correlation is added, the order of the states is not changed. However, CI calculations describe high spin states better than low spin states; the $^1\Sigma_g^+$ is not as well described as $^7\Sigma_u^+$. Considering the difficulty in describing multiple two-electron 3d bonds, as measured by the errors in D_e, it would not be surprising to find that a better calculation would reverse the order of the states. Thus trends in the correlation treatments also support the $^1\Sigma_g^+$ assignment of the ground state.

Figure 2 shows calculated potential curves for the $^3\Sigma_g^-$, and $^1\Gamma_g$ states of V_2. The dominant configurations for the states considered here are:

$$^3\Sigma_g^- \quad 4s\sigma_g^2 \; 3d\sigma_g^2 \; 3d\pi_{xu}^2 \; 3d\pi_{yu}^2 \; 3d\delta_{xyg}^1 \; 3d\delta_{(x^2-y^2)g}^1$$

$$^3\Delta_g \quad 4s\sigma_g^2 \; 3d\sigma_g^1 \; 3d\pi_{xu}^2 \; 3d\pi_{yu}^2 \; 3d\delta_{xyg}^2 \; 3d\delta_{(x^2-y^2)g}^1$$

The $^3\Sigma_g^-$ state of V_2 has the same triple two-electron 3d bond as in Ti_2 with the remaining two electrons in the 3dδ orbitals. Because the 3dδ orbitals still have small overlaps in the region near R_e (about 3.5 a_o), the lowest state is a triplet state arising by forming two one-electron 3dδ bonds. This occupation also gives rise to $^1\Gamma_g$ and $^1\Sigma_g^+$ states analogous to the corresponding states of the $\pi_u^4\pi_g^2$ occupation of O_2. The R_e and ω_e values obtained from the CASSCF curves [9] are in good agreement with the recent results of Langridge-Smith, Morse, Hansen, Smalley and Merer [32] for the $^3\Sigma_g^-$ ground state of V_2. The presence of a very low-lying state (possibly $^1\Sigma_g^+$) was indicated by the spectral analysis of Merer [33] and is consistent with the very low-lying $^1\Gamma_g$ state ($^1\Sigma_g^+$ is obtained as a second root of the CASSCF and is also very low-lying). Table 2 also shows the calculated spectroscopic constants for the $^3\Delta_g$ state of V_2. This state was studied because the corresponding state is very low-lying for Nb_2 but here the excitation energy $^3\Sigma_g^- \rightarrow {}^3\Delta_g$ is 0.92 eV.

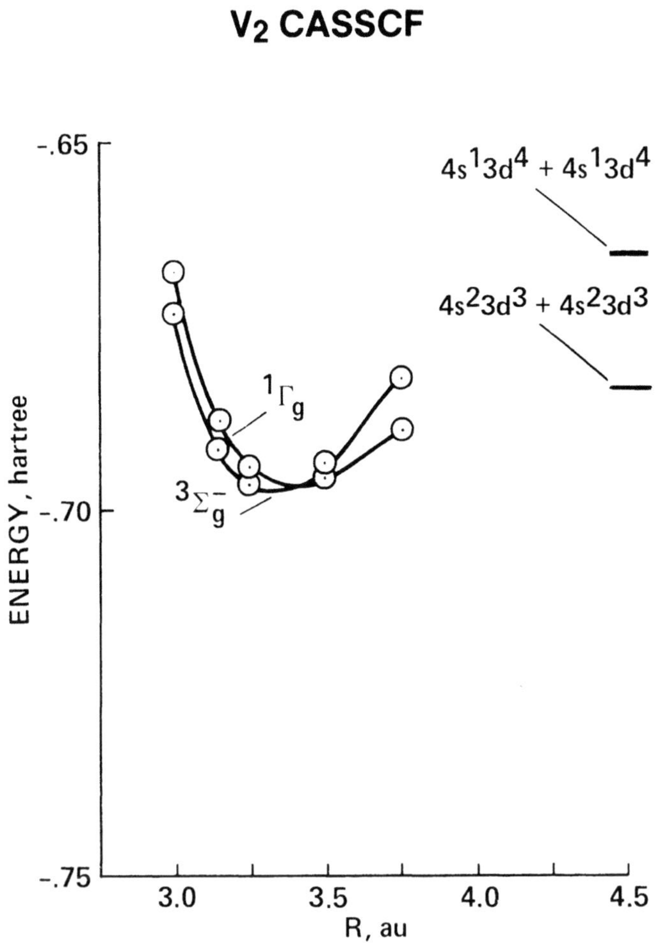

Figure 2. Calculated CASSCF potential curves for selected states of V_2.

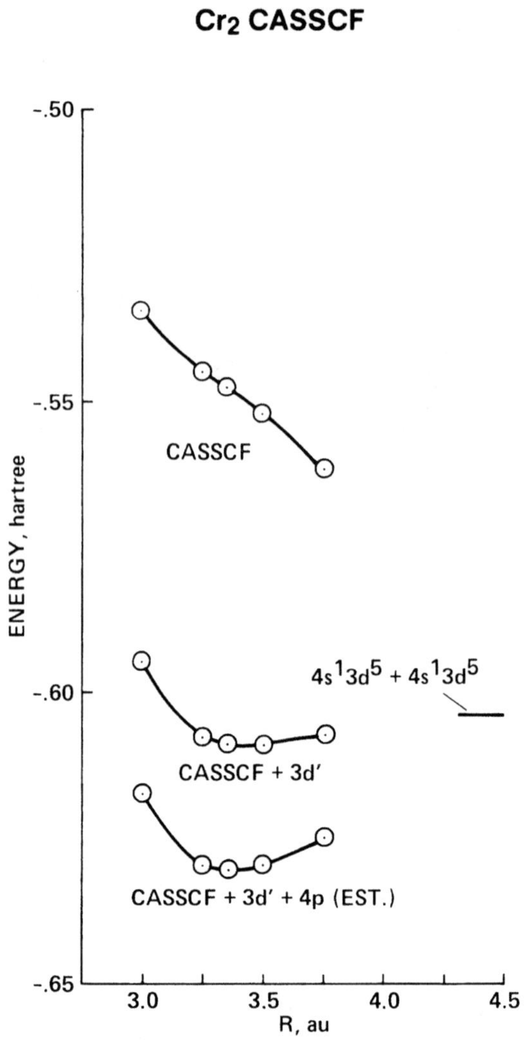

Figure 3. Calculated potential curves for the $^1\Sigma_g^+$ state of Cr_2. The curve labeled CASSCF + 3d′ includes an estimate of 3d′ correlation from the extended CASSCF calculations of Walch (Ref. [12]); while the curve labeled CASSCF + 3d′ + 4p (Est.) also includes an estimate of the 4p correlation based on extended CASSCF calculations on Ti_2 (Ref. [12]). Because the correlation contributions are obtained by summing individual energy contributions the effect of 3d′ is expected to be significantly overestimated (see text). Thus, we expect that a CASSCF calculation in which all the 3d′ orbitals were added simultaneously would show a much reduced binding energy contribution due to 3d′ leading to a potential curve with a more distinct shoulder than the CASSCF curve but no well.

Figure 3 shows calculated potential curves for Cr_2. Here the 3dδ orbitals are doubly occupied which is expected to be unfavorable based on the V_2 result that the 3dδ orbitals were preferentially singly occupied. This is consistent with the weaker bonding in Cr_2, D_e = 1.6 eV [34] as compared to V_2, D_e = 3.0 eV [31]. Because of the weaker bonding in Cr_2 the CASSCF potential curve is not bound [9]. However, the potential curve does exhibit a shoulder near the experimental R_e which is suggestive of an inner well. As discussed in Section 6, a more accurate computational description of the quintuple 3d bond in Cr_2 is a very difficult problem which remains as a major challenge for theory.

Figure 4 shows calculated potential curves for TiV while Table 3 summarizes the calculated spectroscopic constants. The $^4\Sigma^-$, $^4\Pi$ and $^2\Delta$ states may be derived from the V_2 $^3\Sigma_g^-$ configuration by removing a single electron from the 3dσ, 3dπ and 3dδ orbitals, respectively. The ordering of these states may be understood on the basis of orbital overlap and intra atomic coupling [35] arguments. Based on atomic overlap arguments one expects the ordering $^2\Delta < {}^4\Sigma^- < {}^4\Pi$

Table 3. Calculated spectroscopic constants for the TiV molecule

State	R_e(Å)	ω_e(cm^{-1})	D_e(eV)[a]
$^4\Sigma^-$	1.86	495.	0.80
$^4\Pi$	1.95	410.	0.15
$^2\Delta$	1.89	472.	0.43

[a] Calculated with respect to Ti($4s^1 3d^3$) + V($4s^1 3d^4$).

However, while all these states dissociate to pure V ^{6}D and the $^4\Sigma^-$ and $^4\Pi$ states dissociate to pure Ti ^{5}F, the $^2\Delta$ state dissociates to a mixture of 20% Ti ^{5}F and 80% Ti ^{5}P. It is this latter intra atomic coupling effect which leads to the observed ordering of states with

$$^4\Sigma^- < {}^2\Delta < {}^4\Pi$$

The calculated $^4\Sigma^-$ ground state is consistent with the experimental results of Van Zee and Weltner which show an ESR spectrum con-

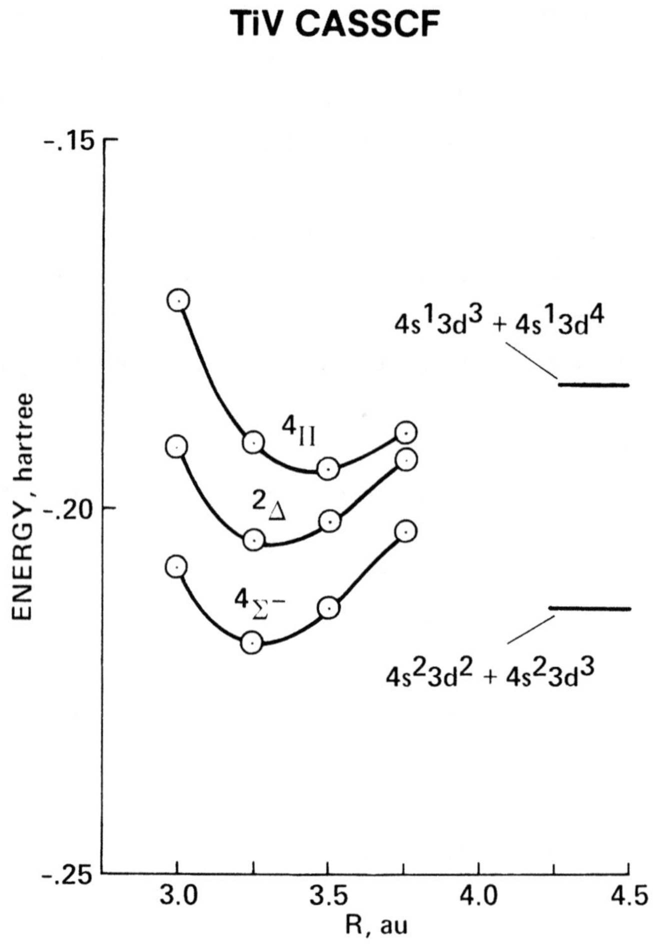

Figure 4. Calculated CASSCF potential curves for selected states of TiV.

sistent with $^4\Sigma^-$ [36]. The calculated bond length of TiV is intermediate between the bond lengths of Ti_2 and V_2.

Table 2 also shows results for the $^1\Sigma_g^+$ state of Cu_2. The electronic configuration here is:

$$^1\Sigma_g^+ \; 4s\sigma_g^2 \; 3d\sigma_g^2 \; 3d\sigma_u^2 \; 3d\pi_u^4 \; 3d\pi_g^4 \; 3d\delta_u^4 \; 3d\delta_g^4$$

This configuration arises from the $4s^13d^{10} + 4s^13d^{10}$ asymptote. The bonding here appears to involve mainly the 4s electrons with the 3d electrons remaining essentially atomic like [37,38]. However, this viewpoint has been challenged by Pauling who argues in favor of $4s^14p^13d^9$ hybrids [39]. A central concern for the Cu_2 calculations has been the calculation of the bond length. Here we find that correlation of the 3d electrons shortens the bond by 0.19 a_o leading to a final bond length of 4.35 a_o which is 0.15 a_o longer than experiment. It has been argued [37] that the remaining discrepancy is largely due to relativistic effects since twice the relativistic contraction of the 4s orbital for the $4s^13d^{10}$ state is 0.13 a_o. Since that time calculations have shown that only about 70% of the atomic core-valence contraction in the alkali dimers is observed in the molecular systems [40]. Martin [41] using an all electron treatment which includes relativistic effects as a perturbation and Laskowski et al. [42] who have added relativistic effects through an effective core potential approach, both compute the contraction to be 0.10 a_o, or about 70% of the atomic contraction. In both approaches [41,42] correlation does not affect this conclusion. More recently Martin and Werner [43] have considered the effect of higher excitations with CEPA, in addition to the relativistic effects, and find a bond length in good agreement with experiment. It has been observed [44] that a POLCI treatment leads to a better D_e for Cu_2 than a SDCI treatment. However, it is not clear that such approximations can be used for those systems where there is d-d bonding.

5. SECOND ROW TRANSITION METAL DIMERS

Figure 5 shows calculated potential curves for selected states of

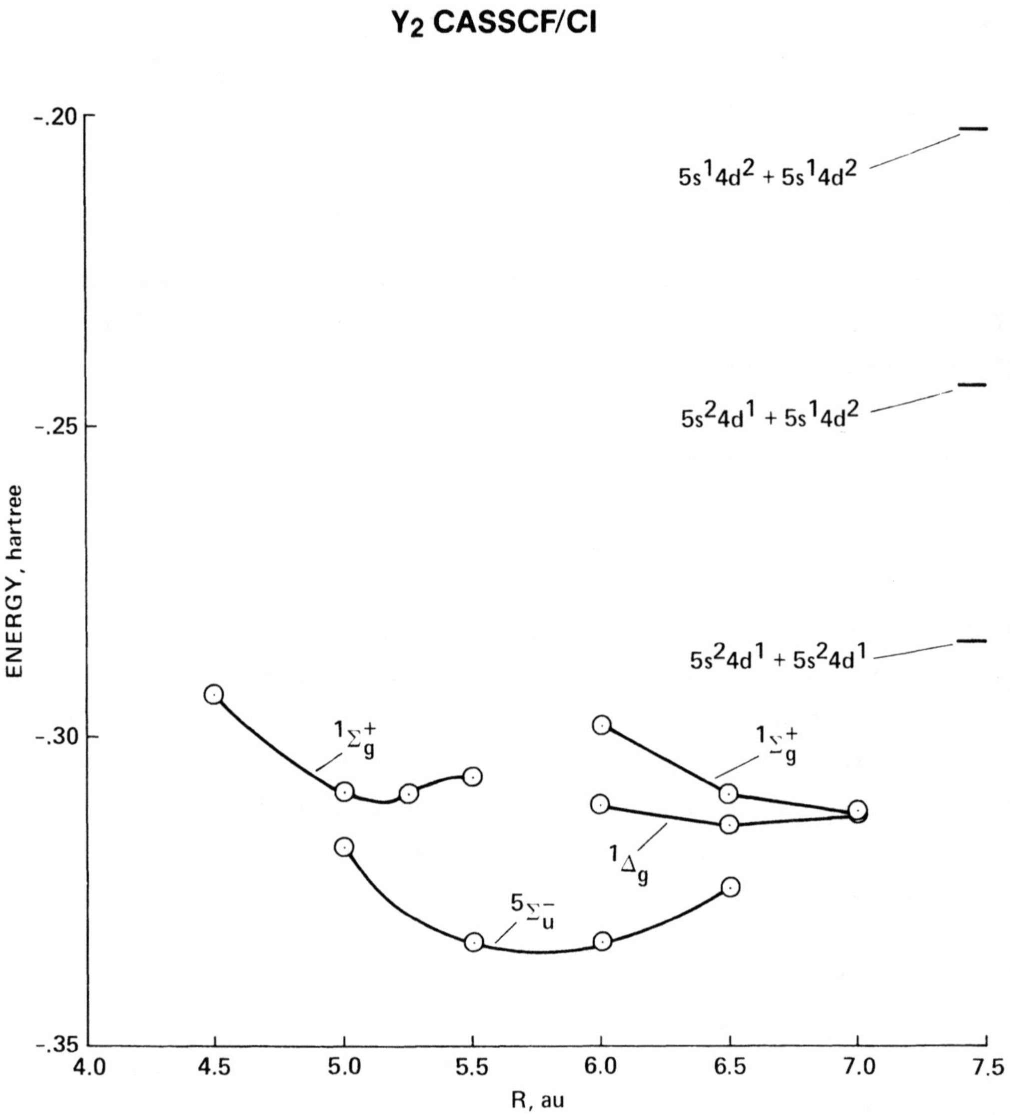

Figure 5. Calculated CASSCF/CI potential curves for selected states of Y_2.

Y_2. The dominant configurations for the states considered here are:

$^1\Sigma_g^+$ $5s\sigma_g^2$ $4d\pi_{xu}^2$ $4d\pi_{yu}^2$

$^5\Sigma_u^-$ $5s\sigma_g^2$ $5s\sigma_u^1$ $4d\sigma_g^1$ $4d\pi_{xu}^1$ $4d\pi_{yu}^1$

$^1\Delta_g$ $5s\sigma_g^2$ $5s\sigma_u^2$ $(4d\pi_{xu}^2 - 4d\pi_{yu}^2)$

$^1\Sigma_g^+$ $5s\sigma_g^2$ $5s\sigma_u^2$ $4d\sigma_g^2$

The $^5\Sigma_u^-$ state here is directly related to the $^5\Sigma_u^-$ state of Sc_2 and arises from the $5s^24d^1 + 5s^14d^2$ atomic asymptote. The long R $^1\Sigma_g^+$ and $^1\Delta_g$ states arise from the $5s^24d^1 + 5s^24d^1$ asymptote. The long R $^1\Sigma_g^+$ state is directly related to the Sc_2 $^1\Sigma_g^+$ state, while the $^1\Delta_g$ state was not studied for Sc_2 but arises from the same configuration (π_u^2) as the $^3\Sigma_g^-$ state of Sc_2. A new state for Y_2 which was high in energy for Sc_2 is the short R $^1\Sigma_g^+$ state which arises from the $5s^14d^2 + 5s^14d^2$ atomic symptote. That this state is low-lying for Y_2 but not for Sc_2 reflects the significantly greater 4d bonding in the second transition row. An unusual feature of the short R $^1\Sigma_g^+$ potential curve is an apparent avoided crossing; the state which is crossing the short R $^1\Sigma_g^+$ state is a $^1\Sigma_g^+$ state derived from the same configuration as the $^1\Delta_g$ state. This $^1\Sigma_g^+$ state should lie slightly above the $^1\Delta_g$ curve. An interesting feature of the Y_2 potential curves as compared to the Sc_2 potential curves is the more comparable R_e values for states arising out of the three possible asymptotes. This result is consistent with the more comparable (n+1)s and nd orbital sizes for the second transition row as compared to the first transition row. Table 4 gives calculated spectroscopic constants for the second transition row dimers. No experimental data exist for Y_2. However, Weltner does not observe an ESR spectrum for Y_2 as was observed for Sc_2 [45]. One possible explanation for this is that the

Table 4. Calculated spectroscopic constants for second row transition metal dimers

Molecule	R_e(Å)	ω_e(cm^{-1})	D_e(eV)	T_e(eV)
Y_2 $^5\Sigma_u^-$				
CI	3.03	171.	2.44	0.00
Y_2 $^1\Sigma_g^+$				
CASSCF	2.73	205.	1.74	
CI	2.74	206.	2.93	0.87
Nb_2 $^3\Sigma_g^-$				
CASSCF	2.10	448.	2.24	0.00
Nb_2 $^1\Gamma_g$				
CASSCF	2.11	427.	2.15	0.09
Nb_2 $^3\Delta_g$				
CASSCF	2.01	501.	2.12	0.12
Nb_2 $^3\Phi_g$				
CASSCF	2.19	340.	1.28	0.96
Mo_2 $^1\Sigma_g^+$				
CASSCF	1.99	399.	0.77	
Exp.	1.93[a]	477.[a]	4.2 [b]	

[a] Ref. [47]
[b] Ref. [48]

$^1\Sigma_g^+$ state is actually the ground state for Y_2 even though the calculations place it 0.87 eV above the $^5\Sigma_u^-$ state. Here, as discussed above for Ti_2, we expect more extensive CI calculations to favor the $^1\Sigma_g^+$ state over the $^5\Sigma_u^-$ state. As was done for the first transition row, the T_e values are obtained by computing the binding energy for each state with respect to its corresponding asymptote and then positioning the asymptotes at the experimental separations. Thus we are including a correction for the error in the Y_2 asymptotic energies due to relativistic effects. Further experiments would be necessary to decide definitely whether $^1\Sigma_g^+$ is the ground state of Y_2. Here the most useful experimental result would be a measurement of R_e since R_e differs significantly between the short R $^1\Sigma_g^+$ and the $^5\Sigma_u^-$ state whereas the vibrational frequencies are not substantially different.

Figure 6 shows calculated potential curves for Nb_2 while the calculated spectroscopic constants are given in Table 4. The dominant configurations for the states considered here are:

$$^3\Sigma_g^- \quad 5s\sigma_g^2\ 4d\sigma_g^2\ 4d\pi_{xu}^2\ 4d\pi_{yu}^2\ 4d\delta_{xyg}^1\ 4d\delta_{(x^2-y^2)g}^1$$

$$^3\Delta_g \quad 5s\sigma_g^2\ 4d\sigma_g^1\ 4d\pi_{xu}^2\ 4d\pi_{yu}^2\ 4d\delta_{xyg}^2\ 4d\delta_{(x^2-y^2)g}^1$$

$$^3\Phi_g \quad 5s\sigma_g^2\ 4d\sigma_g^2\ 4d\pi_u^3\ 4d\delta_g^3$$

Here the $^3\Sigma_g^-$, $^1\Gamma_g$, $^1\Sigma_g^+$ and $^3\Delta_g$ states are directly related to the corresponding states of V_2. The $^3\Delta_g$ and $^3\Phi_g$ states arise by exciting an electron from $4d\sigma$ and $4d\pi$, respectively, into $4d\delta$. From Figure 6 one sees that the electronic structure of Nb_2 seems very similar to that V_2 in that the lowest calculated state is $^3\Sigma_g^-$ and there is a low-lying $^1\Gamma_g$ state. There are, however, two additional low-lying states $^3\Delta_g$ and $^3\Phi_g$. Here the $^3\Delta_g$ state is very close in energy to the $^3\Sigma_g^-$ state while the $^3\Phi_g$ state is 0.96 eV higher. The ordering of these states is consistent with the ordering of the $^4\Sigma^-$ and $^4\Pi$ states

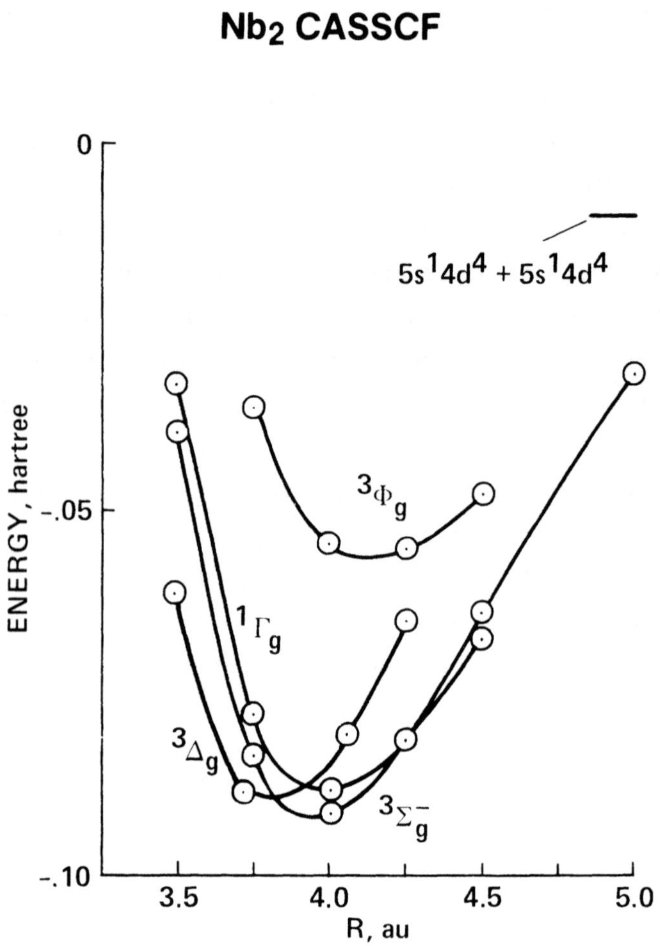

Figure 6. Calculated CASSCF potential curves for selected states of Nb_2.

of TiV. The increased stability of $^3\Delta_g$ for Nb_2 as compared to V_2 implies stronger 4dδ bonding in Nb_2 as compared to the 3dδ bonding in V_2. This is also consistent with the stronger bonding observed in Nb_2 (calculated D_e = 2.24 eV) as compared to V_2 (calculated D_e = 0.60 eV).

Experimentally very little is known about the electronic structure of Nb_2. Smalley has obtained the optical spectrum of Nb_2 but the spectrum shows little resolved structure and is too complex to enable analysis [46]. The calculations also show this increased complexity in the Nb_2 spectrum due to the presence of a low-lying $^3\Delta_g$ state in addition to the $^1\Gamma_g$ and $^1\Sigma_g^+$ states which were also present for V_2. Also for Nb relativistic effects are much larger than for V and may increase the coupling of states. As for Y_2 some experimental information would be very useful here, especially a determination of R_e which could help to distinguish between the $^3\Delta_g$ and $^3\Sigma_g^-$ states as candidates for the ground state of Nb_2.

In order to compare Cr_2 and Mo_2 the ground $^1\Sigma_g^+$ state of Mo_2 was considered at the CASSCF level. This state has the configuration

$$^1\Sigma_g^+ \quad 5s\sigma_g^2 \; 4d\sigma_g^2 \; 4d\pi_{xu}^2 \; 4d\pi_{yu}^2 \; 4d\delta_{xyg}^2 \; 4d\delta_{(x^2-y^2)g}^2$$

which is directly related to the ground state configuration of Cr_2. Mo_2 differs from Cr_2 in that the CASSCF curve is bound due to increased nd bonding for the second transition row. The calculated spectroscopic constants here are (experimental values [47,48] in parenthesis) R_e = 1.99Å (1.93Å), ω_e = 399 cm^{-1} (477 cm^{-1}) and D_e = 0.77 eV (4.2±0.2 eV).

6. THE DESCRIPTION OF ELECTRON CORRELATION IN Cr_2

As indicated in Section 4, Cr_2 might be viewed as a failure of the CASSCF procedure since only a shoulder is obtained in the region where one expects an inner well (3d bonding region). However, the CASSCF procedure with this choice of active space is only a first order description of Cr_2 which allows the spin recoupling of the 4s and 3d electrons from the large R high spin atomic coupling to the region near R_e where the system can be described formally as a hextuple bond, but does not include all of the molecular correlation effects. The particular features of Cr_2 which make the description of the region near R_e so difficult are i) that we are dealing with a quintuple bond and ii) that the 3d overlaps are smaller even near R_e than would be the case for, say, 2p orbitals in a first row molecule. The result is that Cr_2 is a molecule which requires a complex MCSCF zero order description involving very high orders of excitation with respect to a single configuration SCF reference.

In order to quantify this somewhat we carried out small valence CI calculations among the active orbitals from the CASSCF calculation (for R = 3.25 a_o). Here we allowed single and double excitations from various subsets of the most important configurations in the CASSCF wavefunction. The first calculation, which used as reference configurations the 16 configurations which are doubles or less from SCF and thus included up to quadruples with respect to SCF, led to a valence energy 2.42 eV above the CASSCF energy. We then added all the quadruples which are important in the CASSCF (16 more configurations) so that we are including excitations through hextuples and obtained an energy 0.70 eV above the CASSCF energy. Finally we added the hextuple excitations (six more configurations) which are important in the CASSCF so that we are now including through octuple excitations and the resulting energy was still 0.42 eV above the CASSCF energy. From the analysis of the configurations which need to be included to get a reasonable valence energy we see that the reference configurations must include quadruple excitations among the 3dδ orbitals and double excitations among the 3dσ and 3dπ orbitals but that the 4s bonding orbital may remain inactive (i.e., doubly occupied in all reference configurations). This result is in line with the relative overlaps of these orbitals and suggests that the 3dδ orbitals are still in the small overlap region even near R_e. One also expects that V_2 would be easier to describe than Cr_2 since the 3dδ orbitals are singly occupied.

The CASSCF procedure used does not include all of the important molecular correlation effects. For the TM atoms the most important additional correlation effects are derived from the 4p, 3d′ (here 3d′ is a tight diffuse correlating orbital for the 3d) and 4f orbitals and we expect the most important additional molecular correlation terms to involve these orbitals. One approach to introducing these missing correlation effects would be the direct use of multi configuration singles and doubles CI. However, if we use the CASSCF wavefunction (3088 configurations) as a reference, the CI expansion is 57 million configurations for a [8s6p4d2f] basis set. However, given the difficulty in reproducing the CASSCF energy for smaller numbers of reference configurations it is clear that any CI calculation with fewer references than the CASSCF wavefunction will have to be carried out with caution.

Given these difficulties, several alternative methods have been tried for introducing the missing molecular correlation in approximate ways. By far, the most extensive study of additional correlation effects and additivity of correlation effects for Cr_2 is that of Walch [12]. In the extended CASSCF calculations of Walch, the 3d′ shell and 4p shell are added to the CASSCF calculation. This approach would be rigorous if one could include all the additional active orbitals in one calculation. Unfortunately, this is not possible even with orbital occupation constraints and a procedure was used in which the extra active orbitals were added by symmetry block and the separately obtained energy contributions were summed to obtain an estimate of the extra molecular correlation. The problem with the method is that it assumes that the atomic correlation is constant with R, however, "atomic correlation" is reduced as the bonds form. For Ti_2 calculations can be carried out in which all the 3d′ orbitals are included simultaneously and it is found that the extended CASSCF procedure overestimates the effect of 3d′ by a factor of four due mostly to the variation with R of the 3d′ atomic correlation terms. On the other hand comparisons of extended CASSCF calculations and CI calculations for N_2 and Ti_2 indicate that the approach of summing the individual contributions does not give a potential curve that is deeper than the

CI curve and that the overestimation of the effect of 3d´ does not affect the R_e or ω_e significantly. Thus, we conclude that the reasonable potential curve obtained for Cr_2 by Walch (see Figure 3) must arise at least in part by cancellation of errors, i.e. the overestimation of the effect of 3d´ is being compensated by the omission of other molecular correlation effects such as correlation involving 4f. The calculated spectroscopic constants are (experimental values [34, 49-51] in parenthesis) R_e = 1.78Å (1.68Å), ω_e = 383 cm^{-1} (480 cm^{-1}), and D_e = 0.71 eV (1.56 eV). These results, while in part due to cancellation of errors, do indicate the importance of 3d´ and 4p for more extensive CI calculations.

Another approach due to Das and Jaffe [13], is in principle equivalent to the extend CASSCF approach used by Walch. These workers use symmetrically orthogonalized atomic orbitals (called PLO's) in an attempt to greatly reduce the size of the MCSCF expansion. Thus, Das and Jaffe did not attempt to reproduce the valence correlation in the CASSCF calculations [9], but rather computed additional correlation effects arising from the 3d´, 4p, 4f, and 3p shells and then added them to our CASSCF potential curve. They predict that 3d´ and also 3p shell excitations make large contributions to the binding energy. Their 3d´ and 4p correlation effects are much smaller than those determined by Walch and do not result in a well for Cr_2 after the other correlation terms are added. Due to limitations in the CI expansion and the small basis used, their conclusions while interesting are somewhat tentative.

A computationally very different approach due to Goodgame and Goddard [14] assumes that the missing correlation serves mainly to correct the location of the ionic atomic asymptotes. Those authors attempt to include these effects by empirical modification of the integrals to correct the atomic ionization potentials and electron affinities to agree with experiment. This method does lead to a reasonable potential curve for Cr_2 although the bond length is somewhat too short which suggests that this method over corrects to some extent.

The main value of these approximate approaches is expected to be to lead to a better understanding of the additional correlation effects needed to compute a reliable potential curve for Cr_2. In the opinion of the present authors the most promising theoretical method for actually doing the large CI calculations necessary to solve the Cr_2 problem is the externally contracted CI method [52]. These calculations will still be difficult due to the need for an extensive reference wavefunction leading to a few million configurations. However, it is probable that these calculations will be carried out within the next few years.

7. CONCLUSIONS

The CASSCF procedure is found to generally provide reliable R_e and ω_e values. However, D_e values are systematically underestimated at the CASSCF level; the worst case is Cr_2 where the CASSCF curve is not bound although a shoulder is observed in the region near the experimental R_e. The CASSCF procedure provides a consistent set of calculations for these molecules from which qualitative trends and a picture of the electronic structure may be derived.

The CASSCF/CI procedure has lead to confirmation of the $^5\Sigma_u^-$ ground state of Sc_2 and the $^4\Sigma^-$ ground state of TiV. Numerous predictions remain untested. Among these are the prediction that Ti_2 has a $^1\Sigma_g^+$ ground state with a short bond length and essentially all the computed results for Y_2 and Nb_2. The computed results for Nb_2 are especially interesting since the experimental spectrum for Nb_2 is so complex that it has not been analyzed. Experiment has confirmed the prediction of the ground state symmetry and bond length of V_2.

In addition to specific predictions, these studies yield a simple qualitative picture of the bonding in the transition metal dimers. The $4s^2 3d^m + 4s^2 3d^m$ asymptote leads to Van der Waals bonding at large R due to the $4s \rightarrow 4p$ near degeneracy effect. Here the 3d electrons are only weakly coupled. These weakly bound states have not been observed in experiment, and probably are not of chemical interest. If the two atoms have the occupation $4s^1 3d^{m+1}$, strong two-electron d-d bonds (i.e. two electrons per bond orbital as in H_2, a bond order of one) and a short bond length results. Such bonding has been observed for $Ti_2\ ^1\Sigma_g^+$, $V_2\ ^3\Sigma_g^-$, and $Cr_2\ ^1\Sigma_g^+$. When the excitation energy is very large the cost of promoting both atoms to the occupation; $4s^1 3d^{m+1}$ is so large that bonding arises from the mixed asymptote; $4s^1 3d^{m+1}$ + $4s^2 3d^m$. Such bonding has been found for $Sc_2\ ^5\Sigma_u^-$ and $Ti_2\ ^7\Sigma_u^+$. In these states there are three 4s electrons, or a bond order of one half. These states also have long bond lengths, perhaps because of the 4s repulsion, which favor one-electron bonds. Such one-electron bonds (i.e. one electron per bond orbital as in H_2^+, a bond order of one half) naturally retain the atomic high spin coupling to maximize the exchange interaction.

The strengths of the nd bonds are determined largely by overlap considerations. Since the relative overlaps of nd orbitals are $nd\sigma \approx nd\pi > nd\delta$, the $nd\sigma$ and $nd\pi$ orbitals are filled before the $nd\delta$ orbitals. For the first transition row, the $d\delta$ overlap is sufficiently small that two-electron $d\delta$ bonds do not contribute strongly to the binding. For example V_2 has both $d\sigma$ and $d\pi$ two-electron bonds and two $d\delta$ one-electron bonds. In Cr_2 where the two additional electrons are added to the $d\delta$ orbital, instead of increased binding, the D_e is smaller for Cr_2 than V_2. This arises because the small overlap of the $d\delta$ orbitals leads to such weak bonds that they do not compensate for the atomic d-d exchange terms which are lost to form the bond. The loss of these atomic d-d exchange terms is quite important in describing weak bonds. The problems in treating Cr_2 correctly are related to this balance of weak two electron $d\delta$ bonds and the atomic d-d exchange interactions. Since this problem is not present in the other dimers considered, our CASSCF treatment fails only for Cr_2.

There are several differences between the first and second transition row: i) $\langle r_{5s}\rangle/\langle r_{4d}\rangle$ is smaller than $\langle r_{4s}\rangle/\langle r_{3d}\rangle$ for comparable locations in the first row ii) the exchange interactions are smaller for 4d than for 3d and iii) the $5s^2 4d^m \rightarrow 5s^1 4d^{m+1}$ excitation energy is smaller than the $4s^2 3d^m \rightarrow 4s^1 3d^{m+1}$ excitation energy. These features lead to stronger 4d bonding than 3d, with a notable increase in the strength of the dδ bond and a reduced significance of the mixed asymptote. There are several examples of these changes. Y_2 has a low-lying $^1\Sigma_g^+$ state, which is a very likely candidate for the ground state, arising from $5s^1 4d^2 + 5s^1 4d^2$. Mo_2 has a larger D_e than Cr_2 and the CASSCF treatment is now able to describe this bonding. Nb_2 has a low-lying $^3\Delta_g$ state ($4d\sigma_g^1 4d\pi_u^4 4d\delta_g^3$) which arises by moving a 4dσ electron into a 4dδ orbital, while the corresponding state is 0.92 eV up for V_2.

References:

1. W. Weltner, Jr. and R.J. VanZee, Ann. Rev. Phys. Chem., 35, 291 (1984).
2. See articles in "Recent Developments and Applications of Multi Configuration Hartree-Fock Methods," NRCC Proceedings No. 10, M. Dupuis, editor, NRCC, 1981.
3. P.E.M. Siegbahn, A. Heiberg, B.O. Roos and B. Levy, Physica Scripta, 21, 323 (1980); B.O. Roos, P.R. Taylor, P.E.M. Siegbahn, Chem. Phys. 48, 157 (1980); P.E.M. Siegbahn, J. Almlof, A. Heiberg and B.O. Roos, J. Chem. Phys. 74, 2381 (1981).
4. H. Lischka, R. Shepard, F.B. Brown and I. Shavitt, Int. J. Quantum Chem. Symp. 15, 91 (1981); P.E.M. Siegbahn, J. Chem. Phys. 72, 1647 (1980); B. Liu and M. Yoshimine, J. Chem. Phys. 74, 612 (1981); P. Saxe, D. Fox, H.F. Schaefer III and N.C. Handy, J. Chem. Phys. 77, 5584 (1982).
5. W.A. Goddard, III, T.H. Dunning, Jr., W.J. Hunt and P.J. Hay, Acc. Chem. Res., 6, 368 (1973); T.H. Dunning, Jr., in "Advanced Theories and Computational Approaches to the Electronic Structure of Molecules," edited by C.E. Dykstra, Reidel Publishing (1983), p. 67.
6. M.R.A. Blomberg, U.B. Brandemark and P.E.M. Siegbahn, preprint communicated to authors.
7. A.E. Stevens, C.S. Feigerle and W.C. Lineberger, J. Amer. Chem. Soc., 104, 5026 (1982).
8. M.M. Goodgame and W.A. Goddard, III, Phys. Rev. Lett. 48, 135 (1982); J. Phys. Chem. 85, 215 (1981).
9. S.P. Walch, C.W. Bauschlicher, Jr., B.O. Roos and C.J. Nelin, Chem. Phys. Lett., 103, 175 (1983).
10. I. Shim and K. Gingerich, J. Chem. Phys., 77, 2490 (1982); J. Chem. Phys. 78, 5693 (1983).
11. F.A. Cotton and I. Shim, J. Amer. Chem. Soc. 104, 7025 (1982).
12. S.P. Walch, to be published.
13. G.P. Das and R.L. Jaff, Chem. Phys. Lett., 109, 206 (1984).

14. M.M. Goodgame and W.A. Goddard III, paper presented at the Sixth West Coast Theory Conference, Los Alamos, New Mexico, April 1984.
15. The experimental atomic separations are taken from Ref. [16], while the orbital sizes are from numerical Hartree-Fock calculations.
16. C.E. Moore, Atomic energy levels, Natl. Bur. Stand. (US) circ. 467 (1949).
17. S.P. Walch and C.W. Bauschlicher, Jr., Chem. Phys. Lett. 94, 290 (1983).
18. R.L. Martin and P.J. Hay, J. Chem. Phys. 75, 4539 (1981).
19. P.J. Hay, J. Chem. Phys. 66, 4377 (1977).
20. A.J.H. Wachters, J. Chem. Phys. 58, 4452 (1973).
21. A.D. McLean and B. Liu, Chem. Phys. Lett. 101, 144 (1983).
22. S. Huzinaga, J. Chem. Phys. 66, 4245 (1977).
23. S.P. Walch, C.W. Bauschlicher, Jr. and C.J. Nelin, J. Chem. Phys. 79, 3600 (1983).
24. J. Almlof, MOLECULE, a gaussian integral program.
25. P.E.M. Seigbahn, C.W. Bauschlicher, Jr., B. Roos, A. Hieberg, P.R. Taylor and J. Almof, SWEDEN, a vectorized SCF MCSCF direct CI.
26. R.C. Raffenetti, BIGGMOLI, Program No. 328, Quantum Chemistry Program Exchange, Indiana University, Bloomington (1977), with modifications to allow use of the full D_{2h} symmetry.
27. J. Drowart and R.E. Honig, J. Phys. Chem. 61, 6801 (1957).
28. L.B. Knight, R.J. VanZee and W. Weltner, Jr., Chem. Phys. Lett. 94, 296 (1983).
29. D.P. DiLella, W. Limm, R.H. Lipson, M. Moskovits and K.V. Taylor, J. Chem. Phys. 77, 5263 (1982).
30. S.P. Walch and C.W. Bauschlicher, Jr., J. Chem. Phys. 79, 3590 (1983).
31. A. Kant and S.H. Lin, J. Chem. Phys. 51, 1644 (1965).
32. P.R.R. Langridge-Smith, M.D. Morse, G.P. Hansen, R.E. Smalley and A.J. Merer, J. Chem. Phys. 80, 593 (1984).
33. A. Merer, private communication.
34. A. Kant and B. Strauss, J. Chem. Phys. 45, 3161 (1966).
35. S.P. Walch and C.W. Bauschlicher, Jr., J. Chem. Phys. 78, 4597 (1983).
36. R.J. VanZee and W. Weltner, Jr., Chem. Phys. Lett. 107, 173 (1984).
37. C.W. Bauschlicher, Jr., S.P. Walch and P.E.M. Siegbahn, J. Chem. Phys. 76, 6015 (1982).
38. C.W. Bauschlicher, Jr., S.P. Walch and P.E.M. Siegbahn, J. Chem. Phys. 78, 3347 (1983).
39. L. Pauling, J. Chem. Phys. 78, 3346 (1983).
40. H. Partridge, C.W. Bauschlicher, Jr., S.P. Walch and B. Liu, J. Chem. Phys. 79, 1866 (1983).
41. R. Martin, J. Chem. Phys. 78, 5840 (1983).
42. B. Laskowski, C.W. Bauschlicher, Jr. and S.R. Langhoff, to be published.
43. R. Martin and H.J. Werner, private communication.
44. C.W. Bauschlicher, Jr., Chem. Phys. Letters 97, 204 (1983).

45. L.B. Knight, R.W. Woodward, R.J. VanZee and W. Weltner, Jr., J. Chem. Phys. 79, 5820 (1983).
46. R.E. Smalley, Paper presented at the 1984 ACS Annual Meeting, Philadelphia, Pennsylvania, August, 1984.
47. Y.M. Efremov, A.N. Samoilova, V.B. Kozkukhowsky and L.V. Gurvich, J. Mol. Spect. 73, 430 (1978).
48. S.K. Gupta, R.M. Atkins and K.A. Gingerich, Inorg. Chem. 17, 3211 (1978).
49. D.L. Michalopoulos, M.E. Geusic, S.G. Hansen, D.E. Powers and R.E. Smalley, J. Phys. Chem. 86, 3914 (1982).
50. V.E. Bondybey and J.H. English, Chem. Phys. Lett. 94, 443 (1983).
51. S.J. Riley, E.K. Parks, L.G. Pobo and S. Wexler, J. Chem. Phys. 79, 2577 (1983).
52. P.E.M. Siegbahn, Int. J. Quantum Chem. 23, 1869 (1983).

SUPERSONIC CLUSTER BEAMS: AN ALTERNATIVE APPROACH TO SURFACE SCIENCE

R.E. Smalley
Rice Quantum Institute and
Department of Chemistry
Rice University
Houston, Texas 77251

ABSTRACT. Rapid advances in the art of generating and probing supersonic metal cluster beams are beginning to show the outlines of a new approach to surface science. The new approach involves the study of small, bare clusters in the high vacuum of a molecular beam apparatus in much the same way that the more conventional type of surface science uses perfect single crystals in an ultra high vacuum surface machine. In both cases the object of study is a highly idealized model of the real polycrystalline surfaces of practical importance, but in the case of the new cluster beam approach, this model has the advantage of a far more immediate connection to high level *ab initio* theory. As an example of this new molecular cluster approach to surface science, dissociative chemisorption experiments are described for D_2 and N_2 on small clusters of a variety of transition metals.

One of the most difficult problems with the study of surface science as it is currently practiced is that there is a rather weak coupling between theory and experiment. For the most part, the surface experiments most likely to probe the fundamental details of the science are carried out on macroscopic single crystals at surface sites with little or no symmetry. Theory, on the other hand, has great difficulty in coping with macroscopic arrays of atoms without extensive symmetry. Drastic simplifications must be made,

R. J. Bartlett (ed.), Comparison of Ab Initio Quantum Chemistry with Experiment for Small Molecules, 53–65.

such as considering only a several-layer-thick two dimensional slab at a fairly low level of theory, or just a few atoms treated at a very high level of theory in a small cluster chosen to mimic the presumed active site under study on the real surface. In either case, direct comparison between theory and experiment is rarely a high-tension experience for the scientists involved.

Major advances in the understanding of nature have traditionally occurred when theory and experiment reached a stage of real intellectual tension -- a time when clean-cut, definitive experiments were possible in a subject area ripe for an equally clean and sure theoretical approach. At such a time disagreements between theory and experiment have more than casual significance: they can change minds and force the creation of dramatically new concepts. In this respect, surface science is still quite a young field. Most experiments are of a descriptive nature, and most theory is only intended to be suggestive.

Within the past several years a rapid series of advances has occurred in the art and science of studying bare metal clusters in the rarified environment of a molecular beam [1-4]. Increasingly, it is becoming clear that these bare cluster beams can provide a new view of surface science, an alternative to the more traditional approach of studying bulk single crystal surfaces, and one which is nearly perfect for intimate coupling to theory. In a sense, these metal cluster beam techniques are a means of getting away from the macroscopic, low symmetry aspects of surfaces while still retaining the heart of what's new in physics and chemistry of surface science -- it brings surface science experiments into the microscopic, molecular realm ideal for theory.

This realization that metal clusters may have a vital impact as models of surfaces is hardly new. Muetterties [5], Messmer [6] and Ozin [7] have been particularly avid champions of this cause, and the core idea is likely to be considerably older [8]. What is new is that the requisite beam techniques are only just now approaching the level of power necessary to bring the idea to life.

As an example of this new sort of cluster surface science, this paper will review a recent development which allows the surface reactivity of bare clusters to be studied for the first time in the gas phase as a function of cluster size. Previously, the dominant technique for such studies involved the use of rare-gas matrices with spectral probes of the chemisorbed reaction products [7, 9-13]. This matrix isolation technique has great advantages (particularly in the use of infrared and various spin resonance probes) and a great deal has been learned. However, it has proved to be virtually impossible to extend its application beyond the study of reactive atoms and a few diatomics. There just isn't any sure way to distinguish clearly among metal clusters with more than a few atoms when these clusters are imbedded in a matrix. Their spectra are too similar.

A second fundamental limitation has been that metal cluster reactions in cryogenic matrices must occur at low temperatures, or be photochemically induced. As a consequence, metal cluster studies in matrices have never really lived up to their original promise as an entry point to cluster surface science. This is by no means a failure of the practitioners in the matrix isolation field; it is simply an unavoidable weakness of the technique.

In contrast, the supersonic metal cluster beam is an excellent environment to study the surface reactions of small clusters. Clusters in the size range between 2 and 100 atoms can now be routinely generated for virtually any element in the periodic table, and cooled to near absolute zero in a supersonic beam [1]. Identification of the cluster size and constitution is readily obtained by near-threshold direct one-photon laser ionization with time-of-flight mass spectrometric (TOFMS) detection [14]. In order to provide the desired reactive environment for the clusters, a special fast-flow reactor can be fitted to the supersonic nozzle so that the reactions occur over a known time at near room temperature in a bath of roughly 100 torr of helium buffer gas [15]. Figure 1 shows a schematic of such a metal cluster fast-flow reactor. Once

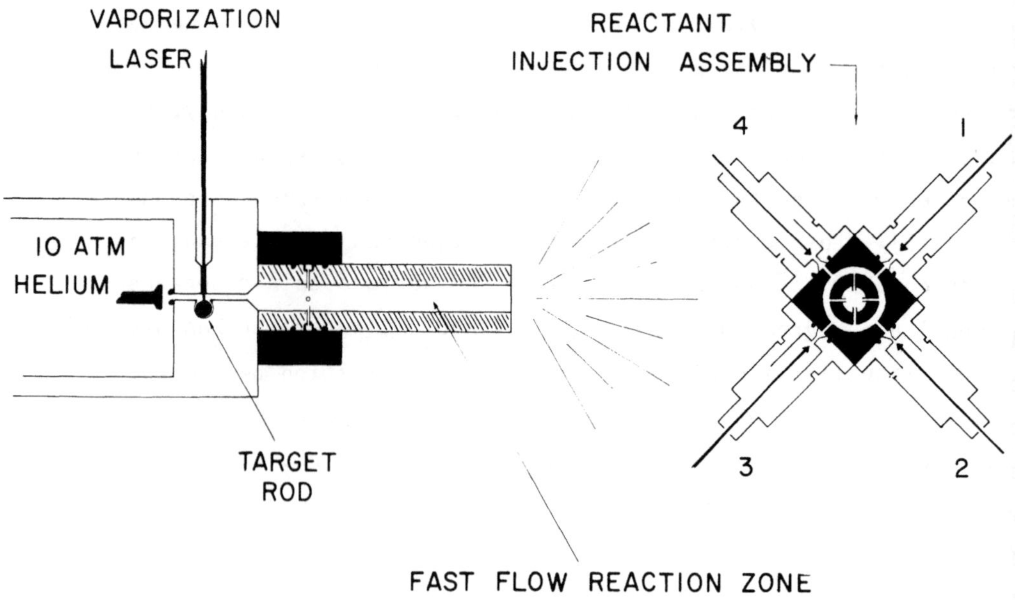

Figure 1. Schematic of gas-phase metal cluster reaction device. Metal clusters are generated by focussing a 30-40 mj, 6 ns duration green laser beam (Nd:YAG 2nd harmonic) to a 1 mm dia. spot on a pure target rod of the desired metal. Helium carrier gas is pulsed over this rod such that the density above the rod at the time of vaporization is 1-2 atm. Metal clusters in the 2-100 atom size range are formed and thermalized in the near-sonic flow of this helium carrier as it passes down a 0.2 cm diameter, 1.8 cm long tube (shown here in cross-section). This helium + metal cluster mixture then flows through a 1 cm dia., 10 cm long reaction tube into which reactants are injected through four needles. As seen in the end view, these four needles are fed from an annular ring which, in turn, receives reactant gas from any of four independent pulsed solenoid valves.

the reactants and reaction products flow to the end of the reaction tube, there is still enough helium density to provide a very intense supersonic expansion so that the contents of the reaction tube are formed into a very cold beam, ideal for detailed laser probes of the extent of reaction and the nature of the reaction products themselves -- all as a function of cluster size.

These cluster reactivity experiments with the fast-flow reactor offer an excellent example of the direct parallel that is developing between UHV single crystal surface science and the new study of bare clusters in high vacuum beam machines. In UHV single crystal experiments, one of the most productive means of studying surface reactivity is to first very carefully prepare a clean, well-characterized surface in UHV, and then translate the crystal into a high pressure reaction vessel where the chemistry is done. Analysis of the results of this high pressure chemistry is then accomplished by returning the crystal to the UHV chamber again, where the various powerful new surface spectroscopies are used to see what was left on the surface.

In the cluster beam experiments roughly the same process is followed. The clusters are prepared by condensation of laser-vaporized metal atoms in the roughly 2 atm of helium carrier gas flowing over the metal target rod shown in the diagram of Figure 1. The cluster-laden helium carrier gas then flows down the reaction tube and freely expands out the exit end to form a supersonic beam. Direct photoionization with an F_2 excimer laser is then used downstream to determine the nature of the metal clusters.

As shown in the Figure, there are a set of four needles near the upstream end of the reaction tube, and these can be fed from any one of the four independent pulsed valves. The cluster beam equivalent of the first step (preparation and characterization) in UHV single crystal reaction studies is to run a control pulse where only pure helium is injected through the needles into the reaction tube. The equivalent of exposing the single crystal surface to reactants at high pressure is to follow this control pulse with one where the desired reactant is injected through the needles into the fast-flow reactor. Clusters with the chemisorbed species attached then travel through the reaction tube and are transported into the high vacuum TOFMS detection chamber by the supersonic molecular beam. Since this process takes only a millisecond (as opposed to the minutes required in UHV experiments) the vacuum requirements in the cluster beams experiments are far more modest than what is routinely necessary in single crystal work. In practice, the cluster nozzle is fired at a rate of 10 pulses per second with control (no reactant) pulses alternating with pulses with the reactant present. The TOFMS cluster beam signal is accumulated under computer control, keeping the data from the control and reaction shots in two separate arrays.

As a result, it is possible to mimic almost perfectly the high pressure reaction experiments done with single crystals. For the reaction tube design of Figure 1, the effective total gas density is in the 50-100 torr range, and the clusters are exposed to this pressure at near room temperature (actually 320K) for a contact time of 150-200 microseconds. Slightly different reactor designs could be made to change these parameters through rather broad limits.

An example of the application of this new technique is shown in Figure 2, which displays the control TOFMS cluster distribution for 4-29 atom cobalt clusters along with experiments performed on alternate shots with D_2 reactant added at two different concentrations [16,17]. Even a quick glance shows there is a dramatically abrupt onset of reactivity at Co_{10}. This high reactivity continues throughout the size range between 10 and 18, followed by a short range of comparatively unreactive clusters (Co_{19} through Co_{22}). Careful examination of the bottom trace shows that at high D_2 concentration, the cobalt clusters in the 10-18 range have effectively been titrated by the D_2, picking up a maximum number of D_2 molecules ranging from five for the smaller clusters to 7 D_2 molecules for Co_{18}.

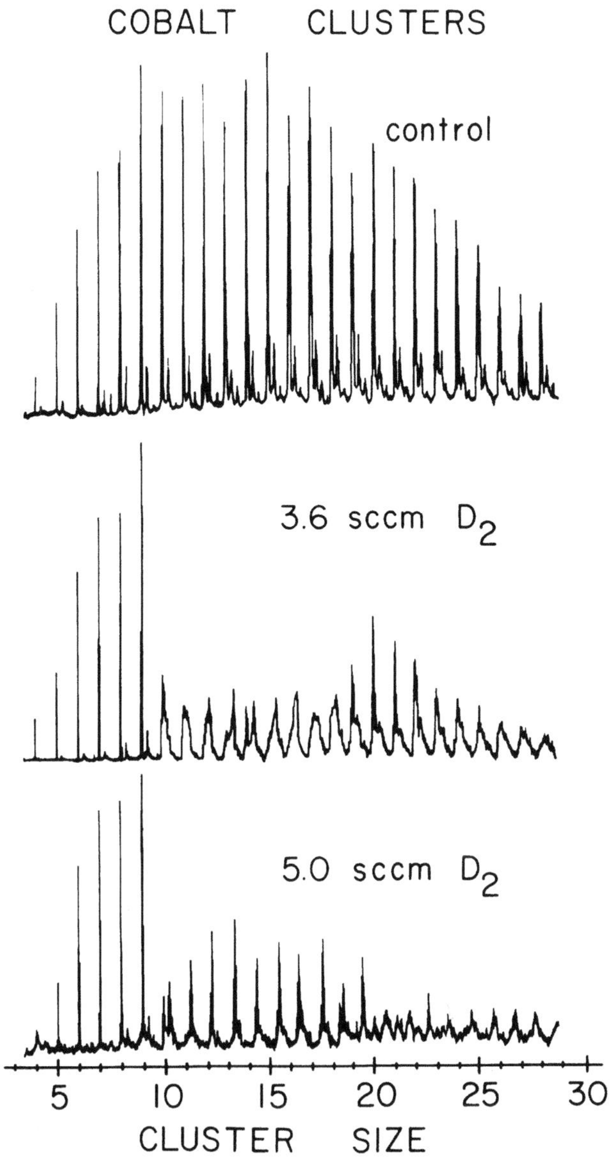

Figure 2. Chemisorption study of D_2 on cobalt clusters. The control mass spectrum was performed with only pure helium injected in as the reactant gas. The lower two mass spectra were taken with 3.6 sccm and 5.0 sccm flow of injected D_2 reactant, respectively. The sharp peaks seen in the bottom-most trace for clusters with more than 10 atoms are all due to cobalt clusters with more than one molecule of D_2 chemisorbed.

Figure 3 shows the results of a similar experiment carried out with N_2 on the cobalt clusters [18]. Notice that the reactivity variations are even more pronounced as a function of cluster size than they were with D_2, but the overall reactivity pattern is still basically the same. This similarity of reactivity patterns is quite graphically brought out in Figure 4 which shows plots of the measured relative reactivity as a function of cluster size for these D_2 and N_2 reactions on the surface of cobalt clusters.

To date quite a variety of transition metal clusters have been examined with this new fast-flow reactor technique for both D_2 and N_2 reactions, and we have yet to encounter the same reactivity pattern twice. Each metal is different [17]. Niobium shows even more pronounced reactivity fluctuations than shown here for cobalt clusters, but in the case of niobium, it is clusters 8, 10, and 16 that are particularly unreactive. Iron shows something of a "W" shaped reactivity pattern with minima at Fe_6 and Fe_{17} and a central maximum at Fe_{10}. Nickel, unfortunately, shows only a mild monotonic rise in reactivity toward D_2 chemisorption as a function of cluster size. As expected from bulk copper surface studies [18], copper clusters were found to be totally unreactive under these room temperature conditions.

Both in the case of D_2 and for the N_2 chemisorption as well, the fact that the cluster reactions are occuring at 320K in a 50-100 torr reaction tube for 150-200 microseconds indicates rather firmly that we are dealing with a dissociative chemisorption event. One of the classic techniques of single crystal UHV chemisorption studies is fast thermal desorption spectroscopy (TDS) which permits a rough estimate of the surface binding energies of various chemisorbed species. In the case of hydrogen and nitrogen, there are no clearly identified molecular chemisorption states known through TDS techniques to survive heating to room temperature on any transition metal surface. This is, of course, only circumstantial evidence, but it is particularly firm in the case of hydrogen, and it will have to suffice until infrared spectral techniques are developed

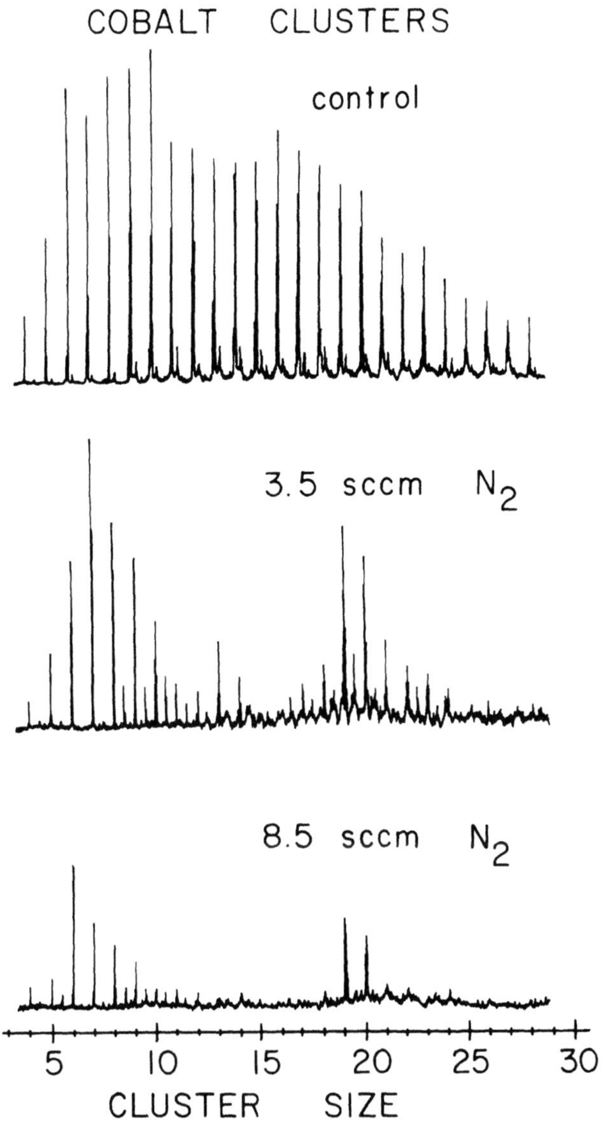

Figure 3. Chemisorption study of N_2 on cobalt clusters. Careful examination of the mass spectra (particularly the middle trace taken with 3.5 sccm flow of N_2) reveals features due to reaction products with one or more N_2 molecules chemisorbed on the cobalt clusters. The higher concentration trace at the bottom is a particularly graphic demonstration of the comparative inertness of Co_{19} and Co_2.

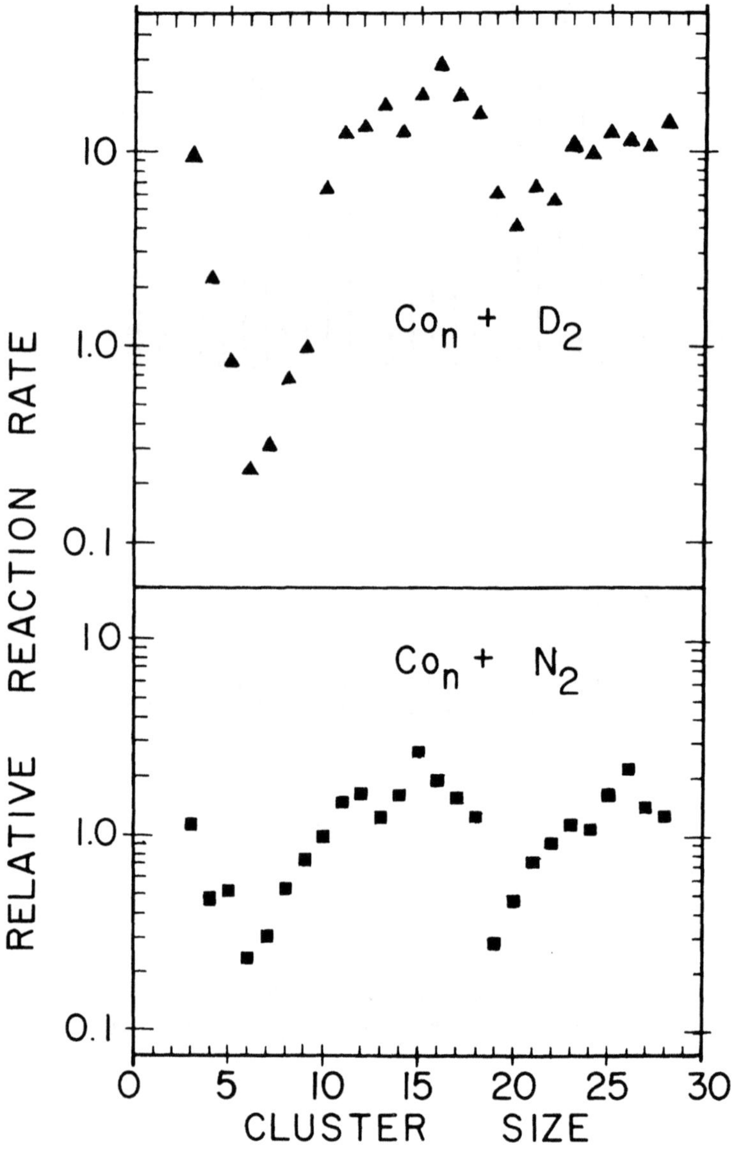

Figure 4. Relative reaction rates of cobalt clusters with D_2 and N_2. Note the striking similarity in the reactivity patterns for these two reactants. The estimated error for these relative rates is ±20% except for rates greater than 10 or less than 0.1 compared to the reference rate; for these extreme rates the measurements are only of semi- quantitative significance.

for the cluster beam and the actual positions of the chemisorbed atoms are determined. Isotope scrambling experiments could be performed, but the observation of scrambling would not necessarily rule out a weak bond remaining between the chemisorbed hydrogen or nitrogen atoms.

One of the most troublesome of the possible complications to this new cluster beam approach to surface science is the question of structure. Since the clusters are generated by extremely rapid nucleation of a highly supersaturated metal vapor at nearly room temperature, one can easily imagine there could be many configurations of the larger clusters frozen in during the formation stages. Certainly in the limit of very large clusters, one will encounter a dendritic growth mechanism where the internal energy liberated by the addition of each successive atom is insufficient to heat the cluster enough to anneal the most favored geometry. Each new atom simply sticks where it hits.

On the other hand, the very small clusters are expected to be quite fluxional at room temperature, and they may very well find it quite easy to relax to the most favored (closest packed?) geometry as they cool. One vital question is how large a cluster one can work with before this ready self-annealing process is no longer sufficient. In this connection, observation of such highly size-dependent properties as shown in Figures 2-4 for these cobalt clusters is rather comforting. In the long run, however, it will be critical to have reliable spectral measures of the structure(s) of these clusters as well as the binding energies and structures of the chemisorbed species on their surfaces.

In summary, new developments on the experimental front in the generation and use of metal cluster beams are beginning to flesh out a new experimental approach to surface science. Although the bare outlines of this new field are just emerging, a variety of tantalizing results are already available and it is reasonable to expect the field will become exceedingly rich within the next few years. If experimental developments continue without encountering a major block, the future of cluster beams could be quite exciting.

This new intrinsically microscopic, molecular approach to surface science has the potential of just the sort of high-tension relationship with <u>ab initio</u> theory that is so vital to the progress of science.

ACKNOWLEDGEMENT

The research described here is the work of many superb students and postdoctoral associates whose names are cited in the relevant references. Research on bare metal clusters in the author's laboratory is supported by the Department of Energy, Division of Chemical Sciences, together with the Robert A. Welch Foundation and the Exxon Education Foundation. Principal support for studies of nonmetal adducts of metal clusters is provided by the National Science Foundation. Acknowledgement is also made to the Donors of the Petroleum Research Foundation for partial support of this research.

References:

1. (a) T.G. Dietz, M.A. Duncan, D.E. Powers and R.E. Smalley, J. Chem. Phys. 74, 6511 (1981).
(b) D.E. Powers, S.G. Hansen, M.E. Geusic, A.C. Puiu, J.B. Hopkins, T.G. Dietz, M.A. Duncan, P.R.R. Langridge-Smith and R.E. Smalley, J. Phys. Chem. 86, 2556 (1982).
(c) J.B. Hopkins, P.R.R. Langridge-Smith, M.D. Morse and R.E. Smalley, J. Chem. Phys. 78, 1627 (1983).
(d) M.D. Morse and R.E. Smalley, Ber. Bunsenges. Phys. Chem. 88, 228 (1984).
2. (a) V.E. Bondybey and J.H. English, J. Chem. Phys. 76, 2165 (1982).
(b) J.L. Gole, J.H. English and V.E. Bondybey, J. Phys. Chem. 86, 2560 (1982).
(c) V.E. Bondybey and J.H. English, J. Chem. Phys. 80, 568 (1984).
3. (a) S.J. Riley, E.K. Parks, C.R. Mao, L.G. Pobo and S. Wexler, J. Phys. Chem. 86, 3911 (1982).
(b) S.J. Riley, E.K. Parks, L.G. Pobo and S. Wexler, J. Chem. Phys. 79, 2577 (1983).
(c) S.J. Riley, E.K. Parks, G.C. Nieman, L.G. Pobo and S. Wexler, J. Chem. Phys. 80, 1360 (1984).
4. (a) E.A. Rohlfing, D.M. Cox and A. Kaldor, Chem. Phys. Lett. 99, 161 (1983).

(b) E.A. Rohlfing, D.M. Cox and A. Kaldor, J. Chem. Phys. 81, 3322 (1984).
(c) E.A. Rohlfing, D.M. Cox and A. Kaldor, J. Phys. Chem. 88, 4497 (1984).
5. (a) E.L. Muetterties, Bull. Soc. Chem. Belg. 84, 959 (1975).
(b) E. Shustorovich, R.C. Baetzold and E.L. Muetterties, J. Phys. Chem. 87, 1100 (1983).
6. (a) R.P. Messmer in The Physical Basis for Heterogeneous Catalysis, E. Drauglis and R.I. Jaffee, eds. (Plenum, New York, 1975).
(b) K.H. Johnson and R.P. Messmer, Int. J. Quantum Chem. Symp 10, 147 (1976).
7. (a) G.A. Ozin Catal. Rev. -Sci. Eng. 16, 191 (1977).
(b) G.A. Ozin, Symp. Faraday Soc. 14, 7 (1980).
(c) G.A. Ozin and C. Gracie, J. Phys. Chem. 88, 643 (1984).
(d) G.A. Ozin and J.G. McCaffrey, J. Phys. Chem. 88, 645 (1984).
8. R. Ugo, Catal. Rev. -Sci. Eng. 11, 255 (1975).
9. K.J. Klabunde, Chemistry of Free Atoms and Particles, (Academic, New York, 1980).
10. W. Weltner, Jr. and R.J. van Zee, Ann. Rev. Phys. Chem. 35, 291 (1984).
11. D.A. Garland and D.M. Lindsay, J. Chem. Phys. 80, 4761 (1984).
12. (a) C. Cosse, M. Fouassier, T. Mejean, M. Tranquille, D.P. DiLella and M. Moskovits, J. Chem. Phys. 73, 6076 (1980).
(b) M. Moskovits and D.P. DiLella, in Metal Bonding and Interactions in High Temperature Systems, ACS Symposium Series 179, ed. J.L. Gole and W.C. Stwalley (Am. Chem. Soc., Washington, D.C., 1982), p. 153.
13. (a) Z.K. Ismail, R.H. Hauge, L. Fredin, J.W. Kaufman and J.L. Margrave, J. Chem. Phys. 77, 1617 (1982).
(b) Z.H. Kafafi, R.H. Hauge, L. Fredin, W.E. Billups and J.L. Margrave, J.C.S. Chem. Commun., 1230 (1983).
14. T.G. Dietz, M.A. Duncan, M.G. Liverman and R.E. Smalley, J. Chem. Phys. 73, 4816 (1980).
15. M.E. Geusic, M.D. Morse, S.C. O'Brien and R.E. Smalley, Rev. Sci. Inst., submitted for publication.
16. M.E. Geusic, M.D. Morse and R.E. Smalley, J. Chem. Phys. 82, 590 (1985).
17. M.D. Morse, M.E. Geusic, J.R. Heath and R.E. Smalley, J. Chem. Phys. 83, xxxx (1985).
18. (a) M. Balooch, M.J. Cardillo, D.R. Miller and R.E. Stickney, Surf. Sci. 46, 358 (1974).
(b) M. Balooch and R.E. Stickney, Surf. Sci. 44, 310 (1974).

THEORETICAL CHARACTERIZATION OF CHEMICAL REACTIONS OF IMPORTANCE IN THE OXIDATION OF HYDROCARBONS: REACTIONS OF ACETYLENE WITH HYDROGEN AND OXYGEN ATOMS*

Thom. H. Dunning, Jr., Lawrence B. Harding,
Albert F. Wagner, George C. Schatz and Joel M. Bowman
Theoretical Chemistry Group
Chemistry Division, Argonne National Laboratory
Argonne, Illinois 60439

1. INTRODUCTION

For more than six years the Theoretical Chemistry Group at Argonne National Laboratory has been interested in the energetics and dynamics of the elementary chemical reactions involved in the oxidation of hydrogen and simple hydrocarbon fuels, e.g., methane, acetylene, and formaldehyde. Hydrocarbon oxidation is a complex free radical chain reaction involving many highly reactive species whose chemistry is difficult to characterize in the laboratory. Theoretical studies of the elementary chemical reactions involved in these processes thus affords many opportunities for contributing to the solution of an important societal problem, namely, the inefficient use of hydrocarbon fuels. Further, to the extent that the chemical processes involved in flames are a microcosm of chemistry, the knowledge gained in these studies can be expected to contribute to our understanding of chemical reactivity in general.

Over the past six years our Group has studied a number of elementary chemical reactions involved in the oxidation of hydrogen and simple hydrocarbon fuels, including $O+H_2$ [1,4], $OH+H_2$ [2,4], $H+O_2$ [3,4], $O+CH_4$ [5], $H+CH_4$ [6], $O+C_2H_2$ [7], $H+C_2H_2$ [8], H+CO [9], $H+H_2CO$ [10], $C+H_2$ [11], H+HCN [12], OH+CO [13], and H+HCO [14].

*Work performed under the auspices of the Office of Basic Energy Sciences, Division of Chemical Sciences, U.S. Department of Energy, under Contract W-31-109-Eng-38.

R. J. Bartlett (ed.), Comparison of Ab Initio Quantum Chemistry with Experiment for Small Molecules, 67–94.

These studies were primarily concerned with the mechanisms, energetics, and rates of the reactions, the effects of temperature and pressure, kinetic isotope effects, and the effect of vibrational excitation of the reagents. We will limit our discussion here to our studies of the reactions of acetylene with hydrogen and oxygen atoms. These studies illustrate the basic approach involved.

Acetylene is important both as a practical hydrocarbon fuel and as a model system for studying soot formation. In addition, acetylene is formed in many hydrocarbon flames, including flames of methane, and thus the reactions of acetylene are an integral part of the overall mechanism for the oxidation of most hydrocarbon fuels. The development of a detailed mechanism for the oxidation of acetylene was the subject of a recent study by Miller and coworkers [15]. From modeling studies of both flame and shock tube experiments they conclude that the reactions

$$O + C_2H_2 \rightarrow CH_2 + CO \qquad (1a)$$

$$O + C_2H_2 \rightarrow H + HCCO \qquad (1b)$$

"are the most important fuel consuming reactions under lean, stoichimetric, and even slightly rich conditions." Further, they noted that in spite of numerous studies "(t)he branching ratio between the two sets of products is still a source of uncertainty." We will report here theoretical studies of reactions (1) emphasizing the role that theoretical studies have played, and are continuing to play, in elucidating the rate and mechanism of this reaction.

A detailed mechanism for the formation of carbonaceous species, e.g., soot, in non-aromatic hydrocarbon flames has not yet been advanced. However, many investigators [16] have proposed that the initial condensation step involves the polymerization of acetylene. Evidence for this comes from the observation that acetylene and polyacetylenes precede the formation of soot in many fuel-rich flames. To obtain information on the reactions involved in the polymerization of acetylene, a number of investigators have studied the pyrolysis of acetylene. Tanzawa and Gardiner [17] found that two different initiation steps were important in the pyrolysis of acetylene: an addition step

$$H + C_2H_2 \; (+M) \rightarrow C_2H_3 \; (+M) \qquad (2a)$$

at low temperature and an abstraction step

$$H + C_2H_2 \rightarrow H_2 + C_2H \qquad (2b)$$

at high temperatures. Condensation then occurs by subsequent reactions of the vinyl (C_2H_3) or ethynyl (C_2H) radicals with acetylene. There have been a number of direct studies of the kinetics of the addition reaction, both as a function of temperature and pressure, providing detailed data on the rate of the addition reaction under the conditions of interest in flames. For the abstraction reaction, or rather its reverse,

$$H_2 + C_2H \rightarrow H + C_2H_2 \qquad (-2b)$$

only a limited number of studies have been reported and these were at room temperature. Thus, rate data are not available on reaction (2b) at the temperatures of interest. We will report here the results of theoretical studies of both the addition and abstraction reactions (2a, -2b) with an emphasis on the quantitative determination of the rate constants for the reactions.

As the reactions of hydrogen atoms with acetylene poses the simpler problem, we shall first discuss reactions (2). The reaction of oxygen atoms with acetylene, which may involve more than one potential energy surface, will be discussed in Section 3.

2. REACTIONS OF HYDROGEN ATOMS WITH ACETYLENE

A summary of the energetics of the abstraction and exchange routes in the reaction of hydrogen atoms with acetylene, (2a) and

(2b), is given in Table 1. As can be seen, the addition reaction is nearly 30 kcal/mol exoergic, while the abstraction reaction is predicted to be nearly 30 kcal/mol endoergic. As expected from Hammond's rule [19], the saddle point for the addition reaction is "early," i.e., it most closely resembles the reactants, while that for the abstraction reaction is "late," i.e., it most closely resembles the products; see Figure 1.

Table 1. Energetics of the abstraction and addition pathways in the reaction of hydrogen atoms with acetylene, (2a) and (2b), from calculation [8] and experiment [18]. In kcal/mol.

	Addition		Abstraction	
	Calcd	Exptl	Calcd	Exptl
ΔE_{rxn}	-36.8		31.0	
ΔE_{zpe}	7.7		-1.5	
ΔH_0	-29.1	-34-38	29.5	15-33

2.1 The Abstraction Reaction: $H + C_2H_2 \rightarrow H_2 + C_2H$

Let us now consider the abstraction reaction in more detail. Because of the extreme endoergicity of this reaction, ~30 kcal/mol, experimental studies of this reaction have instead considered its reverse

$$H_2 + C_2H \rightarrow H + C_2H_2 \qquad (-2b)$$

The rate constants for reaction (-2b) calculated from transition state theory (with the simple Wigner tunneling correction) are plotted in Figure 2. The rate constants at room temperature measured by Lange and Wagner [20] and Laufer and Bass [21] are also plotted in this figure (the much higher rate for this reaction

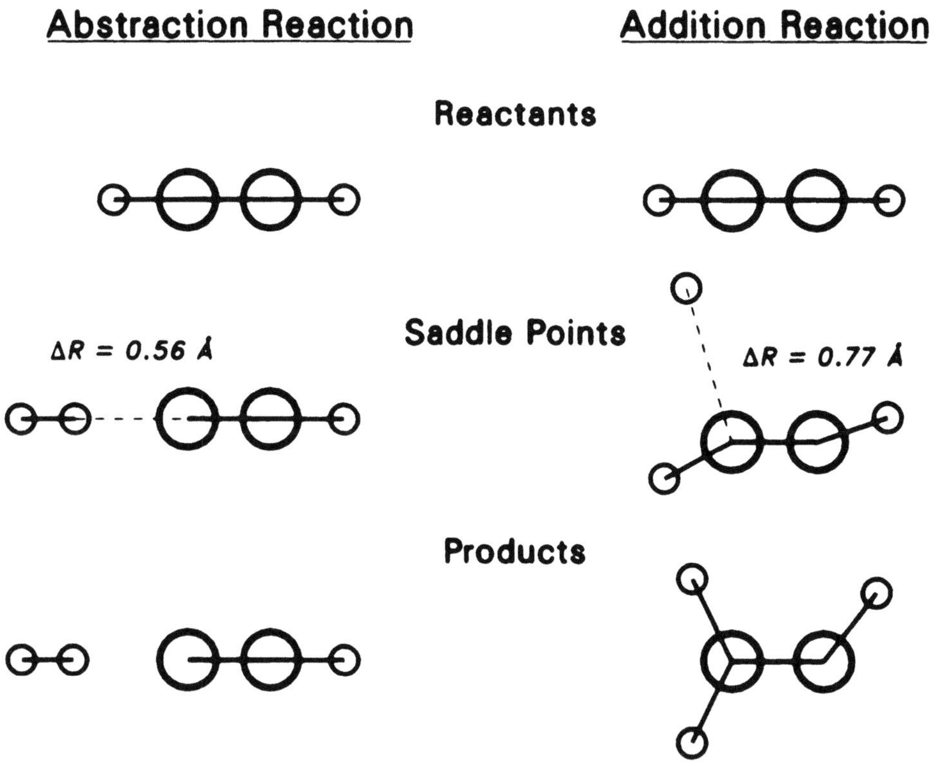

Figure 1. Molecular structures for the reactants, saddle point and products for the abstraction and addition pathways in the reaction of hydrogen atoms with acetylene. Extensions of the critical (dashed) bonds, $\Delta R = R_{sp} - R_e$, are indicated where R_{sp} is the bond distance at the saddle point and R_e is the bond distance in the reactant (abstraction) or product (addition).

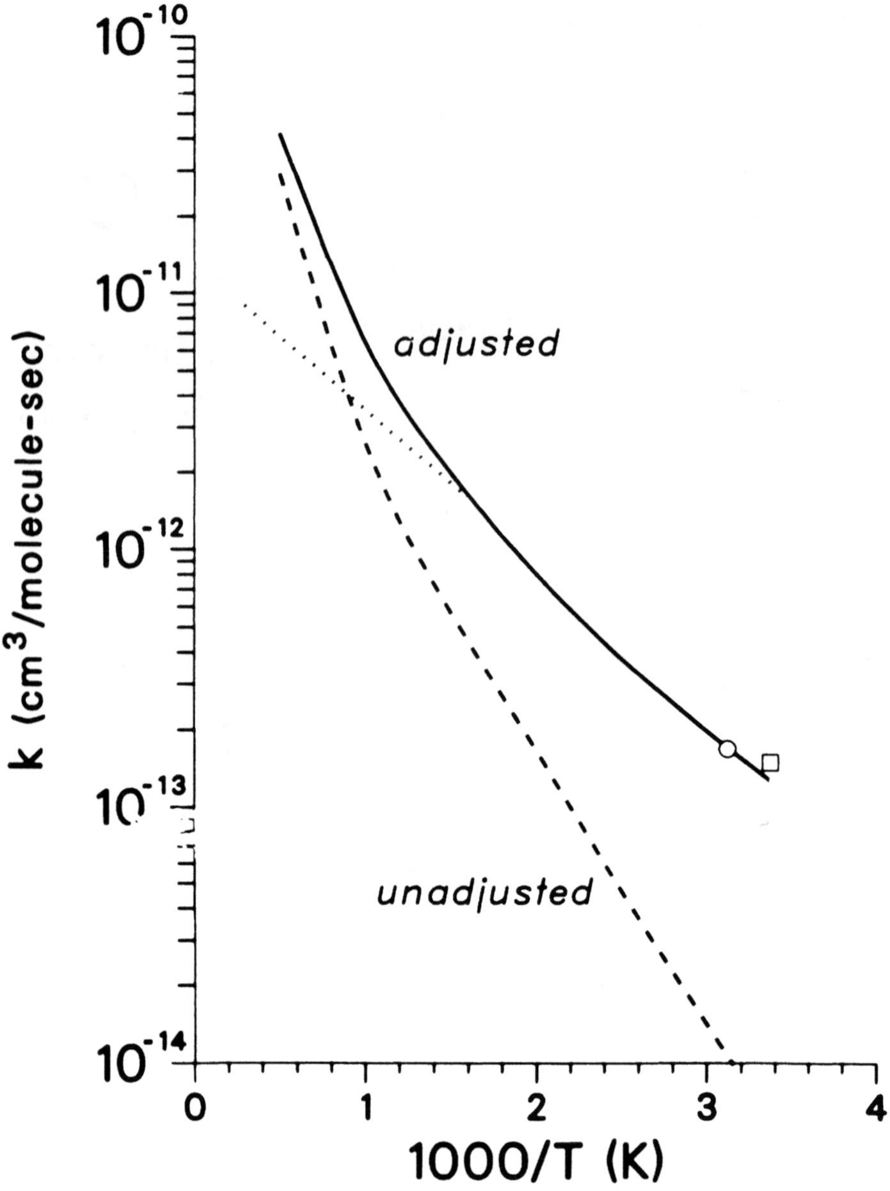

Figure 2. Rate constant for the $H_2 + C_2H \rightarrow H + C_2H_2$ abstraction reaction. The dashed curve (--) is the calculated rate for the calculated barrier height ($\Delta E_b = 4.0$ kcal/mol); the solid curve (-) is the calculated rate for the barrier height adjusted to agree with the room temperature rates of Lange & Wagner [20] and Laufer & Bass [21] ($\Delta E_b = 2.3$ kcal/mol). The rates derived by Gardiner & coworkers [17] at high temperatures have also been plotted.

reported by Wittig and coworkers [22a] has been shown to involve the low-lying excited state of C_2H [22b]). No other direct measurements of the rate of this reaction exist, although Tanzawa and Gardiner [17] reported an estimate of the rate of this reaction at high temperatures obtained from their studies of the thermal decomposition of acetylene; this estimate is also plotted in Figure 2.

The dashed line in Figure 2 is for the calculated barrier height of 4.0 kcal/mol. As expected, use of this barrier in the transition state theory calculations (with the tunneling correction) yields a rate constant much smaller than that obtained from experiment; the calculated barrier is clearly too large. Experience has shown [1-13] that for reactions such as those referenced here the calculated barrier heights are too large by 1-3 kcal/mol. In fact, if we lower the calculated barrier by just 1.7 kcal/mol, to 2.3 kcal/mol, the calculated rate constant at room temperature can be brought into agreement with the reported room temperature rates [20,21].

As can be seen, the calculations predict that the rate constant of this reaction is strongly dependent on temperature, much more so than expected from typical Arrhenius behavior. The cause for this can be seen in the vibrational energy level correlation diagram for this reaction which is plotted in Figure 3. First we note that with the exception of ω_{HH}, the vibrational frequencies of C_2H evolve smoothly into those of C_2H_2; this is as expected as there are few changes in the C_2H moiety upon formation of C_2H_2 (see Figure 1). The HH stretching frequency, ω_{HH}, changes markedly during the course of the reaction as this mode is strongly coupled with the reaction coordinate; the HH stretching mode in H_2 eventually evolves into the asymmetric CH stretching mode in C_2H_2.

The complex formed by the addition of a diatomic molecule, with one vibrational mode, to a linear triatomic molecule, with three vibrational modes (one of which, the bending mode, is doubly degenerate), in a collinear configuration has five additional

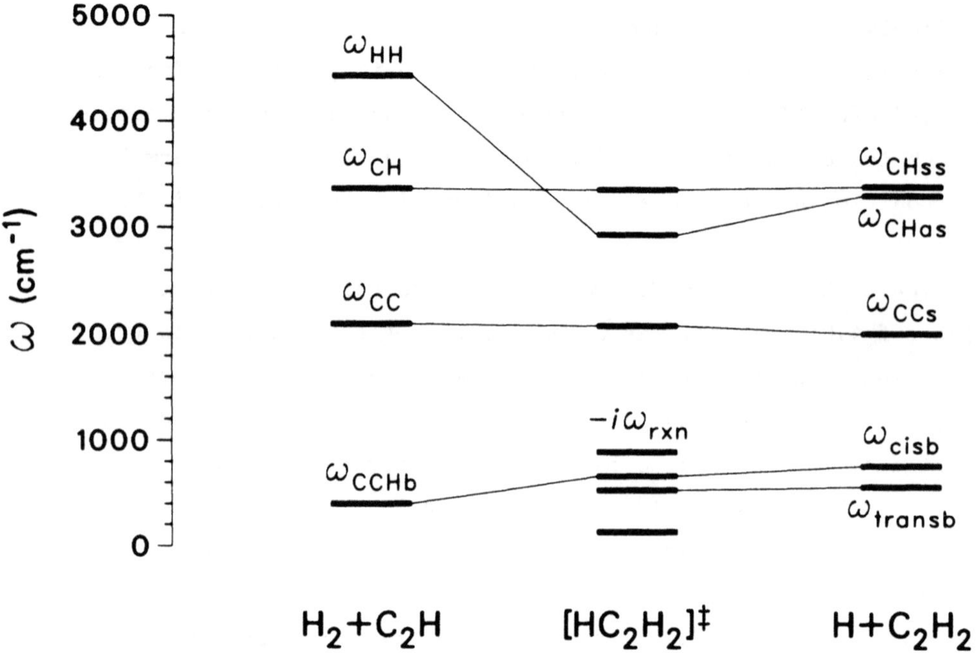

Figure 3. Correlation diagram for the vibrational energy levels for the H_2 + C_2H → H + C_2H_2 abstraction reaction. The frequency associated with the reaction coordinate at the saddle point is plotted as $-i\omega_{rxn}$.

vibrational modes. For the transition state of reaction (-2b) the five additional modes are a non-degenerate reaction coordinate and two doubly degenerate bending modes. As can be seen in Figure 3, one of these modes is of very low frequency, 139 cm^{-1}. It is this low frequency mode which is responsible for the markedly non-Arrhenius behavior of the rate constant of reaction (-2b). The nuclear displacement vectors for one of the components of this mode are plotted in Figure 4; this mode involves an H_2-C_2H wagging motion.

To see how this low frequency mode leads to the strong temperature dependence of the reaction rate constant of the abstraction reaction, we plot in Figure 5 the temperature dependence of the harmonic vibrational partition function for the H_2-wagging mode as well as the reduced rate constant, i.e., the rate constant with the H_2-wagging mode partition function factored out

$$k_{red} = k/Q^2_{H_2\text{-}wag} \qquad (3)$$

(note that the partition function is squared because the H_2-wagging mode is doubly degenerate). As is evident in this figure, the non-Arrhenius behavior of the abstraction rate constant is largely a result of the strong temperature dependence of the partition function of the low frequency H_2-wagging mode.

The low frequency of the transition state vibrational mode is due to the "earliness" of the saddle point for the highly exoergic abstraction reaction. Because of the trend suggested by Hammond's rule [19], we expect strong non-Arrhenius behavior of the rate constants for many highly exoergic reactions. This behavior has, in fact, also been observed in the H_2 + CN abstraction reaction, ΔE_{rxn} = 25.3 kcal/mol, by Bair and coworkers [12]; for this reaction the theoretical prediction has been confirmed by the high temperature measurements of Szekely et al [23].

The rate constant for the abstraction reaction does not agree well with that determined by Tanzawa and Gardiner [17] from their analysis of the thermal decomposition of acetylene - their rate constant is substantially smaller than the one calculated here and displays simple Arrhenius behavior. Such deconvolution analyses

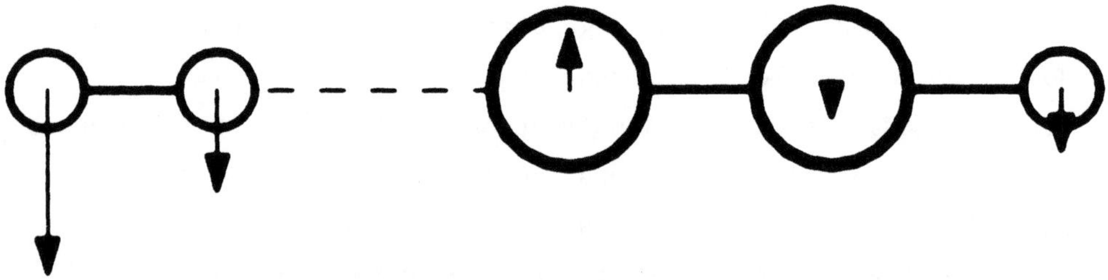

Figure 4. Normal mode displacements for the low frequency (ω=139cm^{-1}) H_2-wagging mode at the saddle point in the H_2 + C_2H → H + C_2H_2 abstraction reaction.

are, of course, fraught with difficulties. In addition, Tanzawa and Gardiner [17] were unaware of the strong temperature dependence expected for the rate constant of the abstraction reaction.

In conclusion then, we believe that the adjusted rate constant plotted in Figure 5 for the abstraction reaction is the best presently available. This rate constant has been used in a recent modeling study of the oxidation of acetylene by Miller and coworkers [15].

2.2 The Addition Reaction: $H + C_2H_2 \rightarrow C_2H_3$

The vibrational modes for the addition reaction

$$H + C_2H_2 \rightarrow C_2H_3 \tag{2a}$$

are plotted in Figures 6 (in-plane modes) and 7 (out-of-plane modes). As can be seen, the vibrational modes of C_2H_2 change smoothly into those of the transition state and then into those of the products (although some of the correlations because of mixing of the modes are not without ambiguity). The reaction coordinate, see Figure 6, consists largely of hydrogen motion, primarily a direct approach of the incoming hydrogen atom to the nearest carbon atom. One point of interest is the coupled motion of the two hydrogen atoms on the C_2H_2 moiety in the reaction coordinate. The hydrogen atom on the carbon atom being attacked rotates away from the incoming hydrogen atom; this is as would be expected from simple electron repulsion arguments. Simultaneously with this motion, the hydrogen atom on the far carbon atom rotates toward the incoming hydrogen atom. The motion of the second hydrogen atom is due, in part, to the repulsion of the two CH bond pairs and, in part, to the transfer of the singly occupied orbital from the incoming hydrogen atom to the far carbon atom (simultaneously, of course, with the transfer of the bonding orbitals from the CC to the CH region).

In the acetylene molecule there are 3N-5 = 3(4)-5 = 7 vibrational modes, while at the transition state of the addition reaction and in the product, the vinyl radical, there are 3N-6 = 3(5)-6 = 9 vibrational modes. Thus, addition of an atom to a

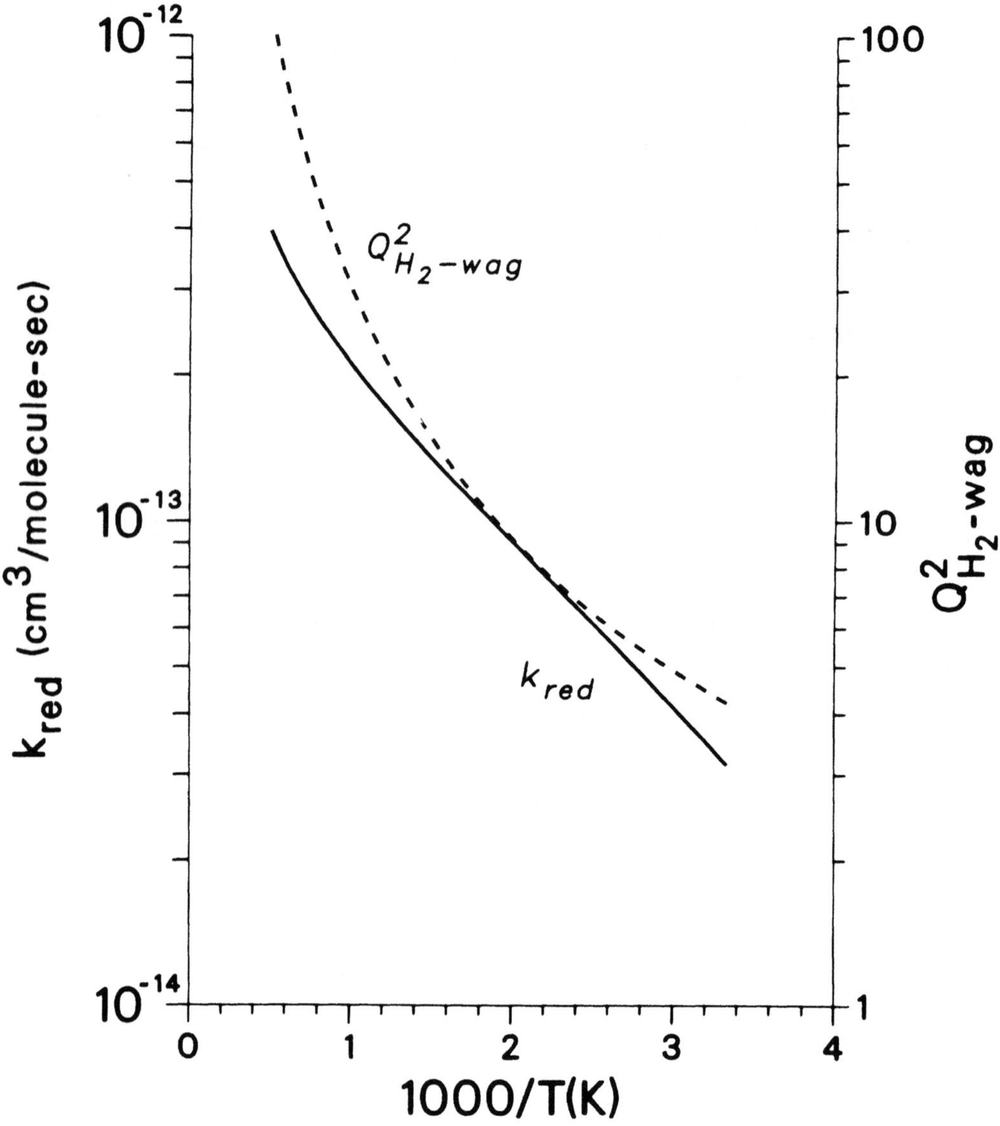

Figure 5. Temperature dependence of the harmonic partition function of the low-frequency H_2-wagging mode and of the reduced rate constant for the $H_2 + C_2H \rightarrow H + C_2H_2$ abstraction reaction.

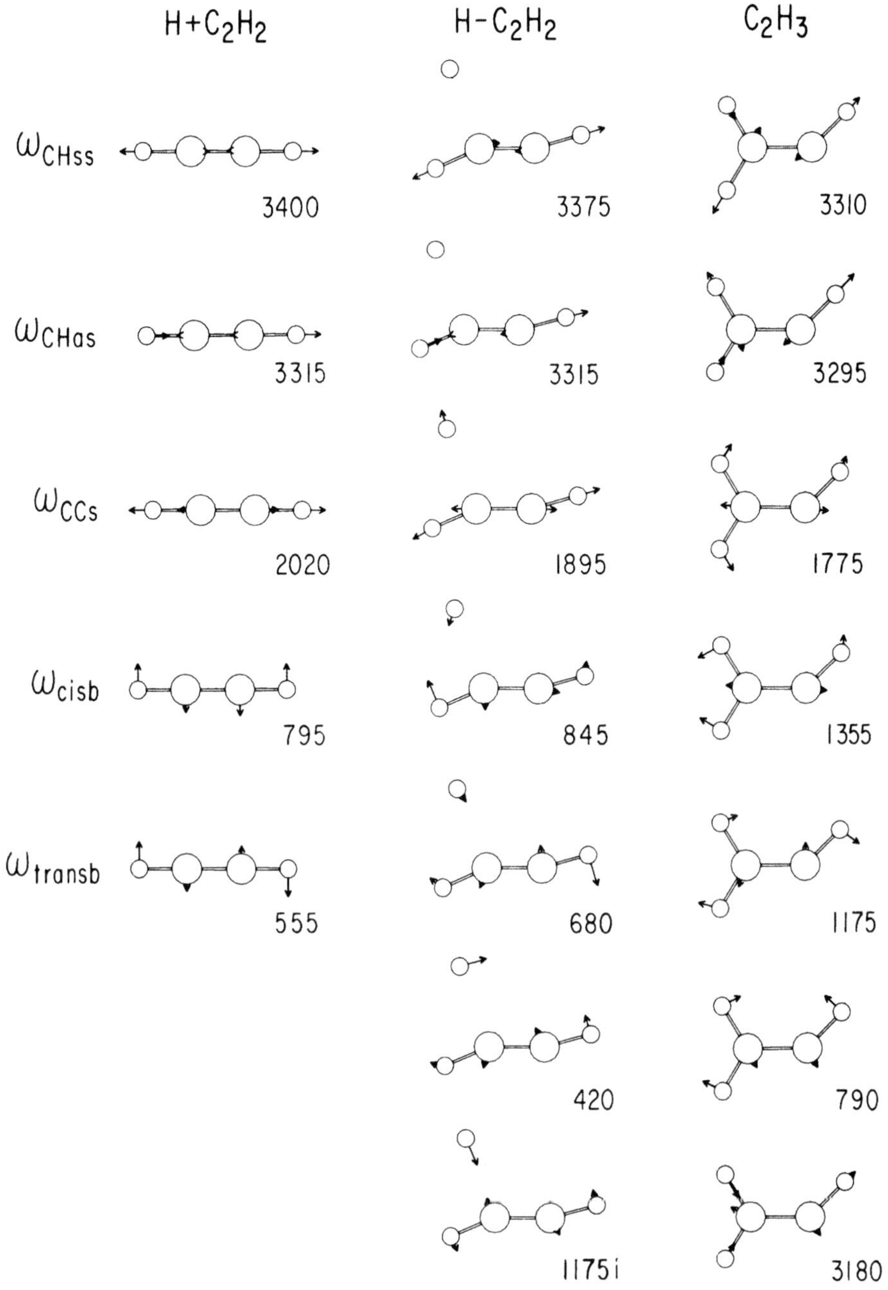

Figure 6. In-plane (a') normal modes for the reactants, saddle point and products for the $H + C_2H_2 \rightarrow C_2H_3$ addition reaction.

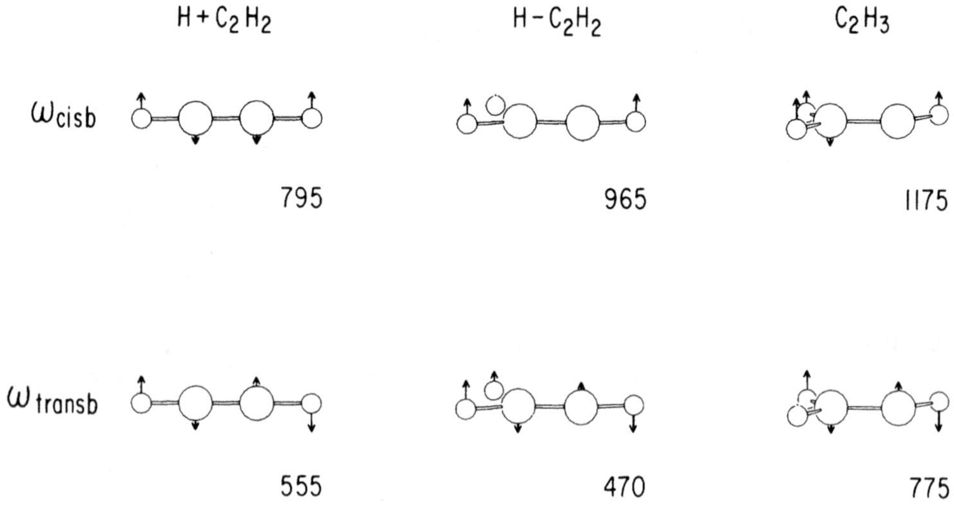

Figure 7. Out-of-plane (a") normal modes for the reactants, saddle point and products for the $H + C_2H_2 \rightarrow C_2H_3$ addition reaction.

linear molecule to form a non-linear product results in the addition of only two, not three, vibrational modes. The "missing" mode is of a" symmetry; this is the cause of the unusual behavior observed in the out-of-plane trans-bending mode upon addition of the hydrogen atom (see Figure 7). At large $H\text{-}C_2H_2$ separations this mode is simply the trans-bending mode of C_2H_2 with a calculated frequency of 555 cm^{-1}. As the system approaches the transition state, amplitude is transferred from the hydrogen atoms of the C_2H_2 moiety to the incoming hydrogen atom; this results in a slight (15%) decrease in the frequency of this mode. Finally, in the product, the vinyl radical, the transfer of amplitude from the acetylenic hydrogen to the incoming hydrogen is essentially complete and the frequency has increased to 775 cm^{-1}.

The calculated barrier, ΔE_b, and reaction exoergicity, ΔE_{rxn}, for the addition reaction (without zero point corrections) are listed in Table 2. The POL-CI entry is a published result [8] determined with a wavefunction comparable to that used in the abstraction studies [1-13]. The following four entries in Table 2 are preliminary results of more sophisticated calculations [25]. The last entry in Table 2 is obtained from an analysis of the high pressure experimental measurements (see below).

Table 2. Summary of large scale configuration interaction calculations on the energetics of the addition reaction, (2b). In kcal/mol.

Method	Basis Set	ΔE_b	ΔE_{rxn}
POL-CI	[3s2p1d/2s*1p]	7.6[8]	-36.8[8]
HF+1+2	[3s2p1d/2s1p]	8.9	-45.0
	[4s3p2d/3s2p]	8.1	-42.0
	[4s3p2d1f/3s2p1d]	8.4	-42.1
HF+1+2+QC	[4s3p2d1f/3s2p1d]	5.9	-42.4
Exptl		~1.7	-42-46

First consider the calculated barrier heights. At the POL-CI level, the error in the calculated barrier height is nearly 6 kcal/mol; this is much larger than that observed in abstraction reactions [1-13], including the H + C_2H_2 abstraction reaction. Further, the results of the HF+1+2 calculations with ever more flexible (and therefore presumably more accurate) basis sets are discouragingly stable; changes of just a few tenths of a kcal/mol, given the magnitude of the error of several kcal/mol; the HF+1+2 calculations give no signficant improvement relative to the POL-CI calculations. To estimate the effect of higher order excitations in the HF+1+2 calculations, the simple formula of Davidson [24] was used. Inclusion of this correction reduces the calculated barrier height by 2.5 kcal/mol. This suggests the need for a multi-configuration zero-order function upon which to base the configuration interaction calculations; this work is in progress. Even with the Davidson correction, however, the error in the calculated barrier height for the addition reaction is more than 3 kcal/mol.

With the exception of the HF+1+2 calculations with the [3s2p1d/2s1p] basis set, the calculated reaction energy defects are also quite stable, with the estimated error ranging from 2-4 kcal/mol (note that the spread in the measured values is 4 kcal/mol). Calculations on the CH molecule at this level [25] are in error by less than 2 kcal/mol. For the reaction exoergicity the Davidson correction has little effect.

There is also a problem with the theoretical description of the kinetics of the addition reaction. A Lindemann-Hinshelwood mechanism is typically used to describe the addition of hydrogen atoms to acetylene

$$H + C_2H_2 \rightarrow C_2H_3^* \qquad (4a)$$

$$C_2H_3^* \rightarrow H + C_2H_2 \qquad (-4a)$$

$$C_2H_3^* + M \rightarrow C_2H_3 + M \quad (4b)$$

At steady state the rate constant for loss of C_2H_2 is

$$k = k_{4a}k_{4b}[M]/(k_{-4a} + k_{4b}[M]) \quad (5)$$

At high pressures, i.e., as $[M] \rightarrow \infty$, the rate of loss of C_2H_2 is simply k_{4a}. That is, at sufficiently high pressures all of the complexes which cross the transition state are stabilized and the rate of the reaction is equal to the rate at which the transition state for complex formation is crossed, k_{4a}.

The rate of reaction (4a) calculated from transition state theory with the Wigner tunneling correction is compared to the measured high pressure rate constants [26-29] in Figure 8. In calculating the rate constant we have adjusted the barrier height for the addition reaction to agree with that inferred from experiment (as per the discussion of the abstraction reaction above). As can be seen, the calculated pre-exponential factor for the addition reaction is approximately a factor of 25 larger than that determined from experiment, although it should be noted that some uncertainty exists in the experimental data. If the analysis leading to (5) is correct and transition state theory is applicable to (4a), then this discrepancy could be attributed to errors in the calculated vibrational frequencies at the transition stage. There is, however, little evidence to indicate substantial errors in the calculated transition state frequencies from the other theoretical studies [1-13] carried out to date. Further, Harding and coworkers [8] investigated the sensitivity of the pre-exponential factor to changes in the frequencies and concluded that reasonable changes in these frequencies would not improve the agreement with the experimental results.

There are, of course, many possible sources for this error besides that mentioned above. Transition state theory could fail to accurately represent the high pressure limit because of substan-

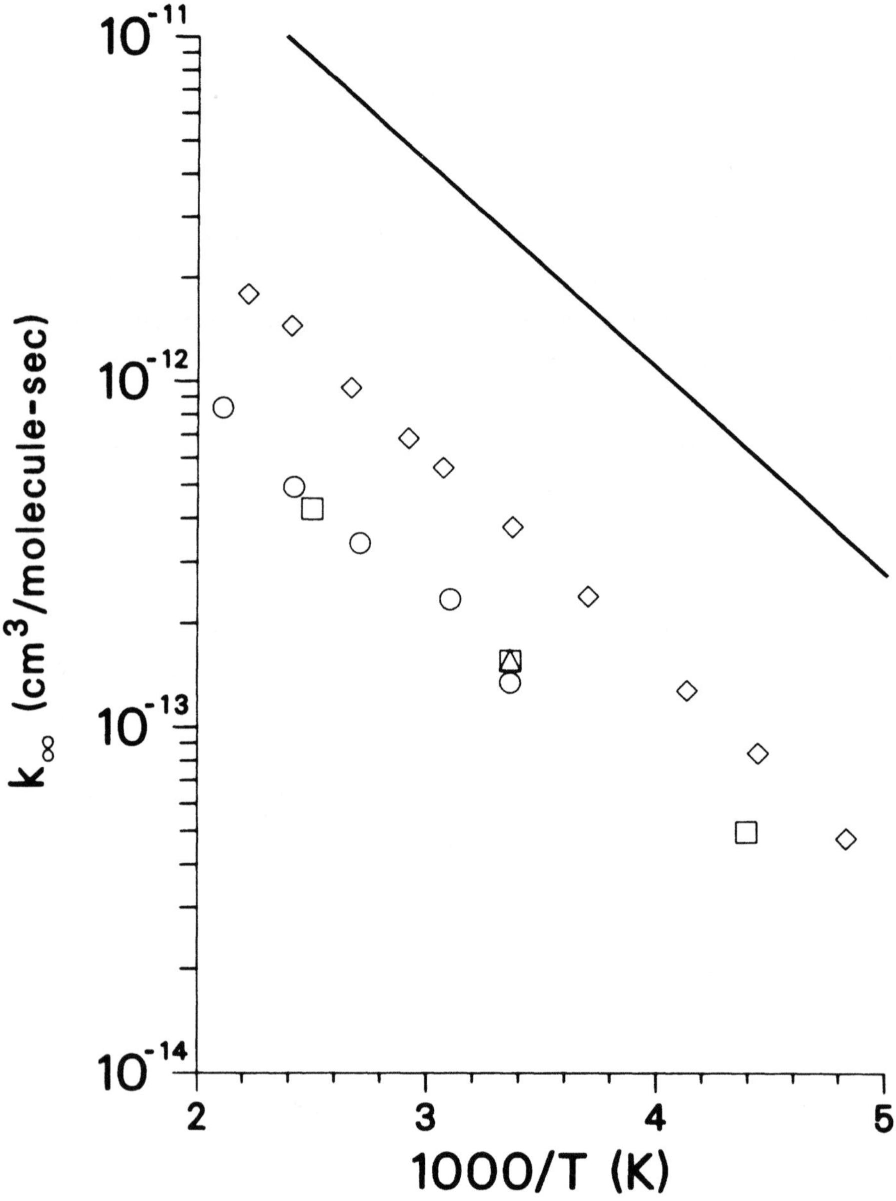

Figure 8. Calculated (–) and measured (Δ [26], [] [27], O[28], ◇ [29]) high pressure rate constants for the H + $C_2H_2 \rightarrow C_2H_3$ addition reaction. In the calculations the barrier height has been adjusted to give agreement with the measured activation energy (ΔE_b = 1.7 kcal/mol).

tial recrossing effects or inaccurate representations of quantum (e.g., tunneling) effects. Or, the experimental measurements may refer only to an apparent high pressure limit, in reality a plateau in the rate constant as a function of pressure that ultimately ends with an ascent to the true high pressure limit at much higher pressures. The theoretical calculations have not found evidence of this, but critical assumptions in the other rate constants k_{-4a} and k_{4b}, such as statistical decay of C_2H_3*, would prevent the calculation of rate constant plateaus [8]. Finally, the Lindemann-Hinschelwood mechanism simply may not form a suitable basis for the interpretation of the experimental data, e.g., collision-induced dissociation of the complex is ignored; a master equation approach may well be required.

3. REACTIONS OF OXYGEN ATOMS WITH ACETYLENE

Since the early studies of Fenimore and Jones [30], the major routes for the reaction of oxygen atoms with acetylene were identified as

$$O + C_2H_2 \rightarrow \begin{cases} H + HCCO & (1a) \\ CH_2 + CO & (1b) \end{cases}$$

The initial step of the reaction was postulated to involve addition of the atomic oxygen to acetylene to form a 'hot' OC_2H_2 complex which then either eliminated a hydrogen atom, (1a), or isomerized [(1,2)-hydrogen migration] and then decomposed, (1b). Neither the simple addition pathway nor the abstraction pathway were found to be of importance in the temperature-pressure regimes of interest here.

Probably the best known aspect of this reaction is its overall rate at low temperatures (250-600 K). The latest NASA review [31] assigns an uncertainty of only 30% to the room temperature rate constant based on the results of numerous studies [32-34].

Measurements up to 600 K [33-37] are also in relatively close agreement although there is some dispute about the degree of curvature in the Arrhenius plot. There is only one direct measurement of the rate constant of (1) at high temperatures, that from the shock tube studies of Lohr and Roth [38].

In order to calculate the overall rate for this reaction, we are currently examining both the barrier to addition of oxygen and the barriers to hydrogen migration and elimination that control the branching ratio between routes (1a) and (1b) with a highly correlated wavefunction (Hartree-Fock plus all single and double excitations). This work is an improvement over an earlier study [7] which involved a less correlated wavefunction and did not examine the addition barriers. The preliminary results of the new studies [25] indicate that oxygen atom addition proceeds on two triplet surfaces ($^3A''$ and $^3A'$) with a barrier of about 10.5 kcal/mol on the lower ($^3A''$) surface. On this surface the barriers for hydrogen elimination (to form H + HCCO) or migration-dissociation (to form $^3CH_2$ + CO) are nearly equal and are approximately 8 kcal/mol below the addition barrier.

The barrier to complex formation on the higher energy ($^3A'$) surface is only 3 kcal/mol above that on the $^3A''$ surface. This surface does not directly lead to products. It may, however, interact strongly with either the lower triplet surface or with the strongly bound singlet surface correlating with ketene [39]. The $^3A'$ surface is buried within the vibration-rotation manifolds of the $^3A''$ and $^1A''$ surfaces. In addition, there is a crossing of the singlet ($^1A''$) and triplet ($^3A'$) surfaces in the entrance channel. Thus, collisions of the long-lived $OC_2H_2^*$ complex may lead to either internal conversion to the lower $^3A''$ state or to intersystem crossing to the $^1A''$ state. In either case, products are expected to be readily produced, although in the case of the CH_2 + CO channel to $^1CH_2$ and not $^3CH_2$.

Because the barriers to both hydrogen migration and hydrogen elimination on the $^3A''$ surface are substantially below the oxygen addition barrier, essentially any metastable complex that forms is expected to decay to products. Thus the overall rate of the reaction is simply the complex formation rate. That rate, determined from transition state theory with the calculated transition state properties but with an adjusted addition barrier, is compared to the experimental values in Figure 9. The adjusted addition barrier selected to reproduce the consensus value of the measured rate constants at room temperature, is 2 kcal/mole. As in the addition reaction discussed in the previous section, the calculated addition barrier is much too high; the error is 8.5 kcal/mole for the oxygen addition reaction compared to an error of 7.2 kcal/mol for the hydrogen addition reaction (see Table 2). Two theoretical rates are shown in Figure 9, the lower one only for reaction on the $^3A''$ surface and the upper one for reaction on both the $^3A''$ and $^3A'$ surfaces (with the barriers separated by the calculated value of 3 kcal/mol). The upper curve would apply only if surface hopping opened up a route to the formation of products for reaction on the $^3A'$ surface.

As Figure 9 shows, below 600 K the agreement with experiment is good, although the calculated values are slightly below most of the measured values. Above 1500 K there are two experimental values given, both from Lohr and Roth [38]. They measured separately both oxygen loss, to obtain a total rate constant, and hydrogen gain, to obtain a rate constant for reaction (1a). Both measurements are subject to an experimental uncertainty of about 20% and to interference from poorly characterized secondary reactions. The comparison between theory and experiment would suggest that the measured total rate constant is too high and should in fact be about equal to the measured value of the rate for hydrogen production.

Measurements of the branching ratio of reaction (1) have a long and controversial history. From an analysis of the final products of the reaction of oxygen atoms with acetylene Arrington et

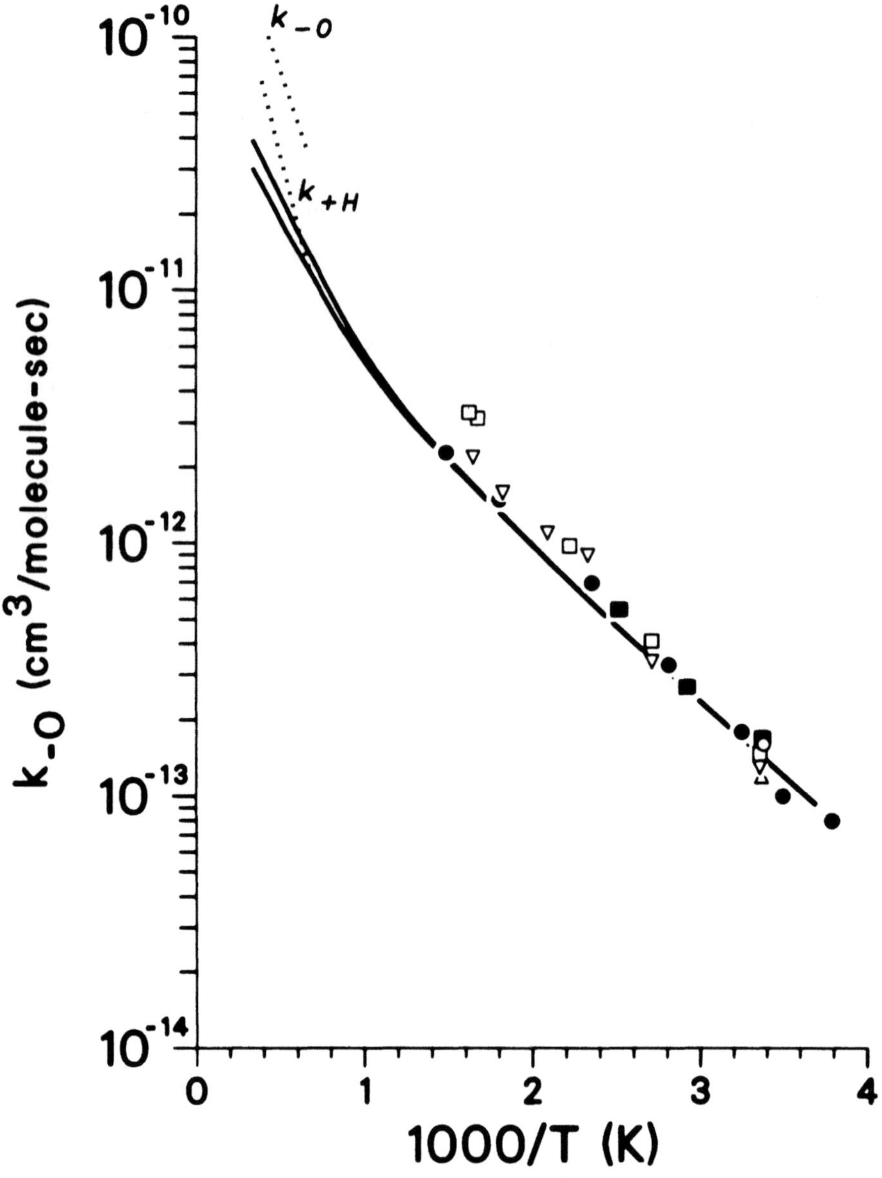

Figure 9. Rate Constant for the loss of oxygen atoms in the O + C_2H_2 addition reaction (in Ref. [32] acetylene loss was measured). Experimental values are: Δ[33], O[32], ∇[34], ■[35], ●[36]. The calculated rate constants are given by the solid curves (see discussion in text); the dotted curves are from the experimental studies of Lohr & Roth [38].

al. [40] concluded that reaction (1b) was the major route for the reaction, i.e., channel (1a) was simply not needed to explain their results. This interpretation was followed by subsequent workers [36,41]. However, Williamson [42], in studies of the O + C_2D_2 reaction in the presence of H_2, reported the observation of substantial amounts of HD, indicating that (1a) was also an important channel. This work was further supported by the observation of Jones and Bayes [32], using photoionization mass spectrometric techniques, of both HCCO and CH_2 in the oxygen-acetylene reaction. From these results Jones and Bayes concluded that H + HCCO was a major reaction channel, accounting for 45-80% of the initial products formed. Shortly thereafter, Gutman and coworkers [43] using the same technique independently concluded that both reactions, (1a) and (1b), were major channels. Later, however, Blumenberg et al. [44] reported studies which indicated that 95% of the reactions between oxygen atoms and acetylene lead to CH_2 + CO.

The most recent experimental studies on this question, besides those of Lohr and Roth [38], are molecular beam studies reported by Lee and coworkers [45] (see also Clemo et al. [46]) and fluorescence studies under jet conditions carried out by Aleksandrov et al [34]. The measurements of Aleksandrov et al. indicate that at room temperature hydrogen elimination constitutes only about 5% of the total reaction while by 608 K it accounts for about 20%. The measurements of Lee and coworkers are at 4 and 8 kcal/mole relative translational energy with the initial vibrational and rotational distribution of acetylene at room temperature. They obtained branching ratios for hydrogen production of 0.22 and 0.70, respectively. Lohr and Roth [38] also find that the branching ratio for hydrogen production increases with temperature, from 0.35 at 1500 K to 0.54 at 2500 K.

The angular distributions measured by Lee and coworkers [45] (and also by Clemo et al. [46]) indicate the collision involves formation of a metastable OC_2H_2* complex that lives for at least a rotational period before decaying into products. If we then make the assumption that all of the initial energy is statistically redistributed during the lifetime of the complex, the molecular beam experiments at 4 and 8 kcal/mole translational energy produce complexes with an energy content equivalent to those produced by a thermal experiment at about 460 and 670 K, respectively. At these two "temperatures" Lee and coworkers find that the branching ratios for hydrogen elimination, (1a), are approximately 0.22 and 0.70, respectively. At the temperature equivalent to the lower energy molecular beam experiment (460 K) the measurements of Aleksandrov et al. [34] predict a value of approximately 0.12 for the branching ratio. Given the uncertainty quoted by Lee and coworkers for their value ($\pm$0.09) and a comparable uncertainty in the value of Aleksandrov et al., the two different measurements are not inconsistent. However, at the equivalent temperature of the higher energy molecular beam experiment (670 K), the value of Lee and coworkers is more than three times higher than the highest value reported by Aleksandrov et al. which was measured at a temperature only 60 K lower (608 K). Thus the molecular beam measurements, converted to the equivalent thermal measurements assuming statistical randomization of the energy in the complex, predict a much more rapid rise in the branching ratio for hydrogen elimination with temperature than does the direct thermal measurements of Aleksandrov et al. [34].

If, on the other hand, we convert the translational energies in the molecular beam experiments to effective temperatures using $3/2RT = KE$, then the effective temperatures of the molecular beam experiments are approximately 1300 K and 2700 K. At these temperatures the measurements of Lohr and Roth [38] would predict branching ratios of 0.30 and 0.56 (both temperatures are within 200 K of the range studied), in fair agreement with the branching ratios reported by Lee and coworkers [45], namely, 0.22 and 0.70. However, as noted above the present calculations call into question the measurements of the total rate constant for reaction (1) reported by Lohr and Roth and, if the total rate is in error, the branching ratio inferred from this data (k_{+H}/k_{-O} was measured, not k_{+H}/k_{+CH_2}) is questionable.

We are currently re-examining the question of the branching ratio for reaction (1) and also the observation from the molecular beam studies that the OC_2H_2 complex must live at least a rotational period. Our initial study [7] found that hydrogen migration had both a tighter transition state and a 15 kcal/mole higher barrier than hydrogen elimination; we thus concluded that the elimination pathway, (1a), would dominate at all temperatures. Now, as mentioned above, preliminary results from the present calculations indicate that both the elimination and migration barriers are at about the same energy and prediction of the branching ratio is no longer obvious. Direct calculation of the branching ratio and collision complex lifetimes with RRKM theories are in progress.

References:

1. S.P. Walch, T.H. Dunning, Jr., R.C. Raffenetti and F.W. Bobrowicz, "A Theoretical Study of the Potential Energy Surface for $O(^3P)+H_2$," J. Chem. Phys. 72, 406 (1980); S.P. Walch, A.F. Wagner, T.H. Dunning, Jr. and G.C. Schatz, "Theoretical Studies of the $O+H_2$ Reaction," J. Chem. Phys. 72, 2894 (1980); G.C. Schatz, A.F. Wagner, S.P. Walch and J.M. Bowman, "A Comparative Study of the Reaction Dynamics of Several Potential Energy Surfaces of $O(^3P)+H_2 \rightarrow OH+H$. I," J. Chem. Phys. 74, 4984 (1981); A. F. Wagner, G.C. Schatz and J.M. Bowman, "The Evaluation of Fitting Functions for the Representation of an $O(^3P)+H_2$ Potential Energy Surface. I," J. Chem. Phys. 74, 4960 (1981); K.T. Lee, J.M. Bowman, A.F. Wagner and G.C. Schatz, "A Comparative Study of the Reaction Dynamics of the $O(^3P)+H_2 \rightarrow OH+H$ Reaction on Several Potential Energy Surfaces. III. Collinear Exact Quantum Transmission Coefficient Correction to Transition State Theory," J. Chem. Phys. 76, 3583 (1982).
2. S.P. Walch and T.H. Dunning, Jr., "A Theoretical Study of the Potential Energy Surface for $OH+H_2$," J. Chem. Phys. 72, 1303 (1980).
3. T.H. Dunning, Jr., S.P. Walch and M.M. Goodgame, "Theoretical Characterization of the Potential Energy Curve for Hydrogen Atom Addition to Molecular Oxygen," J. Chem. Phys. 74, 3482 (1981).
4. For a review of theoretical studies of the above three reactions see: T.H. Dunning, Jr., S.P. Walch and A.F. Wagner, "Theoretical Studies of Selected Reactions in the Hydrogen-

Oxygen System" in Potential Energy Surfaces and Dynamics Calculations, Ed. D.G. Truhlar, (Plenum Publishing Corporation, New York, 1981), pp. 329-357.

5. S.P. Walch and T.H. Dunning, Jr., "Calculated Barrier to Hydrogen Atom Abstraction from CH_4 by $O(^3P)$," J. Chem. Phys. 72, 3221 (1980).
6. S.P. Walch, "Calculated Barriers to Abstraction and Exchange for CH_4+H," J. Chem. Phys. 72, 4932 (1980); G.C. Schatz, S.P. Walch and A.F. Wagner, "Ab Initio Calculation of Transition State Normal Mode Properties and Rate Constants for the H(T)+$CH_4(CD_4)$ Abstraction and Exchange Reactions," J. Chem. Phys. 73, 4536 (1980); G.C. Schatz, A.F. Wagner and T.H. Dunning, Jr., "A Theoretical Study of Deuterium Isotope Effects in the Reactions H_2+CH_3 and H+CH_4," J. Phys. Chem. 88, 221 (1984).
7. L.B. Harding, "Theoretical Studies of the Reaction of Atomic Oxygen (^3P) with Acetylene," J. Phys. Chem. 85, 10 (1981).
8. L.B. Harding, G.C. Schatz and R.A. Chiles, "An Ab Initio Determination of the Rate Constant for $H_2+C_2H \rightarrow H+C_2H_2$," J. Chem. Phys. 76, 5172 (1982); L.B. Harding, A.F. Wagner, J.M. Bowman, G.C. Schatz and K. Christoffel, "Ab Initio Calculation of the Transition-State Properties and Addition Rate Constants for H+C_2H_2 and Selected Isotopic Analogues," J. Phys. Chem. 86, 4312 (1982).
9. T.H. Dunning, Jr., "Theoretical Characterization of the Potential Energy Surface of the Ground State of the HCO System," J. Chem. Phys. 73, 2304 (1980).
10. L.B. Harding and G.C. Schatz, "An Ab Initio Determination of the Rate Constant for $H+H_2CO \rightarrow H_2+HCO$," J. Chem. Phys. 76, 4296 (1982).
11. L.B. Harding, "Theoretical Studies of the Potential Energy Surface for the Reaction $C(^3P)+H_2 \rightarrow CH_2(^3P_1)$," J. Phys. Chem. 87, 441 (1983).
12. R.A. Bair and T.H. Dunning, Jr., "Theoretical Studies of the Reactions of HCN with Atomic Hydrogen," J. Chem. Phys. 82, 2280 (1985); R.A. Bair and A.F. Wagner, "An Ab Initio Determination of the Rate Constant for $H_2+CN \rightarrow H+HCN$," Int. J. Chem. Kinet., to be submitted.
13. L.B. Harding and A.F. Wagner, unpublished calculations.
14. L.B. Harding and A.F. Wagner, unpublished calculations.
15. J.A. Miller, R.E. Mitchell, M.D. Smooke and R.J. Kee, "Toward a Comprehensive Chemical Kinetic Mechanism for the Oxidation of Acetylene: Comparison of Model Predictions with Results from Flame and Shock Tube Experiments," Nineteenth Symposium (International) on Combustion, The Combustion Institute, Pittsburgh, 1982, pp. 181-196.
16. See, e.g., the review: H.F. Calcote, "Mechanisms of Soot Nucleation in Flames - A Critical Review," Combust. Flame 42, 215 (1981).
17. T. Tanzawa and W.C. Gardiner, Jr., "Reaction Mechanism of the Homogenous Thermal Decomposition of Acetylene," J. Phys. Chem. 84, 236 (1980); see also, Y. Hidaka, C.S. Eubank, W.C.

Gardiner, Jr. and S.M. Hwang, "Shock Tube and Modeling Study of Acetylene Oxidation," J. Phys. Chem. 88, 1006 (1984).
18. S. Benson, Thermochemical Kinetics, Wiley, New York, 1976.
19. G.S. Hammond, "A Correlation of Reaction Rates," J. Am. Chem. Soc. 77, 334 (1955).
20. W. Lange and H.G. Wagner, "Massenspektrometrische Untersochungen uber Erzeugung und Reaktionen von C_2H-Radikalen," Ber. Bunsenges. Phys. Chem. 79, 165 (1975); in German.
21. A.H. Laufer and A.M. Bass, "Photochemistry of Acetylene. Biomolecular Rate Constant for the Formation of Butadiyne and Reactions of Ethynyl Radicals," J. Phys. Chem. 83, 310 (1979).
22. (a) A.M. Renlund, F. Shokoohi, H. Reisler and C. Wittig, "Gas-Phase Reactions of $C_2H(X^2\Sigma^+)$ with O_2, H_2, and CH_4 Studied via Time-Resolved Product Emissions," Chem. Phys. Lett. 84, 293 (1981). (b) C. Wittig, private communication.
23. A. Szekely, R.K. Hanson and C.T. Bowman, "High Temperature Determination of the Rate Coefficient for the Reaction $H_2+CN \rightarrow H+HCN$," Intern. J. Chem. Kinetics 15, 915 (1983).
24. S.R. Langhoff and E.R. Davidson, "Configuration Interaction Calculations on the Nitrogen Molecule," Int. J. Quantum Chem. 8, 61 (1974); E.R. Davidson and D.W. Silver, "Size Consistency in the Dilute Helium Gas Electronic Structure," Chem. Phys. Lett. 52, 403 (1977).
25. L.B. Harding, unpublished.
26. D.G. Keil, K.P. Lynch, J.A. Cowfer, and J.V. Michael, "An Investigation of Nonequilibrium Kinetic Isotope Effects in Chemically Activated Vinyl Radicals," Int. J. Chem. Kinet. 8, 825 (1976).
27. W.A. Payne and L.J. Stief, "Absolute Rate Constant for the Reaction of Atomic Hydrogen with Acetylene Over an Extended Pressure and Temperature Range," J. Chem. Phys. 64, 1150 (1976).
28. R. Ellul, P. Potzinger, B. Reimann and P. Camilleri, "Arrhenius Parameters for the System $(CH_3)_3Si+D_2 \rightarrow (CH_3)_3SiD+D$. The $(CH_3)_3Si$-D Bond Dissociation Energy," Ber. Bunsenges. Phys. Chem. 85, 407 (1981).
29. K. Sugaware, K. Okazaki and S. Sato, "Kinetic Isotope Effect of Chemically Activated Vinyl Radicals," Bull. Chem. Soc. Japan 54, 1222 (1981).
30. C.P. Fenimore and G.W. Jones, "Destruction of Acetylene in Flames with Oxygen," J. Chem. Phys. 39, 1514 (1963).
31. NASA Panel for Data Evaluation, "Chemical Kinetics and Photochemical Data for Use in Stratospheric Modeling. Evaluation No. 5," JPL Publication 82-57, 1982.
32. I.T.N. Jones and K.D. Bayes, "Free-Radical Formation in the $O+C_2H_2$ Reaction," Fourteenth Symposium (International) on Combustion, The Combustion Institute, 1973, pp. 277-284; I.T.N. Jones and K.D. Bayes, "The Kinetics and Mechanism of the Reaction of Atomic Oxygen with Acetylene," Proc. R. Soc. Lond. A335, 547-562 (1973).
33. (a) A.A. Westenberg and N. de Haas, "Absolute Measurements of

the $O+C_2H_2$ Rate Coefficient," J. Phys. Chem. 73, 1181 (1969); (b) A.A. Westenberg and N. de Haas, "A Flash Photolysis-Resonance Fluorescence Study of the $O+C_2H_2$ and $O+C_2H_3Cl$ Reactions," J. Chem. Phys. 66, 4900 (1977).
34. E.N. Aleksandrov, V.S. Arutyunov and S.N. Kozlov, "Investigation of the Reaction of Atomic Oxygen with Acetylene," Kinetics and Catalysis, 22, 391 (1981).
35. D. Saunders and J. Hiecklen, "Some Reactions of Oxygen Atoms. I. C_2F_4, C_3F_6, C_2H_2, C_2H_4, C_3H_6, 1-C_4H_8, C_2H_6, c-C_3H_6, and C_3H_8," J. Phys. Chem. 70, 1950 (1966).
36. K. Hoyermann, H. Gg. Wagner and J. Wolfrum, "Zur Reaktion $O+C_2H_2 \rightarrow CO+CH_2$," Z. Phys. Chem. Neue Folge, Bd. 63, 193 (1966); in German.
37. J.T. Herron and R.E. Huie, "Rate Constants for the Reactions of Atomic Oxygen (O^3P) With Organic Compounds in the Gas Phase," J. Phys. Chem. Ref. Data 2, 467 (1973).
38. R. Lohr and P. Roth, "Shock Tube Measurements of the Reaction Behavior of Acetylene with O-atoms," Ber. Bunsenges. Phys. Chem. 85, 153 (1981).
39. J. Bargon, K. Tanaka and M. Yoshimine, "Computer Chemistry Studies of Chemical Reactions: The Wolff Rearrangement," in Computational Methods in Chemistry, Ed. J. Bargon, Plenum Press, New York, 1980, pp. 239-274.
40. C.A. Arrington, W. Brennen, G.P. Glass, J.V. Michael and H. Niki, "Reactions of Atomic Oxygen with Acetylene. I. Kinetics and Mechanism," J. Chem. Phys. 43, 525 (1965).
41. J.M. Brown and B.A. Thrush, "E.S.R. Studies of the Reactions of Atomic Oxygen and Hydrogen with Simple Hydrocarbons," Trans. Faraday Soc. 63, 630 (1967).
42. D.G. Williamson, "The Reaction of $O(^3P)$ with Dideuterioacetylene," J. Phys. Chem. 75, 4053 (1971).
43. J.R. Kanofsky, D. Lucas, F. Pruss and D. Gutmann, "Direct Identification of the Reactive Channels in the Reactions of Oxygen Atoms and Hydroxyl Radical with Acetylene and Methylacetylene," J. Phys. Chem. 78, 311 (1974).
44. B. Blumenberg, K. Hoyermann and R. Sievert, "Primary Products in the Reactions of Oxygen Atoms with Simple and Substituted Hydrocarbons," Sixteenth Symposium (International) on Combustion, The Combustion Institute, 1977, pp. 841-852.
45. R.J. Buss, R.J. Baseman, G. He and Y.T. Lee, "Further Investigation of the Reaction of $O(^3P)$ with Acetylene by the Crossed Molecular Beams Method," in Report on the Combustion Contractors' Meeting, Brookhaven National Laboratory, 1983, pp. 176-178.
46. A.R. Clemo, G.L. Duncan and R. Grice, "Reactive Scattering of a Supersonic Oxygen-atom Beam: $O+C_2H_4$, C_2H_2," J. Chem. Soc. Faraday Trans. 2 78, 1231-1238 (1982).

DYNAMICS CALCULATIONS BASED ON AB INITIO POTENTIAL ENERGY SURFACES

Donald G. Truhlar, Franklin B. Brown, David W. Schwenke and Rozeanne Steckler, Department of Chemistry, University of Minnesota, Minneapolis, Minnesota 55455

Bruce C. Garrett, Chemical Dynamics Corporation, Columbus, Ohio 43220

1. INTRODUCTION

One of the most exciting developments in theoretical chemistry in the last few years has been the production of ab initio potential energy surfaces by electronic structure calculations and the use of these surfaces for dynamics calculations. With accurate enough surfaces, these dynamics calculations may yield results that rival the accuracy attainable experimentally. When that is achieved one also benefits from the extra detail available in the theoretical output. For example, the theoretical results may include interesting information about the dependence of cross sections on initial vibrational states in cases where only the initial translational energy has been experimentally varied, or they may yield product rotational distributions in cases where the experimental product-state resolution is only sufficient to distinguish vibrational structure. In other cases, theoretical rates may be calculated for systems on which no experiments have been performed. An even more dramatic example is the ability of theory to provide opacity functions, which are transition probabilities as functions of impact parameter. These functions are absolutely unattainable experimentally. Of course, for many or even most systems of interest the available potential energy surfaces and those that may be calculated with state-of-the-art methods and basis sets are either not of chemical accuracy or are at least not of demonstrated

R. J. Bartlett (ed.), Comparison of Ab Initio Quantum Chemistry with Experiment for Small Molecules, 95–139.

reliability. Sometimes additional errors are also introduced by the dynamics calculations. Thus the field of ab initio potential energy surfaces combined with dynamics calculations is currently in a critical infancy stage involving the testing of various methodologies and attempts to demonstrate for a few prototype systems what can and cannot be accomplished.

In our own research group, we have used ab initio potential energy surfaces to perform dynamics calculations (i.e., calculations of cross sections, rate constants, transition probabilities, or dynamical attributes such as threshold energies) on several systems including the chemical reactions $H + p\text{-}H_2 \rightarrow o\text{-}H_2 + H$ or 3H [1-19], $OH + H_2 \rightarrow H_2O + H$ [8,20-22], $^{35}Cl + H^{37}Cl \rightarrow H^{35}Cl + {}^{37}Cl$ [23], $O + OH \rightarrow O_2 + H$ or HO_2 [24], $O + H_2 \rightarrow OH + H$ [25], $F + H_2 \rightarrow HF + H$ [26,27] $H + H'F \rightarrow HF + H'$ [26], $H + H'Cl \rightarrow HCl + H'$ [28] (including in most cases additional isotopic analogs), and energy-transfer processes in collisions of H with H_2 [29,30], He with HD [31], He with I_2 [32,33], and HF with HF [34]. In keeping with the theme of the present symposium, namely the state of the art of electronic structure calculations and the comparison of results obtained from such calculations to experiment, we present here a review of a selected subset of systems recently studied in our group for which the potential energy surfaces are, in some sense, state-of-the-art. In particular, we discuss $F + H_2$, $H + H'F$, $H + H'Cl$, $H + CH_3$, $He + I_2$, and HF + HF collisions, as well as additional isotopic analogs in some cases. In all these cases we performed at least some electronic structure calculations [26,28,33-37] in our own group; but for HF + HF our dynamics calculations are based entirely on a surface calculated by Binkley and fit by Redmon [38]. It is becoming increasingly clear that, when electronic structure calculations and dynamics calculations are not performed in the same group, at least a close collaboration of potential-energy-surface builders with potential-energy-surface users is highly desirable.

The methods used in the electronic structure calculations that were performed in our group can be divided into two categories:

1) methods employing a single reference (SR) configuration and 2) methods employing multiple (two or more) reference (MR) configurations. If the system under investigation can be described reasonably well by a single-configuration wavefunction, as is the case for the H-F-H' system even at its saddlepoint [26], then SR methods are employed. In particular the orbitals are optimized using the spin-restricted Hartree-Fock self-consistent-field method [39], which will be abbreviated RHF, and the correlation energy is calculated from a configuration interaction (CI) wavefunction that includes all single and double (SD) excitations from one reference configuration [40]; such a CI calculation will be abbreviated SR-CISD, and the combination of methods will be denoted RHF/SR-CISD. When more than one configuration is too important to be considered as a perturbation, e.g., in the dissociation of CH_4 [35], then MR methods are used. In the MR methods, the orbitals are optimized with respect to the set of reference configurations in a multi-configuration self-consistent-field (MCSCF) calculation [41-44], and the CI wavefunction consists of all single and double excitations from the same set of reference configurations. The resulting wavefunction will be denoted as MCSCF/MR-CISD. A special case occurs when the set of reference configurations consists of all spin- and symmetry-allowed occupations of a number of electrons in a pre-specified manifold of orbitals, called the active orbitals. In this case the reference configurations are said to form a complete-active-space (CAS) [41] set, and the resulting wavefunction will be denoted CASSCF/MR-CISD. These methods, RHF/SR-CISD, MCSCF/MR-CISD, and CASSCF/MR-CISD, were used as implemented in the COLUMBUS [45-48] electronic structure codes.

Although the present chapter is written in the form of a review, it does include some new work (both electronic structure and dynamics calculations) not described elsewhere. Since the present chapter is devoted primarily to our own work, we conclude this introduction by giving a few references to recent reviews from which other recent work along similar lines may be traced [49-52].

2. CHEMICAL REACTIONS

2.1 H + FH' → HF + H' and H + ClH' → HCl + H' : *ab initio* predictions of high barriers

Several recent experimental [53-55] and theoretical [56-60] investigations have shed light on the potential energy surface of the H + FH' → HF + H' reaction, especially in the region of the H-F-H' saddlepoint. These investigations have indicated that the barrier height for this thermoneutral exchange reaction is significantly higher than that predicted by many of the standard extended-LEPS-type [61] potential energy surfaces, including the Muckerman surface no. 5 [62] (M5), which is widely used for the alternative reaction channel F + HH' → HF + H', but which has a barrier height for the exchange channel of only 1.8 kcal/mol [63]. In comparison, Bott has concluded that the activation energy, E_a, must be greater than 19 kcal/mol to be consistent with rate constants measured in a shock tube study [54]. Furthermore, by monitoring the chemiluminescence and mass spectrum of the products from the reaction of vibrationally excited HF molecules with D, Bartoszek *et al*. [55] have obtained a set of bounds placing the effective threshold energy for the exchange channel in the range 41-52 kcal/mol. *Ab initio* values of the barrier height derived from calculations at the collinear H-F-H' saddlepoint geometry are presented in the first six rows of Table 1 and these values, which are in the range 44-49 kcal/mol, are in excellent agreement with the bounds of Bartoszek *et al*. [55]. Wadt and Winter [58] have also calculated the unconstrained H-F-H' saddlepoint and their result is given in the last row of Table 1; the energy lowering due to relaxing the collinear constraint is only about 1 kcal/mol. In further investigations of the bend potential of the H-F-H' saddlepoint, they found, for R(H-F) = 2.154 a_0, that the potential energy for C_{2v} H-F-H' is almost constant when the bond angle $\theta_{HFH'}$ is varied from 180 to 120 deg. Thus, the remainder of this section will only discuss the collinear saddlepoint.

We performed several large-scale configuration interaction

Table 1. Summary of _ab initio_ energies E and classical barrier heights $V^{\neq}$ for H - F - H'.

$R(H-F)^a (a_0)$	$E(E_h)$	$V^{\neq}$ (kcal/mol)	Source
Collinear saddle point			
$R(HF)(a_o)$			
2.154	...	49.0	Bender _et al._ [56]
2.160	-100.7750	44.9	Botschwina and Meyer [57]
2.154	-100.6583	48.1	Wadt and Winter [58]
2.154	...	47.7	Voter and Goddard [59]
2.230	-100.6622	48.8	Dunning [60]
2.154	-100.7328	44.4	Table 2 and Ref. [26]
Unconstrained saddle point[b]			
2.041	-100.6602	46.9	Wadt and Winter [58]

[a] R(H-F) = R(F-H')

[b] θ, the H-F-H angle, is 106 deg.

calculations on FH_2 to calibrate two new analytic potential energy surfaces for this system. These new FH_2 surfaces, called surfaces nos. 4 and 5, are based on _ab initio_ calculations for the H-F-H' saddlepoint region and experimental data [64-66] for the F-H-H' saddlepoint region, and they have also been calibrated in the F-H...H exit channel region, with surface no. 4 based in that region on results from additional _ab initio_ calculations [26] that will be discussed in the next section, and surface no. 5 further adjusted semiempirically.

In order to determine which _ab initio_ approach, SR or MR, is needed to characterize the H-F-H' saddle point region accurately, a preliminary set of calculations was performed using both methods with a 6-311G** basis set [26]. A comparison of the results from these calculations indicated that this saddlepoint region is described well by a single dominant configuration and thus the RHF/SR-CISD method was used in the larger-scale calculations. In all calculations, the F 1s orbital was constrained to be doubly occupied and a partial Hartree-Fock interacting space limitation [67] was imposed on the configuration space. Also we calculated a correction for the effect of unlinked quadruple excitations [40] using the standard Davidson correction [68], namely,

$$\Delta E_Q = \Delta E_{SD}(1 - c_0^2) \qquad (1)$$

where ΔE_Q is the estimate of the energy contribution from the unlinked quadruple excitations, ΔE_{SD} is the correlation energy from the RHF/SR-CISD calculation, and c_0 is the coefficient of the reference configuration in the RHF/SR-CISD wavefunction. Results including the Davidson quadruples correction will be denoted RHF/SR-CISD + Q.

In the final calculations used in determining the H-F-H' barrier height, we employed a large (10s6p4d/5s3p)/[8s5p4d/4s3p] contracted gaussian basis set. This basis set was adapted from the smaller basis of Botschwina and Meyer [57] by addition of diffuse p subshells to the F and H atoms (with exponential parameters of 0.0796 and 0.120, respectively) and by replacement of the two d subshells on F with four (with exponential parameters of 3.0, 1.09, 0.40, and 0.14). In this basis set, the SCF/SR-CISD wavefunction in C_{2v} symmetry includes 22760 configurations.

The results of our calculations [26] for two geometries in the vicinity of the H-F-H' saddlepoint are summarized in Table 2. The geometry given in the first row is that of the saddlepoint that was optimized by Wadt and Winter [58] using the PNO-CEPA method with a (12s8p3d1f/6s2p)/[9s6p3d1f/4s2p] gaussian basis set. The geometry given in the second row is obtained by an asymmetric stretch of the first geometry. The calculated barrier height of 44.4 kcal/mol is consistent with the high barrier obtained in the previous *ab initio* calculations [56-60], as shown in Table 1. The results in Table 2 were used in the calibration of surfaces nos. 4 and 5 discussed in the next section. Using the final surface, no. 5, the shape of the H-F-H' exchange barrier may be compared to that for the very well characterized H-H-H' exchange barrier. In a recent paper Liu [69] has presented new calculations for the potential energy surface for H_3 that are believed to converge it to within about 0.1 kcal/mol of the exact surface in the 3-body interaction region. On the basis of these calculations he recommended subtracting 0.15-0.21 kcal/mol from the previous [70] H + H_2 interaction energies. In Figure 1 we compare H + HH' interaction energies, obtained by subtracting 0.15 kcal/mol (which is hardly noticeable on the scale of the plot) from the previous [70] surface, to H + FH' interaction energies calculated from surface no. 5. In both cases the interaction energy is given as a function of the asymmetric stretch coordinate. It can be seen that the two exchange barriers have quite different characters. For H-F-H', the rectilinear asymmetric stretch lowers the interaction energy to a small fraction of its transition state value before a repulsive wall is encountered, whereas for H-H-H' this is not true. The consequences of this difference for the dynamics are unknown.

Table 2. Total and interaction energies for SR-CISD calculations in the vicinity of the saddlepoint for H+FH' → HF+H'.

R(F-H') (a_0)	R(F-H) (a_0)	$\theta_{HFH'}$ (deg)	E(SR-CISD) (E_h)	ΔE(SR-CISD)[a] (kcal/mol)	ΔE(SR-CISD+Q)[a] (kcal/mol)
2.1540	2.1540	180	-100.732837	47.85	44.40
1.8540	2.4540	180	-100.756668	32.89	31.58

[a] ΔE = E(H-F-H') - E(H+H'F) separated from H'F by 20 a_0.

Another exchange reaction that has received much experimental [71-80] and theoretical [57,59,60,81,82] attention is the H + ClH' → HCl + H' reaction and its isotopic analogs. The height of the H-Cl-H' transition state is not nearly as well characterized

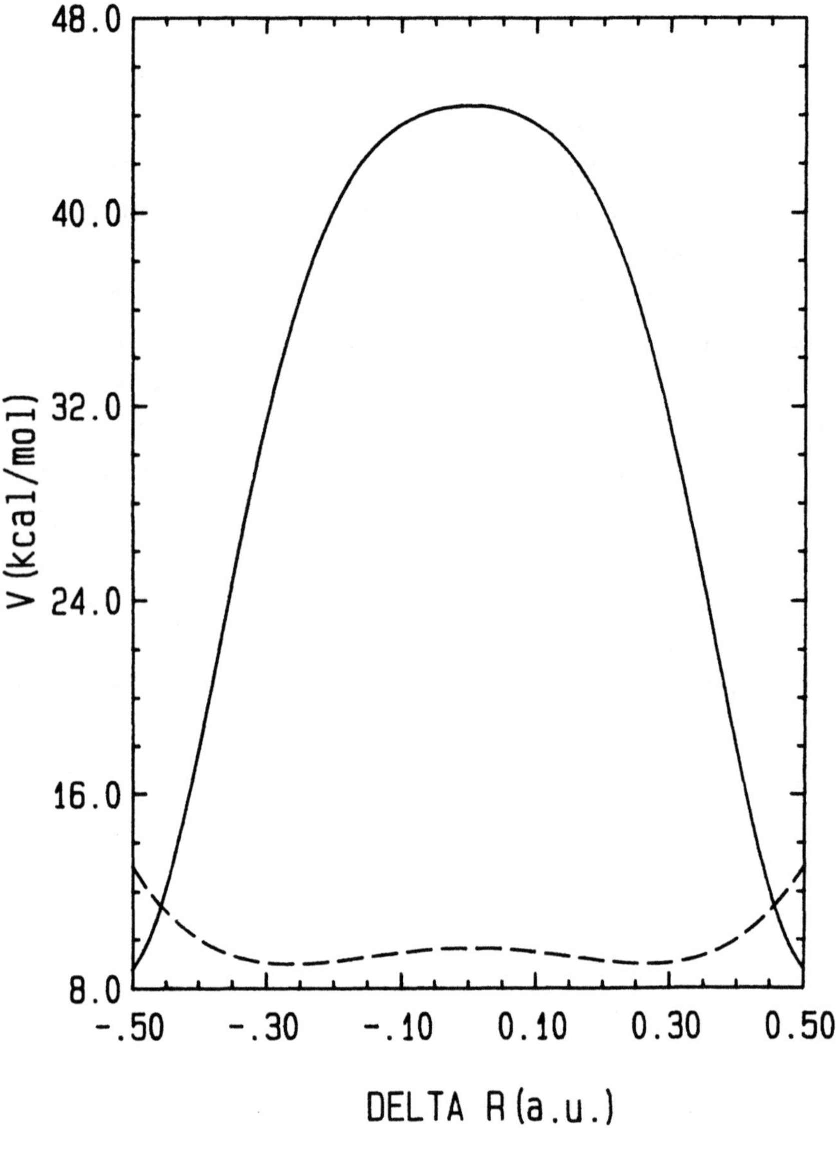

Figure 1. Exchange barrier for H+FH' (solid curve) and H+HH' (dashed curve) along the asymmetric stretch coordinate. The geometries are defined by $R(H-X) = R^{\neq}(H-X) + \Delta R$, $R(X-H') = R^{\neq}(H-X) - \Delta R$, and $R(H-H') = R(H-X) + R(X-H')$, where $R^{\neq}(H-X)$ is 2.154 a_0 for X=F and 1.757 a_0 for X=H. The ordinate is interaction energy, defined as $\Delta E = E(HXH') - E(H + XH')$.

as is that of H-F-H'. The results from several relevant experimental measurements are mutually inconsistent [71-81]. For example, Klein and Veltman [75] have measured the branching ratios for the D(H) + HCl(DCl) system using a discharge flow reactor with a quadrupole mass spectrometer for product detection. Combining their value for the difference in activation energies of the abstraction and exchange reactions with a recent value of 3.2 kcal/mol [79] for E_a of the abstraction reaction, one obtains a value of E_a = 2.1 kcal/mol for the exchange reaction. Endo and Glass [74], employing a discharge flow reactor with EPR detection, obtained $E_a \geqslant 4$ kcal/mol. A more direct experiment, by Miller and Gordon [79], which utilized laser photolysis to generate the D atoms and resonance fluorescence to monitor the concentration of the H, D, and Cl atoms present, yielded no observable exchange at 325K, and this put an approximate lower limit of 7 kcal/mol on E_a.

McDonald and Herschbach [73] observed the exchange reaction D + HCl → DCl + H in a molecular beam apparatus with a nearly Maxwellian speed distribution and a mean relative transitional energy of 9 kcal/mol. For such a distribution, 13% of the collisions have a collision energy greater than 20 kcal/mol and 10% have a collision energy in excess of 22 kcal/mol. Thus, as pointed out by Miller and Gordon [79], this experiment only places an approximate upper bound of about 20-22 kcal/mol on the threshold energy and hence presumably an approximate bound of about the same magnitude on the barrier height. Toennies and coworkers, in a later study, at first reported a confirmation of the McDonald-Herschbach experiment, and obtained an apparent activation energy of 20 ± 6 kcal/mol [77]. Later, however, a more refined and extensive experimental study [78] indicated that the observed H atoms did not arise from the exchange reaction as originally assumed. As a result these experiments do not appear to give any information about the threshold energy for exchange.

Very recently, Wight et al. [80] have observed infrared fluorescence from vibrationally and rotationally excited HCl produced by the reaction of DCl with H atoms with 22 kcal/mol translational energy. This places an upper bound of 22 kcal/mol on the exchange threshold. The bound could not be improved because no precursor molecule is known for which excimer laser photolysis yields H atoms with a lower translational energy.

If we assume that activation energy, threshold energy, and classical barrier height are the same within about 2 kcal/mol, then the two most recent experiments [79,80] may be combined to place the exchange barrier in the range 5–24 kcal/mol, whereas the Wood [72] and Klein-Veltman [75] studies imply lower barriers.

Semiempirical LEPS-type potential energy surfaces that have been calibrated for the Cl + HH' → HCl + H' reaction have a shallow well on the exchange reaction path [83]. *Ab initio* calculations of the barrier height for the exchange reaction have not yet completely converged, but they definitely yield a more consistent picture than the experiments or the semiempirical calculations. Firstly, they all predict a barrier rather than a well for symmetric H-Cl-H' geometries. Botschwina and Meyer [57] have performed PNO-CEPA calculations on this system using a large (13s10p3d1f/6s2p)/[9s7p3d1f/4s2p] basis set and find a barrier of 22.1 kcal/mol. However, they estimate the value of the true barrier height to lie between 10 and 15 kcal/mol based upon the use of a semiempirical methods [57,84] for correcting the calculated value for errors in the correlation energy arising from the use of a truncated basis set and a truncated CI expansion. Dunning [82] has investigated this system by optimizing the geometry of the saddlepoint using the POL-CI method in a (12s9p2d/5s1p)/[4s4p2d/3s1p] basis set. A barrier height of 25.3 kcal/mol was obtained from a GVB+1+2 calculation using this basis set and geometry. Based upon an estimate of the error in the calculated Cl-H-H' barrier height, Dunning concludes that the true H-Cl-H' barrier height lies in the range 14.1-21.7 kcal/mol. Using the generalized resonating valence bond method with a valence double zeta plus polarization basis set, Voter and Goddard [59] have calculated a barrier of 25.5 kcal/mol at Dunning's collinear geometry. Recently, Dunning has re-examined the H-Cl-H' barrier height using the same GVB+1+2 method but with a smaller (11s7p1d/4s2p)/[4s3p1d/2s1p] basis, and he obtained a barrier of 23.5 kcal/mol [60]. Thus all previous *ab initio* calculations predict a high barrier height in the range of 21-26 kcal/mol, although in some cases lower values in the range 10-22 kcal/mol have been estimated by considering various errors in the calculations.

We have recently begun an investigation of H-Cl-H' barrier region using the CASSCF/MR-CISD method [28]. The reference space consists of 28 configurations constructed from the 9 valence electrons occupying the 6 valence orbitals. We have used Botschwina and Meyer's larger, (13s10p3d1f)/[9s7p3d1f], chlorine basis, and we augmented their smaller, (5s2p)/[4s2p], hydrogen basis with a d subshell with an exponential parameter of 1.67. The largest calculation performed on bent C_{2v} H-Cl-H' included 323908 configurations. Using the PNO-CEPA collinear saddlepoint geometry of Botschwina and Meyer [57], for which R(H-Cl) = R(Cl-H') = 2.8384 a_0, we have obtained a total energy of -460.787253 E_h and a barrier height of 20.9 kcal/mol. Interestingly, an RHF/SR-CISD+Q calculation with the same one-electron basis yields a barrier height of 20.7 kcal/mol indicating that the multi-reference CISD calculation includes the most important geometry-dependent quadrupole excitations from the dominant configuration. These values, 20.7 - 20.9 kcal/mol, are slightly lower than the uncorrected values reported earlier [57,59,60,82] and are within the upper bound that may be estimated from the experiment of Wight *et al* [80].

In summary, *ab initio* calculations without semiempirical corrections yield high barrier heights of about 21 kcal/mol or greater. Although this high a value does not contradict the results from recent experiments, further work needs to be done on this system to make the predictions more definitive.

2.2 Entrance- and exit-channel barriers in the $F + H_2 \rightarrow HF + H$ reaction and isotopic analogs

The $F + H_2$ reaction has been widely studied both experimentally and theoretically, largely because of its importance in pumping the HF chemical laser [85-87]. Currently, interest has been heightened by theoretical and experimental evidence of reactive reasonances for this system [64-66,88,89]. Most of the early experiments concentrated on measurements of product state distributions [90-96] and recently there have been precise determinations of the room temperature thermal rate constants [97,98] and measurements of detailed vibrational-state-resolved differential cross sections [64-66, 99].

Although _ab initio_ potential energy data has been available for the $F + H_2$ reaction for over 12 years [100-102], most dynamical studies have used the semiempirical surface no. 5 of Muckerman (M5) [62] because of its convenience and because a fit to the available _ab initio_ results is not definitely more reliable. The M5 surface is an extended LEPS form in which the two Sato parameters were adjusted so that: (a) the room temperature activation energy computed using conventional transition state theory reproduced the experimental value [103] accepted at that time of 1.7 kcal/mol; and (b) the average vibrational energy of the product HF molecule computed using quasiclassical trajectories [62] agreed with experiments [91] for the fraction of the exothermicity deposited in product vibration under thermal conditions.

As experiments on the $F + H_2$ reaction and its isotopic analogs have become more refined and have offered more precise and detailed information, comparison between these data and dynamical calculations on the M5 surface have indicated deficiencies in this surface. First, two recent experiments [97,98] give an activation energy for the $F + H_2$ reaction of 0.9 to 1.1 kcal/mol rather than the 1.7 kcal/mol used to adjust the M5 surface. Secondly, cross sections for production of HF(v=3) and DF(v=4) (v denotes vibrational quantum number) computed using approximate quantal methods [89] on the M5 surface are not appreciable until almost 2 kcal/mol above the energetic threshold. These "delayed thresholds" are not observed experimentally [64-66]. In Figure 2a the heights of the vibrationally adiabatic potential curves are given at critical points of the potential energy surface for the $F + H_2$ and $F + D_2$ reactions. In each case the zero of energy is taken as the infinitely separated reactants in their ground vibrational state. In this diagram one sees that HF(v=3) and DF(v=4) both have energies relatively close to ground-state reactants, hence the delayed threshold effect is seen at energies just above the overall reaction threshold.

Direct comparison with *ab initio* data also indicates errors in the M5 surface. The potential energy change as F-H-H is bent from collinear geometries in the entrance channel is much steeper for the M5 surface than seen in *ab initio* calculations [100]. Also, as discussed in the previous section, the H-F-H' exchange barrier on the M5 surface is much lower than indicated by *ab initio* calculations [26,56-60].

We have used variational transition state theory (VTST) methods [2,6,8,104-108] first to learn which regions of the surface are most critical for determining thermal rate data (including activation energies) and are responsible for the magnitudes or absence of delayed thresholds, and, second, to characterize those portions of the potential energy surface. In VTST the dynamical bottleneck to chemical reaction at low temperatures is located at the maximum in the ground-state adiabatic potential curve [6]. The adiabatic potential curve is given by

$$V_a(\alpha,s) = V_{MEP}(s) + \varepsilon_{int}(\alpha,s) \qquad (2)$$

where s is the distance along the minimum energy path (MEP), α and $\varepsilon_{int}(\alpha,s)$ are respectively the set of quantum numbers and the internal energy for the degrees of freedom normal to the reaction path, and $V_{MEP}(s)$ is the potential along the MEP. For the $F + H_2$ reaction the barrier on the entrance-channel portion of the ground-state adiabatic potential curve controls the thermal rates and

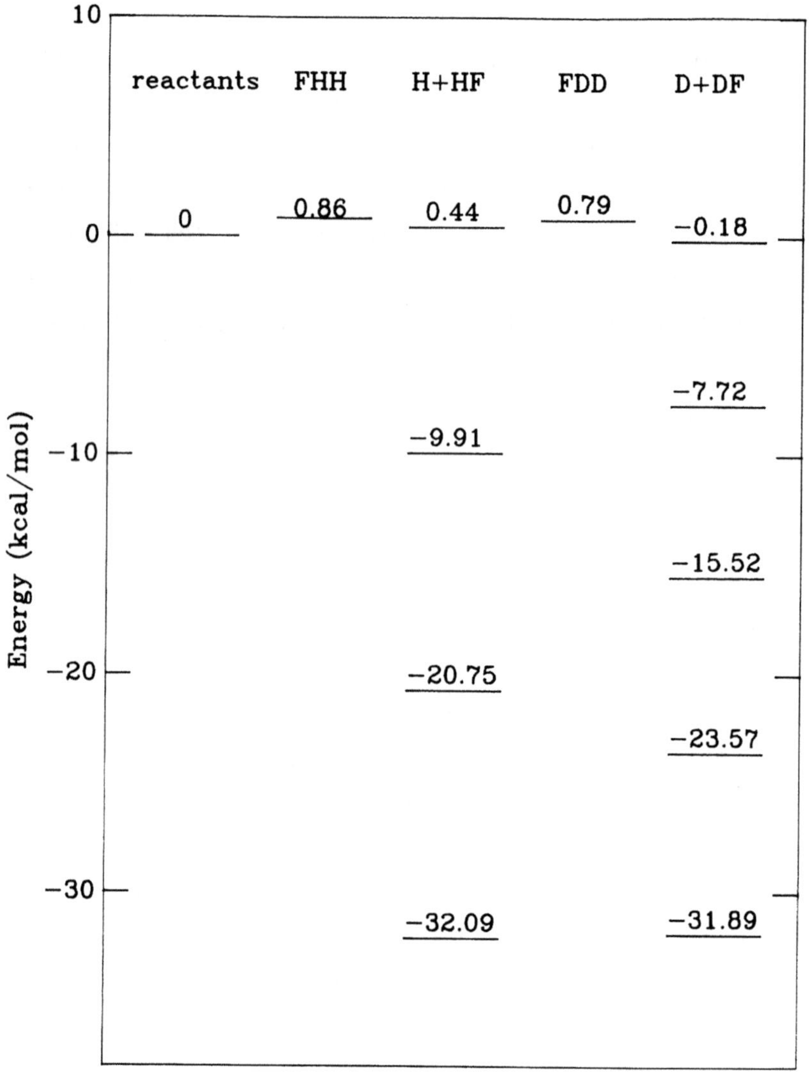

Figure 2a. Energetics of the three-dimensional $F + H_2 \rightarrow H + HF$ and $F + D_2 \rightarrow D + DF$ systems. The energies for each system are relative to reactants in the ground vibrational state for that system; this zero of energy is shown in the first column. The vibrationally adiabatic energies for two saddlepoints, FHH and FDD, are also shown in the figure. The third and fifth columns show the vibrational levels of the products, namely the ground state and the first three excited vibrational states for HF and the ground state and the first four excited vibrational states for DF.

activation energies. This barrier is below the energetic threshold for production of the excited species HF(v=3) on the M5 surface. Thus the entrance-channel barrier does not determine the threshold for production of this excited state product. We have compared [109] barriers on exit-channel portions of excited-state adiabatic potential curves with thresholds for production of excited states obtained from an approximate quantal method (the bend-corrected rotating linear model or BCRLM) [110], and we found good agreement when the stretch vibrational energy levels in Eqn. (2) are computed using the WKB approximation [108]. The exit-channel barrier determines the delayed threshold when it is higher than both the energetic threshold and the entrance-channel barrier. In summary then, if a surface is to be accurate both for overall rate constants and for state-specific production distributions, it is important that the surface be accurate in two regions, namely the region near the maximum of $V_a(\alpha$ = ground state, s) in the entrance channel and also the region where $V_a(\alpha,s)$ has its maxima for HF(v=3) and DF(v=4) in the exit channel.

One approach to fitting these critical portions of the potential energy surface more accurately is to vary the potential in these regions until the results of dynamical calculations agree with the experimental results. Alternatively, the potential in a critical region can be improved by fitting to accurate ab initio potential data. Although the latter approach is more pertinent to the theme of the present discussion, it was used only for the exit channel. First we discuss an attempt [111] we made to use the former approach to reoptimize an extended LEPS surface in the entrance channel region in a manner similar to that used by Muckerman [62]. As the first step, the Sato parameters were adjusted (i) so that the activation energy computed by canonical variational theory [2,8,104-106,108] with a small-curvature semiclassical adiabatic ground state tunneling correction [8,107] (CVT/SCSAG) reproduces the experimental activation energy [97,98] for the F + D_2 reaction in the temperature range 295-373 K; and (ii) so that a model prediction of the average vibrational energy of the DF product agrees with experiment. The reoptimized LEPS potential was used for the collinear part of a new surface and the non-collinear part was refit using ab initio data [100]. The fit was accomplished by replacing the constant HF Sato parameter by one that is a function of the angle between the H_2-to-F vector and the HH axis. Some thermal rate constants for the $F + H_2$ and $F + D_2$ reactions computed using the CVT/SCSAG method on surfaces M5 and the new surface, called no. 2, are presented in Table 3 and are compared with experiment [97,98,112]. Surface no. 2 is seen to be in much better agreement than surface M5 with these experiments for both reactions; however, it still predicts significant, and presumably erroneous, delayed thresholds for HF(v=3) and DF(v=4).

Table 3. Thermal rate constants (units of cm^3 $molecule^{-1}s^{-1}$) for the $F + H_2$ and $F+D_2$ reactions

T(K)	Surface	CVT/SCSAG M5	CVT/SCSAG No. 2	CVT/SCSAG No. 5	experiment
200	$F + H_2$	1.1(-12)[a]	8.7(-12)	1.3(-11)	1.1(-11)[b]
300		3.9(-12)	1.7(-11)	2.5(-11)	2.8(-11)[b]
600		1.5(-11)	3.7(-11)	5.7(-11)	7.3(-11)[b]
200	$F + D_2$	7.9(-13)	5.5(-12)	7.1(-12)	5.7(-12)[c]
300		2.5(-12)	1.1(-11)	1.4(-11)	1.4(-11)[c,d]
600		9.4(-12)	2.3(-11)	3.3(-11)	3.9(-11)[d]

[a] Numbers in parenthesis are powers of ten.
[b] Preferred values from review of reference 112.
[c] Reference 97.
[d] Reference 98.

To try to learn the nature of the potential energy surface in the region responsible for the delayed thresholds, ab initio electronic structure calculations were performed for geometries in the product channel [26]. The details of these calculations have been described in Ref. 26 and Section 2.1. In the product region an SR approach might be expected to be adequate because FH...H is similar to Ne...H, and one electronic configuration dominates. At four geometries, we performed both RHF and CASSCF calculations to test this assumption, and it was confirmed. Thus the SR approach was used for the calibration. Using surface no. 2 as a starting point, the ab initio data was used to fit the HH Sato parameter to a function of the HH distance and the bend angle. The fit was tailored to change the potential in the entrance channel as little as possible. The resulting surface is called surface no. 4. In Figure 2b, we compare the entrance- and exit-channel adiabatic potential curves for the F + H_2 reaction for the M5 and no. 4 surfaces. By using ab initio data to refit the exit-channel region of the potential energy surface, the exit-channel barrier for the production of HF(v=3) has been reduced from 2.5 kcal/mol above the energetic threshold on surface M5 to 1.0 kcal/mol above this energetic threshold, as shown in Figure 2b.

Although surface no. 4 gives a much better description of the state-specific threshold energies than surface M5, the predicted highest-product-state thresholds are still larger than their experimental counterparts, which appear not to be delayed at all above the energetic thresholds. Furthermore, CASSCF/MR-CISD ab initio calculations [26] indicate that improving the reference space and the basis set (by adding an f shell and by changing the exponents on the polarization functions), which increases the number of configurations to 218512, still decreases the potential energies (relative to the HF + H asymptote), but the decrease is very small as the basis is increased. We were able to lower the ab initio surface in the critical region by only 0.2-0.5 kcal/mol, and these small changes are not large enough to account for the discrepancy with experiment. Thus we decided to employ an empirical modification to surface no. 4 to lower the exit-channel barrier even further. The exit-channel region of surface no. 4 was readjusted to fit the threshold energy for the F + HD → HF(v=3) + D reaction,

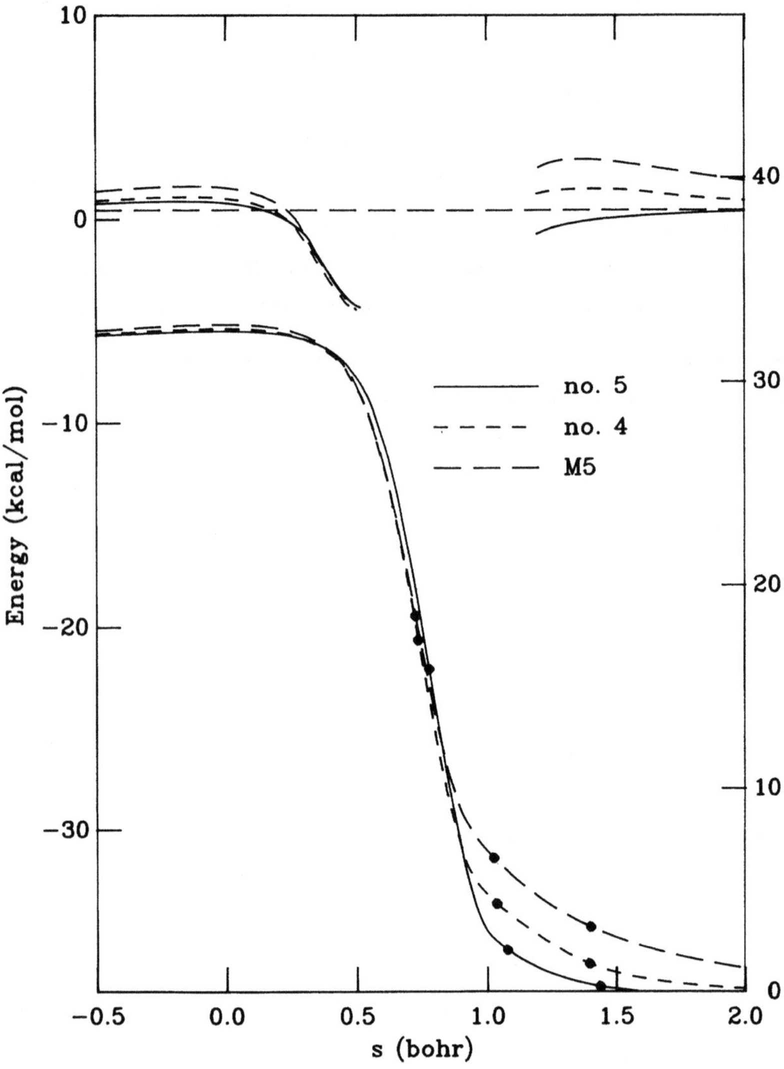

Figure 2b. The classical (V_{MEP}) and vibrationally adiabatic (V_a) energy curves along the MEP for three-dimensional $F + H_2(v=0) \rightarrow HF(v=3) + H$ on surfaces M5, no. 4, and no. 5. The abscissa gives the distance along the MEP with s=0 corresponding to the saddlepoint. The upper two sets of curves show the vibrationally adiabatic entrance-channel (v=0) and exit-channel (v=3) barriers for these surfaces. The lower curves are the potential energy along the MEP with the circles corresponding to H-H bond lengths of 2.0 a_0, 2.5 a_0, and 3.0 a_0. The left-side energy scale is the same as for Figure 2a and the right-side scale is the same as for Figure 1. The dashed horizontal line is the energetic threshold for HF(v=3).

which has the most accurately determined experimental threshold energy [64-66]. The entrance- and exit-channel barriers for the $F + H_2$ reaction on this final surface, no. 5, are also shown in Figure 2b. Surface no. 5 exhibits no delayed threshold for either the $F + H_2 \rightarrow HF(v=3) + H$ or the $F + D_2 \rightarrow DF(v=4) + D$ reaction. The thermal rate constants on surface no. 5 are also shown in Table 3. The entrance-channel region for surface no. 5 is very similar to that for surfaces nos. 2 and 4 and the thermal rate constants do not differ greatly on the three surfaces.

Using a combination of *ab initio* and experimental data a potential energy surface has been constructed which is consistent with the most modern thermal rate data and with the experimental threshold energies for production of vibrationally excited products, as well as with *ab initio* calculations on the H-F-H' exchange barrier. It will be interesting to see the effect surface no. 5 will have upon accurate dynamical calculations of detailed quantities such as differential cross sections and the prediction of reactive resonances.

2.3 $CH_3 + H \rightarrow CH_4$

The dissociation of CH_4 presents an interesting system as the starting point for the development of polyatomic potential energy surfaces. This system can serve as a prototype for both radical recombination and polyatomic dissociation potentials for more complicated organic systems. In a recent study, Duchovic and Hase [113] have demonstrated the sensitivity of calculated rate constants for the recombination reaction $H + CH_3 \rightarrow CH_4$ to the shape of the potential curve in the region where R(C-H) is 3.5-6.0 a_0 (about two to three times the equilibrium value of 2.052 a_0 in methane). In this work, two different potential curves along the reaction coordinate, which corresponds to the making of a C-H bond in methane, were obtained by fitting previous calculations [114] carried out using Møller-Plesset (Many-Body) fourth order (MP4) perturbation theory with 6-31G** basis set to two different functional forms, and these curves were both used to calculate rate constants. The first functional form used was a standard Morse function, and the second form was a "stiff" Morse function in which the constant range parameter was replaced by a polynominal in R(C–H). Figure 3 shows that the main difference in these two potential curves is in the region $3.5\ a_0 < R(C\text{–}H) < 6.0\ a_0$ where the stiff Morse function, which more closely fits the MP4 results, is less attractive than the standard Morse function. Thermal rate constants for the H + CH_3 association reaction at 300 K were calculated by running Monte Carlo classical trajectories on two potential energy surfaces [115] that differ only in the form of the potential curve along the reaction coordinate. It was found that the surface with the stiff Morse function yielded a rate constant ten times smaller than the rate constant from the surface with the standard Morse function, while the experimental [116] rate constant is near the geometric mean of these two values. This shows that the potential in the region $3.5\ a_0 < R(C\text{–}H) < 6.0\ a_0$ plays a very large role in the dynamics of this system.

In a recent study [117] of the symmetric dissociation of H_2O, it was shown that the CASSCF/MR-CISD method gives a more balanced treatment of the correlation energy along the dissociation coordinate than the MP4 method and yields potential energy curves which are more parallel to the true potential. We expect that this may be true for the potential along any dissociation coordinate leading to fragments with free valence. Hence, in order to check the large deviation of the MP4 results [114] from the standard Morse model along the dissociation coordinate of methane, we have investigated [35] this potential curve using the CASSCF/MR-CISD method. For these calculations, we employed a large 6-311++G(df,p) basis set which includes both diffuse and polarization functions on both C and H. CASSCF/MR-CISD calculations, with 63608 configurations, were performed at the same geometries as the MP4 calculations of Duchovic <u>et al</u>. [114] and the results are shown in Figure 3.

In the region where R(C–H) is between 3.5 and 6.0 a_0 our

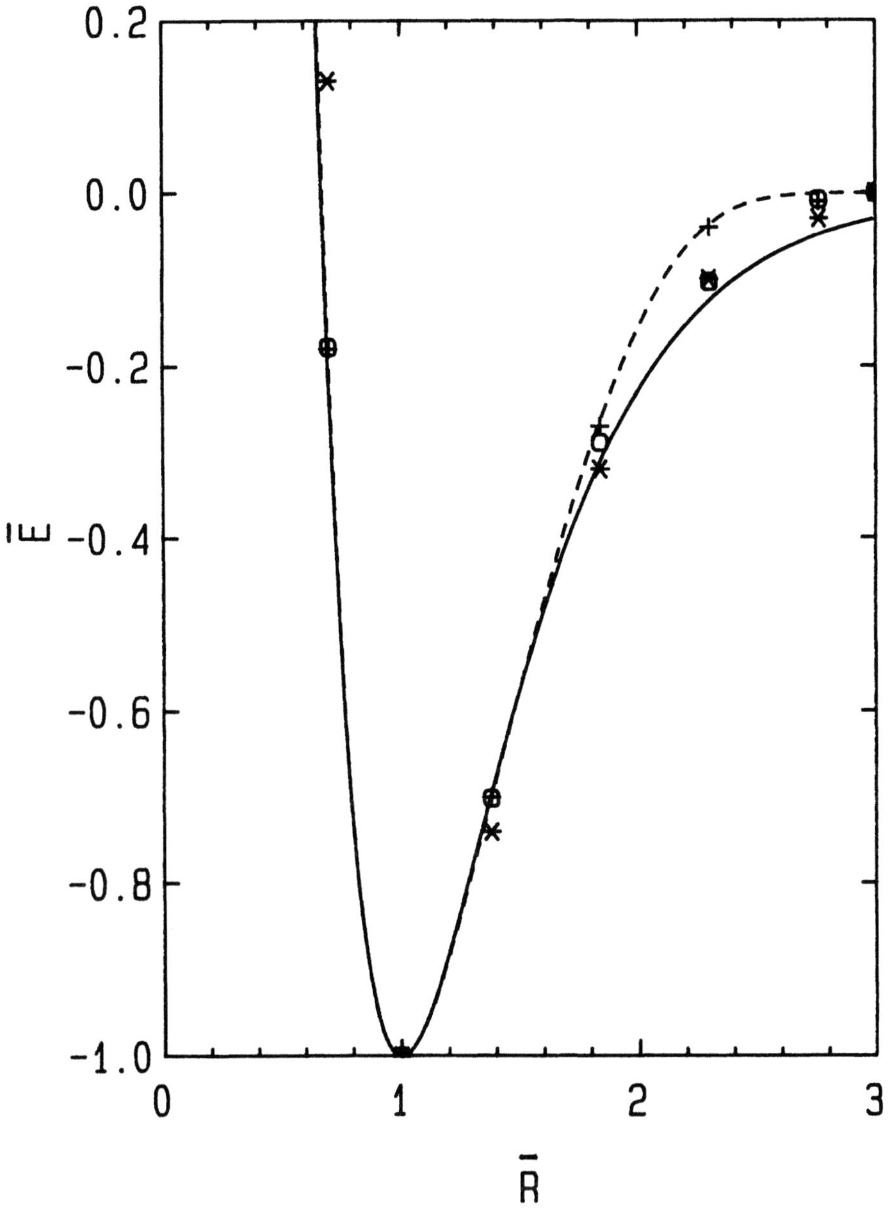

Figure 3. Potential in reduced units, $\bar{E} = [E(R(C\text{-}H)) - E(\infty)]/D$, as a function of reduced C-H stretch coordinate, $\bar{R} = R(C\text{-}H)/R_e$, in CH_4: +, MP4 results from Ref. 114; —, standard Morse function based on MP4 results; ---, stiff Morse function based on MP4 results; *, Padé approximant results from Ref. 114; □, CASSCF/MR-CISD results.

MR-CISD results lie between the stiff Morse potential function, corresponding to the MP4 data, and the standard Morse potential function. Thus, presumably, a calculation of the thermal rate constant at 300 K using a functional fit of the CASSCF/MR-CISD data would yield a rate constant in better agreement with the experimental value. We have found [35] that the "first-order" Lippincott function [118]

$$V = -D \exp[-\beta(R-R_e)^2/2R] \tag{3}$$

and the Varshni III function [119]

$$V = D[\{1 - \frac{R_e}{R} \exp[-\beta(R^2 - R_e^2)]\}^2 - 1] \tag{4}$$

fit the CASSCF/MR-CISD results well, which should not be too surprising since, in fact, these functional forms fit the RKR curves and experimentally derived spectroscopic constants for several diatomics better than the standard Morse curve [35,118,119]. Interestingly, Halonen and Child [120,121] have quite successfully calculated stretch overtone bands using the standard Morse function with the parameters fit to the known stretching vibrational energy levels for the C-H stretch in CH_4 and its C_{3v} isotopic analogs. However, these energy levels are located in the bottom half of the well where the Morse approximation works well, as Figure 3 shows. More recently, Peyerimhoff *et al.* [122] have performed a set of RHF/MR-CISD calculations (MRD-CI in their notation) and found that a standard Morse function gives a "reasonable" fit to these *ab initio* results along the C-H stretch. In contrast, we find that although the potential energy curve along the reaction coordinate of the polyatomic reaction is not unlike the dissociation potential of a typical diatomic system, it is not quantitatively represented by a Morse potential.

The failure of the MP4 method, a single-reference method, in describing H_3C-H in the intermediate bonding region can be attributed to the inability of a single reference function to describe the system adequately in this region. This inadequacy is demonstrated in Table 4 where the square of the coefficient c_0 of the dominant configuration and sum of the squares of the coefficients c_i of the reference configurations in the CASSCF/MR-CISD wavefunction are presented for each geometry. The table shows that, near equilibrium, a single configuration describes the system well, whereas at larger R(C-H), the dominant configuration has a much smaller coefficient. In contrast, however, the table also shows that the total contribution of all the reference configurations is almost constant at all geometries. Thus, we believe that the CASSCF/MR-CISD method treats the potential energy curve for dissociation in a more balanced way than does the MP4 method.

Duchovic <u>et al</u>. used Padé approximants [123] to estimate the contributions of higher order terms in the perturbation series, and we can now compare the CASSCF/MR-CISD results to their results. The comparison is given in Figure 3, which shows that the Padé approximant results are indeed in better agreement with the CASSCF/MR-CISD results than are the MP4 results.

Table 4. Analysis of the MR-CISD wavefunction in terms of the coefficients of the dominant configuration and the reference configurations as a function of R(C-H).

R(C-H)(Å)	c_0^2	$\sum_{i=0}^{10} c_i^2$
0.757	0.957	0.960
1.086	0.938	0.944
1.500	0.916	0.941
2.000	0.850	0.942
2.500	0.737	0.942
3.000	0.617	0.942
10.584	0.454	0.943

3. ENERGY TRANSFER PROCESSES

3.1 He - I_2: cross sections for vibrational excitation

The construction of potential energy surfaces which describe the interaction between two closed-shell atoms and/or molecules presents an enormous challenge to theoretical chemists. Because the interaction energy for such systems at the van der Waals minimum is so tiny compared to the total energy, e.g., 1 meV for HeH_2 [124] and 2.8 meV for HeI_2 [36], accurate *ab initio* descriptions of this region of the potential energy surface are only currently possible for the smallest systems. Meyer *et al.* [124] used very large basis sets and performed configuration interaction calculations which included triple excitations to describe the van der Waals well region in the four-electron HeH_2 system, but a comparable treatment for systems with more occupied orbitals still appears prohibitive.

Even if one is able to describe the various regions of a potential energy surface to a high degree of accuracy using *ab initio* methods, one must also be able to analytically fit the results from these calculations to a functional form which allows for the accurate determination of dynamical properties. For the HeH_2 system, Alexander and Berard [125] have investigated the sensitivity of dynamics calculations to several fits of *ab initio* data from Gordon and Secrest [126] and have shown that the results are quite sensitive to the analytic fit. Duff and one of the present authors [127] further examined this question and suggested that the force exerted by He in the classically allowed region on unstretched H_2 is very useful for explaining the different dynamical results obtained from the different fits.

Recently, Hall *et al.* [128] have reported the first measurement of the energy dependence of a vibrational-rotational excitation cross section for a collisional system that involves only uncharged species; in particular they measured energy-dependent cross sections for $He-I_2$ collisions in the ground electronic state. Two potential energy surfaces based on *ab initio* calculations have now been constructed for $He-I_2$ in the ground electronic state, with special emphasis on those properties of the potential energy surface that are expected to have a strong effect on vibrational excitation [33,36].

The first surface [36] constructed in our group for $He-I_2$ collisions is a pairwise additive (PA) approximation that is based in part on *ab initio* calculations and in part on theoretical and experimental estimates of the long-range forces and the properties of the van der Waals complex. In the *ab initio* calculations, a non-relativistic effective core potential [129] was used to describe the core electrons of I, and the Hartree-Fock and Møller-Plesset third-order perturbation methods were employed using a triple-zeta-plus-polarization basis set. Because of the PA approximation, *ab initio* calculations were only required at T-shaped geometries of the He-I_2 complex. Dynamics calculations [32] based on the rotational infinite-order-sudden approximation [130,131] on this potential yielded semiquantitative agreement with the experimental vibrational excitation cross sections; however the calculations showed significantly more rotational excitation accompanying the vibrationally inelastic events. This was ascribed to incorrect anisotropy of the PA potential. Although these calculations are only in semiquantitative agreement with experiment, they did further quantify our understanding of how the dynamics are related to the potential energy surface. We observed [32,33] that it is possible to correlate the square of the force along the I_2 bond at the classical turning point with vibrational excitation probabilities. This force $F_{int}(R_{tp})$ is $-\left.\frac{\partial V_{int}(R,r,\chi)}{\partial r}\right|_{r=r_e,R=R_{tp}}$, where V_{int} is the interaction energy, R is the length of the vector $\vec{R}$ connecting the center of mass of the I_2 to He, r is the length of the vector $\vec{r}$ connecting the I atoms, χ is the angle between $\vec{R}$ and $\vec{r}$, r_e is the diatomic equilibrium internuclear distance, and R_{tp} is the classical translational turning point, i.e., the root of

$$V_{int}(R,r=r_e,\chi) + \ell(\ell+1)/(2\mu R^2) - E_{rel} = 0 \qquad (5)$$

where E_{rel} is the relative translational energy in the initial state, μ is the reduced mass, and ℓ is the orbital angular momentum. For notational convenience, we define the squared vibrational force product (SVFP) for a general collisional system by

$$SVFP = [\prod_n F_{int}^{(n)}(R_{tp})]^2 \tag{6}$$

where n runs over all (one in this subsection and two in subsection 3.2) oscillators of the target and the projectile, and $F_{int}^{(n)}(R_{tp})$ is the force exerted on the nth oscillator when the radial coordinate is at its turning point and the oscillators are at their equilibrium distance. Figure 4 shows that the correlation between the vibrational excitation probability and SVFP is almost linear and not strongly dependent on the potential energy surface.

Our second HeI_2 potential energy surface [33] is improved relative to the one just discussed in three ways (i) the PA assumption is removed, (ii) results from higher quality _ab initio_ calculations, with up to 53623 configurations, are used in constructing the potential, and (iii) an improved analytic functional form is employed.

The improvements in the _ab initio_ methods are threefold. First, a more accurate effective core potential [132], which incorporates higher-angular-momentum projectors and relativistic effects, has been employed for the core electrons of I. Second, we used the MCSCF/MR-CISD method, which has been described in Section I, rather than the single-reference Møller-Plesset perturbation method. Since two configurations are required to describe the dissociation of I_2, the MCSCF calculation and the multi-configuration reference space include two configurations. Third, we used an improved one-electron basis [33]. The one-electron basis set consisted of a triple zeta valence set plus two p sets on He. It is both larger and better optimized than the basis used in the previous calculations. The exponential parameters for the d subshell and for the bond centered sp shell were adjusted to best describe the I_2 bond length. Although it is not clear whether a good description of the isolated diatom is necessary for calculating the interaction force, we did insure that the calculated value for r_e, equilibrium displacement, and ω_e, the harmonic stretching fre-

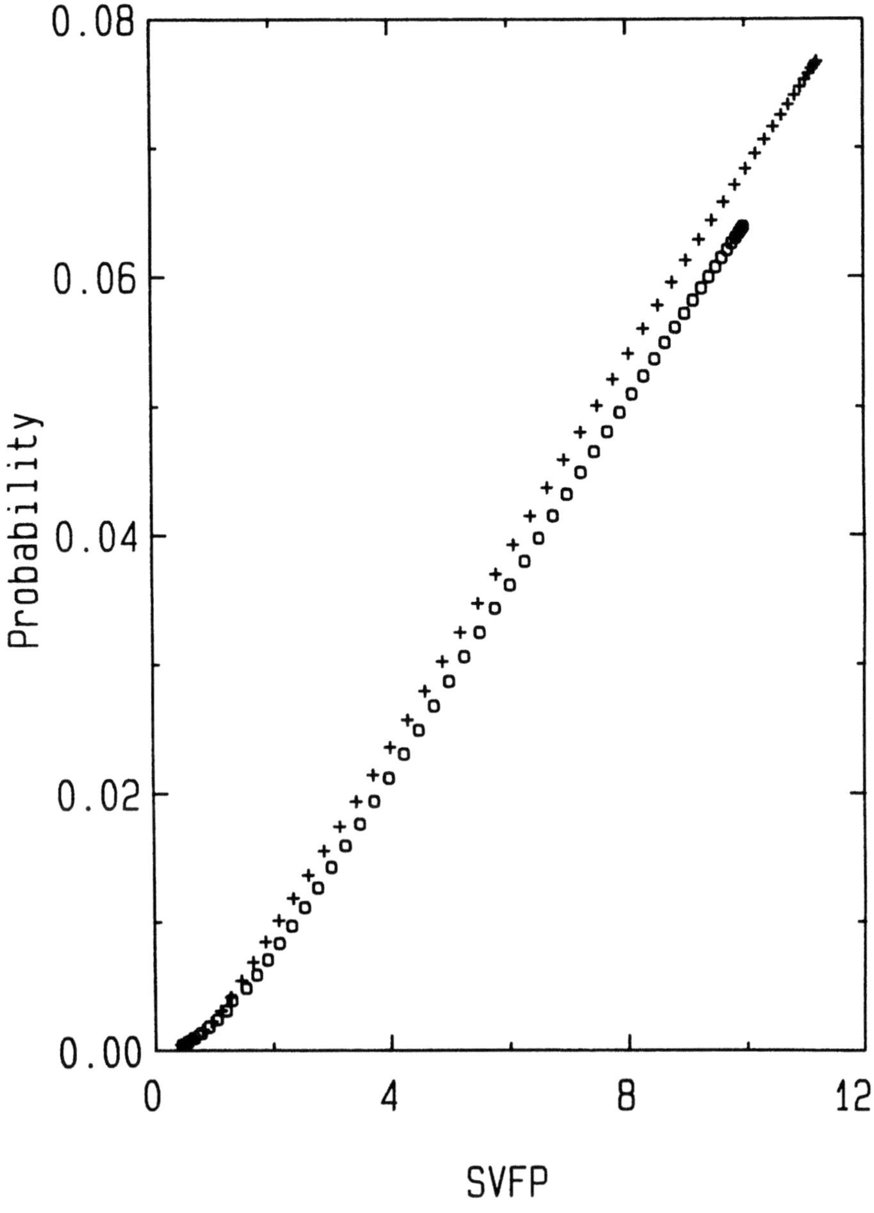

Figure 4. The probability of vibrational excitation from v=0 to v=1 in He - I_2 collisions as a function of the squared vibrational force product (SVFP) in units of 10^{-6} a.u. for E_{rel} = 0.0867 eV and $\chi = 0^o$: +, potential of Ref. 33; □, potental of Ref. 36.

quency, agree well with the experimental values. The exponential parameter for the tight p on He was adjusted to give the lowest energy for isolated He at the CISD level, and the exponential parameter for the diffuse p on He was adjusted to give a reasonable van der Waals well depth and geometry. While the prevailing experimental interpretation [133] indicates that a nonlinear geometry is more stable than a collinear one, our calculations predict that the collinear complex is more stable than the T-shaped complex by 0.09 meV. While this discrepancy is somewhat disappointing, our tests indicate that the force is better converged. In our previous study [33] we investigated the variation of the force with the He basis set and found that, when V_{int} is approximately 100 meV, the force does not vary greatly with basis changes. To test this further, in Table 5 we show the variation of the force with respect to the basis set and the treatment of electron correlation. This shows good agreement between the two CISD calculations, indicating that the calculations are well converged with respect to size of the reference space. Although Clary [134] has treated vibrational energy transfer processes for atom-polyatomic molecule collisions using small-basis-set RHF calculations, Table 5 shows that it is very important that we include polarization functions and correlation effects when estimating the force.

Table 5. Variation of the interaction energy V_{int}, and the force on the I_2 bond, F_{int}, for T-shaped HeI_2 as a function calculation method [a,b].

$R(a_o)$	RHF[c]		RHF[d]		MCSCF[d]		RHF/SR-CISD[d]		MCSCF/MR-CISD[d]	
	V_{int}	F_{int}	V_{int}	F_{int}	V_{int}	F_{int}	V_{int}	F_{int}	V_{int}	F_{int}
4.0	775	215	724	184	723	191	633	170	631	174
5.0	163	24	159	21.4	159	22.7	118	16.6	118	17.3
6.0	32	-0.1	29	-1.7	29	-1.3	12	-3.3	12	-3.0
7.0	6	-1.0	5	-1.4	5	-1.2	-2	-1.8	-2	-1.7

[a] V_{int}, given in meV, and F_{int}, given in meV/a_o, are both evaluated at $\chi=90°$ and r=5.054 a.u., which is r_e for the MCSCF/MR-CISD calculations.

Table 5 (cont.)

---

[b] For χ = 90° and each indicated R, F_{int} is obtained by calculating V_{int} at r=4.854, 5.054 and 5.254 a_0, fitting it to a parabola, and analytically evaluating the derivative at r=5.054 a_0.
[c] Triple zeta basis on I and quadruple zeta basis on He without any bond-centered functions and polarization functions.
[d] Polarized basis of Ref. 33.

Once we have reliable estimates of the force at selected points it is important to employ an analytic representation which is capable of reproducing these forces. We developed a procedure, described in detail elsewhere [33], that involves only linear parameters and that fits exactly the input *ab initio* points at a set of r and χ values for each R, but that does not involve a truncated Legendre expansion of the potential. The avoidance of a truncated Legendre expansion of the potential, which has often been employed, is considered important because a Legendre expansion of the anisotropy may converge slowly in the repulsive region. The new functional form is based on a multiplicative correction to a pairwise additive potential. Truncating the Legendre expansion of the multiplicative correction is much less serious than truncating a Legendre expansion of the potential itself. It should be noted that for some values of R and χ, our analytic representation predicts the force to have the opposite sign than that predicted by the pairwise additive potential that serves as the original factor. Figure 5 shows the values of V_{int} and $-F_{int}$ on the two surfaces for the same collinear orientation as used for Figure 4.

3.2 HF + HF: a challenge for the dynamicist

In this subsection we discuss selected aspects of the quantum mechanical treatment of vibrational-to-vibrational (V-V) energy transfer in the collision of two hydrogen fluoride (HF) molecules. V-V energy transfer has long been a topic of widespread interest for dynamicists because of the possibility of resonant or near-resonant transitions, which have large probabilities and dominate nonequilibrium energy relaxation under many circumstances. An

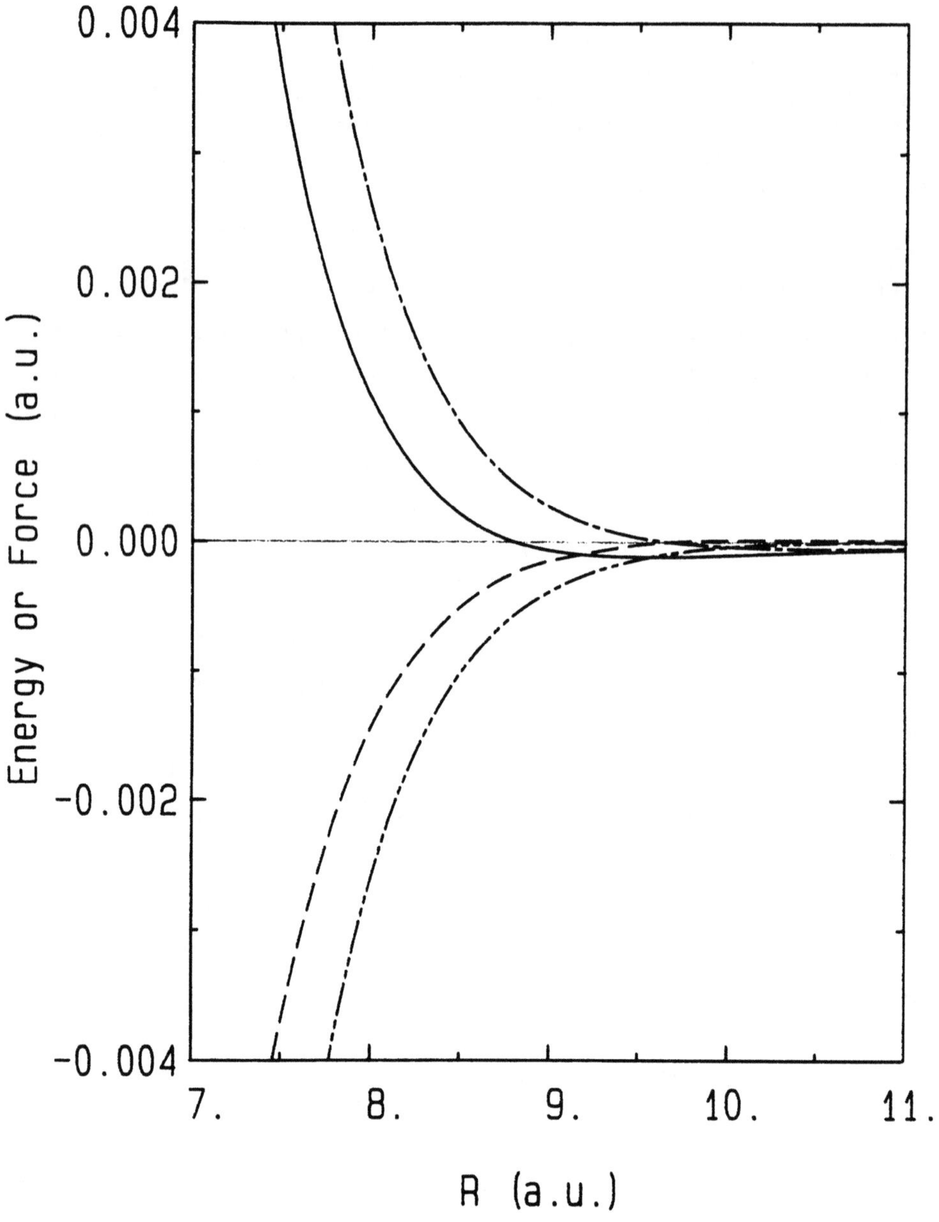

Figure 5. V_{int} and F_{int} for He + I_2 with $r = r_e$ and $\chi = 0^o$: solid curve, V_{int} of Ref. 33; long-short dashed curve, V_{int} from Ref. 36; dashed curve, $-F_{int}$ from Ref. 33; long-short-short dashed curve, $-F_{int}$ of Ref. 36.

important question in interpreting the large transition probabilities often observed for V-V energy transfer is ascertaining the respective roles and relative importance of short-range versus long-range forces. Early calculations on V-V energy transfer were performed using simple interaction potential functions, such as a nearest-neighbor exponential repulsions or truncated multipole series, and were based on simplified treatments of the dynamics, such as perturbation theory. However, to draw reliable conclusions, it is necessary to use both accurate interaction potentials and accurate dynamics. The system that has received most attention is HF - HF, partly because of its technological importance and partly because of the availability of a convenient source of excitation. In the present discussion we restrict ourselves to this one most studied system.

Several HF - HF potentials have been proposed on the basis of *ab initio* calculations [135-141]. Until recently, the most accurate of these were based on the Hartree-Fock, double-zeta-plus-polarization calculations of Yarkony *et al.* [142] of the interaction potential for two HF molecules, both at the diatomic equilibrium internuclear separation, but at various distances and relative orientations with respect to each other. In order to extend these calculations to predict the dependence on vibrational coordinates, Poulsen, Billing, and Steinfeld [136] fit them to a sum of two-center terms (the reliability of this kind of extension is suspect). More recently Binkley and Redmon [38] have produced a new potential energy surface. This is an improvement over all previous analytic representations in that it is based on a set of *ab initio* points that include varying HF distances and also in that it was computed using a larger one-electron basis set (6-311G**) and including electron correlation at the level of fourth-order Møller-Plesset perturbation theory. The *ab initio* points were fit [38] to a functional form made up of the sum of two-body, three-body, and four-body terms (similar to that used previously for H_2O [143]), and it includes a large number of adjustable parameters. This makes the potential expensive to evaluate, but it is probably unavoidable to use a complicated fit in order to accurately represent all of the important regions of the potential.

In performing classical trajectory calculations, the evaluation of the potential and its derivatives can become the most time-consuming part of the calculations, and one strives hard for a functional representation that is economical to evaluate. Dynamics calculations employing quantum mechanics do not involve nearly as large a percentage of the effort in the potential subroutine, though, so in the quantal case there is less emphasis on achieving an economical fit than on achieving a reliable one. In quantal dynamics calculations one expands the angular part of the interaction potential in terms of some complete set of functions which then allows the analytic evaluation of some integrals necessary for the calculation of matrix elements of the potential. The expansion we have used for our calculations is

$$V(\vec{R},\vec{r}_1,\vec{r}_2) = \sum_{q_1 q_2 \mu} v_{q_1 q_2 \mu}(R,r_1,r_2)\, \mathcal{Y}_{q_1 q_2 \mu}(\hat{r}_1,\hat{r}_2) \qquad (7)$$

$$\mathcal{Y}_{q_1 q_2 \mu}(\hat{r}_1,\hat{r}_2) = \frac{4\pi}{[2(1+\delta_{\mu 0})]^{1/2}} [Y_{q_1 \mu}(\hat{r}_1) Y_{q_2 -\mu}(\hat{r}_2)$$

$$+ Y_{q_1 -\mu}(\hat{r}_1) Y_{q_2 \mu}(\hat{r}_2)] \qquad (8)$$

where $\vec{R}$ is the vector connecting the centers of masses of the two molecules, $\vec{r}_i$ is the vector along the bond of molecule i in the coordinate system where the z axis is in the direction of $\vec{R}$, and $Y_{q\mu}$ is a spherical harmonic. This form was chosen because it is completely general and yet it depends only on the six independent quantities $R, r_1, r_2, \theta_1, \theta_2$, and $\phi_1 - \phi_2$, where θ_i and ϕ_i are the inclination and azimuthal angles of $\vec{r}_i$ with respect to the vector $\vec{R}$. The coefficients $v_{q_1 q_2 \mu}$ are determined numerically for each value of R, r_1, and r_2 required for a calculation, and a large number of them are necessary to accurately reproduce $V(\overline{R},\overline{r}_1,\overline{r}_2)$. In our scattering calculations we retain a maximum of 161 terms in Eqn. (7). We keep all terms with $q_1 + q_2 \leq 10$ [note that $\mu \leq \min(q_1,q_2)$] and use 11-point quadrature in each of the three angles θ_1, θ_2, and $\phi_1 - \phi_2$ to determine the 161 coefficients. This requires the evaluation of $V(\overline{R},\overline{r}_1,\overline{r}_2)$ at a total of 1331 different angles for each value of R, r_1, and r_2 required, which is far greater than the number of angles used in generating the analytic representation $V(\overline{R},\overline{r}_1,\overline{r}_2)$. Since our quantum mechanical scattering calculations require the potential at about 1.5×10^4 sets of R, r_1 and r_2, it would require about two million words of memory or mass storage to hold the required potential information in the form of Eqn. (7). Thus we recompute it from the original multi-center fit for each scattering calculation. One important aspect of using an expansion like this is that the work of evaluating the potential function is independent of the number of coupled channels included in the quantal scattering calculation, although only when a large number of coupled channels (>100) are considered does the potential function evaluation take less than half of the total time for a scattering calculation at one energy.

Although there are superficial similarities between the Poulsen-Billing-Steinfield (PBS) [136] surface and the Binkley-Redmon (BR) [38] one, they can give quite different energy transfer probabilities. To illustrate this we have performed some calculations using a model [144] in which the two HF molecules are treated as breathing spheres. Table 6 gives the probabilities computed from these breathing-sphere calculations for the process $HF(v_1=1) + HF(v_2=1) \rightarrow HF(v_1=0 \text{ or } 2) + HF(v_2=2 \text{ or } 0)$. The effective spherical potential for the breathing sphere calculations was obtained first from the PBS surface and then from the BR surface, in each case using the potential corresponding to collinear intermolecular approach with the hydrogen end of molecule 2 approaching the fluorine end of molecule 1 (this is a hydrogen-bonding orientation).

Table 6. V-V energy transfer probabilities for HF-HF scattering in the breathing sphere model at a relative transitional energy of 0.03 eV.

Orbital angular momentum	Poulsen et al. P.E.S.[a]		Binkley-Redmon P.E.S.[b]	
	$P_{11\to02}$	$P_{11\to20}$	$P_{11\to02}$	$P_{11\to20}$
0	6.5(-2)[c]	3.7(-3)	2.8(-2)	7.2(-4)
12	6.5(-2)	3.6(-3)	2.7(-2)	6.4(-4)
24	6.4(-2)	3.3(-3)	2.5(-2)	4.2(-4)
36	6.4(-2)	2.8(-3)	2.1(-2)	1.3(-4)
48	6.2(-2)	2.2(-3)	1.6(-2)	3.6(-5)
60	5.8(-2)	1.5(-3)	1.0(-3)	1.0(-3)
72	5.1(-2)	8.0(-4)	4.3(-3)	5.1(-3)
84	4.0(-2)	2.3(-4)	7.9(-4)	1.2(-2)
96	1(-12)	1(-15)	1(-9)	4(-7)

[a] using the potential energy surface of Ref. 136

[b] using the potential energy surface of Ref. 38.

[c] 6.5(-2) = 6.5×10^{-2}.

The calculations show that the V-V energy transfer probabilities for the breathing sphere calculations in which the potential is based on the PBS surface slowly decrease as the relative orbital angular momentum quantum number increases from zero up to about $\ell = 96$, where they suddenly drop several orders of magnitude. This sudden drop occurs when the centrifugal barrier becomes large enough to prevent (classically) the molecules from getting close enough together to feel the hard core repulsion of the potential. The probabilities from the BR surface are quite different for ℓ in the range 0 to 96: the $P_{11\to20}$ probability decreases as ℓ increases from 0 to 48, but then increases dramatically as ℓ further increases from 60 to 84.

In the previous section it was pointed out that it is possible to correlate the square of the force along the oscillator coordinate at the classical turning point with vibrational excitation probability in He + I_2 collisions. The analog of this correlation for a V-V energy transfer problem is to correlate the square of the product of the forces on each bond with the transition probability. Figure 6 shows a test of this correlation for the HF - HF system by plotting the logarithm of $P_{11 \to 20}$ or $P_{11 \to 02}$ as a function of the logarithm of the square of the product of the forces on each molecule, i.e., versus the logarithm of the quantity defined in Eqn. (6). The correlation is not as linear as was observed for HeI_2, but the points in Figure 6 do for the most part lie in a small set of almost monotonic sequences, indicating that the forces do correlate strongly with the probabilities.

In an attempt to explain the quite different results predicted using potentials based on the the two surfaces, in Figure 7 we have plotted the forces as a function of the center-of-mass separation of the two molecules. Also shown is the interaction potential. For both surfaces the force on the first molecule is positive for large separations and slowly increases as the separation decreases until a maximum is reached at about 6 a_0. At smaller distances the force decreases rapidly and goes to a very large negative number.

The situation is different for the second molecule. The PBS surface gives a force that is positive everywhere and increases as the separation decreases, while the RB surface gives a positive force for both large and small separations with local maxima at about 3.5 and 7.5 a_0 and a local minimum around 5 a_0. We tentatively attribute the qualitatively different opacity-function shapes for the two interaction potentials to this qualitative difference in the behavior of the forces.

It would be interesting to know what accurate _ab initio_ calculations predict for the forces. This would require using a larger one-electron basis and directly calculating _ab initio_ energies at geometries close to where the forces are required, the former to reduce possible basis-set superposition errors (as discussed below) and the latter to ensure that the forces need not be obtained from global fits with possibly large local errors. An alternative way to reduce the additional possible error in the forces due to the fit would be to employ a fitting procedure similar to that discussed

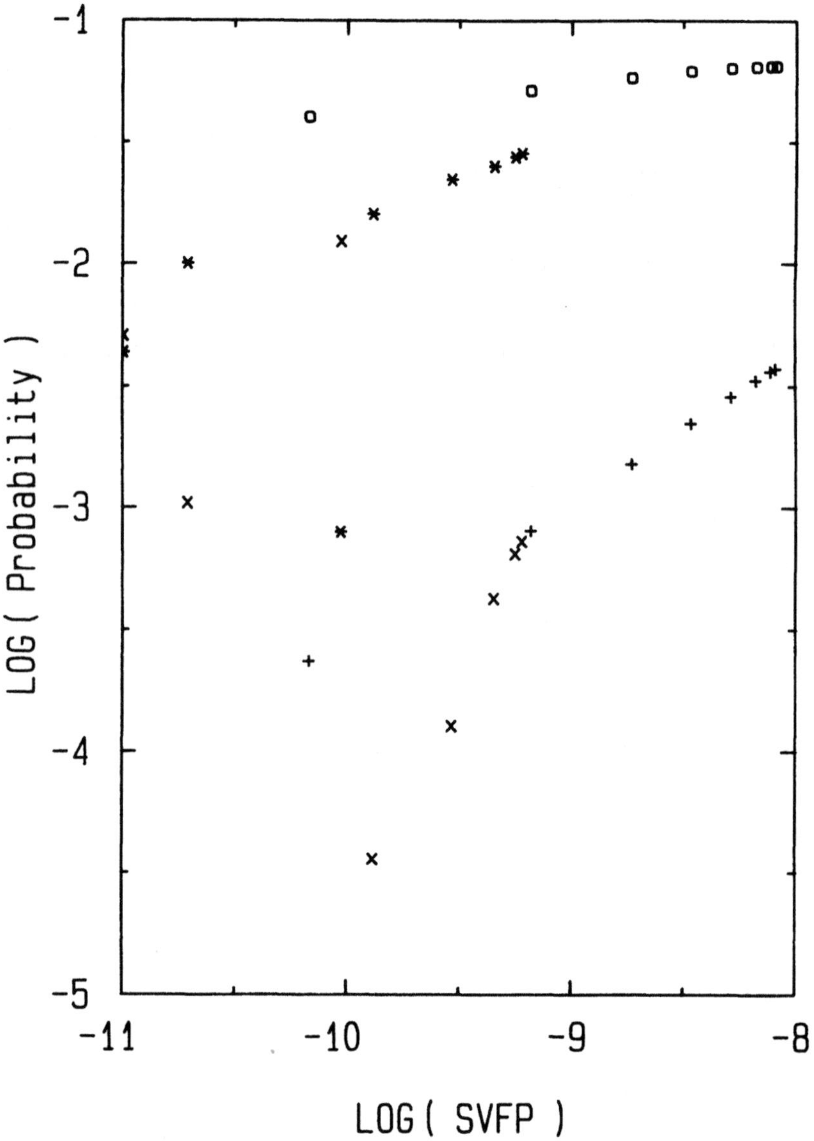

Figure 6. The logarithm of $P_{11\to 20}$ and $P_{11\to 02}$ for HF + HF as a function of the logarithm of the squared vibrational force product (SVFP) in a.u. for E_{rel} = 0.030 eV. This is from a breathing sphere calculation with the potential from the collinear hydrogen-bonding orientation $H_1 - F_1 \cdots H_2 - F_2$ (orientation I of Ref. 144); □, $P_{11\to 02}$ using PBS V_{int}; +, $P_{11\to 20}$ from PBS V_{int}; ×, $P_{11\to 02}$ from RB V_{int}; *, $P_{11\to 20}$ using RB V_{int}.

above for He - I_2; the kind of fit we used for He - I_2, as opposed to a more general multicenter fit, is designed to introduce minimal fitting error into the vibrational forces. It would also be interesting to replace the breathing-sphere calculations by converged quantal dynamics on the fully anisotropic (161-term) BR potential. Preliminary work on these full quantal dynamics calculations indicates that over 10^3 coupled channels must be included for convergence, even with total angular momentum zero [34]. Further work on these large-scale dynamics calculations will be reported subsequently.

To learn more about the requirements on the one-electron basis set for accurate interaction potentials for the HF - HF system, we have done some restricted single-configuration Hartree-Fock calculations with a large number of basis sets for selected geometries of this system [37]. Our calculations used small, medium, large, and very large gaussian basis sets, and we attempted to assess the effect of basis-set superposition error on the calculation of interaction energies. Included in our calculations are the basis sets used by Yarkony et al. [142] and Binkley [38]. For the collinear hydrogen bonding approach, we considered a geometry with $R = 4\ a_0$. The basis of Yarkony et al. [142] gave an energy of 16.2 kcal/mol, and the basis used by Binkley [38] gave an interaction energy of 16.6 kcal/mol, while our best estimate of the Hartree-Fock limit of the interaction energy is 18.5 ± 0.5 kcal/mol. Our large basis sets on which we base this estimate include more diffuse functions than either of the above basis sets, and these diffuse functions are important for reducing basis-set superposition error and for treating polarizability contributions to long-range interactions. Although we have not calculated the force predicted by these basis sets, it would be interesting to do so and include this information in the construction of an even more accurate potential energy surface for HF - HF collisions. We note that the BR fit at the hydrogen-bonding collinear geometry with $r_1 = r_2 = 1.723\ a_0$ and $R = 4.0\ a_0$ yields 13.07 kcal/mol, and the PBS fit at this geometry yields 4.73 kcal/mol. The ab initio calculations

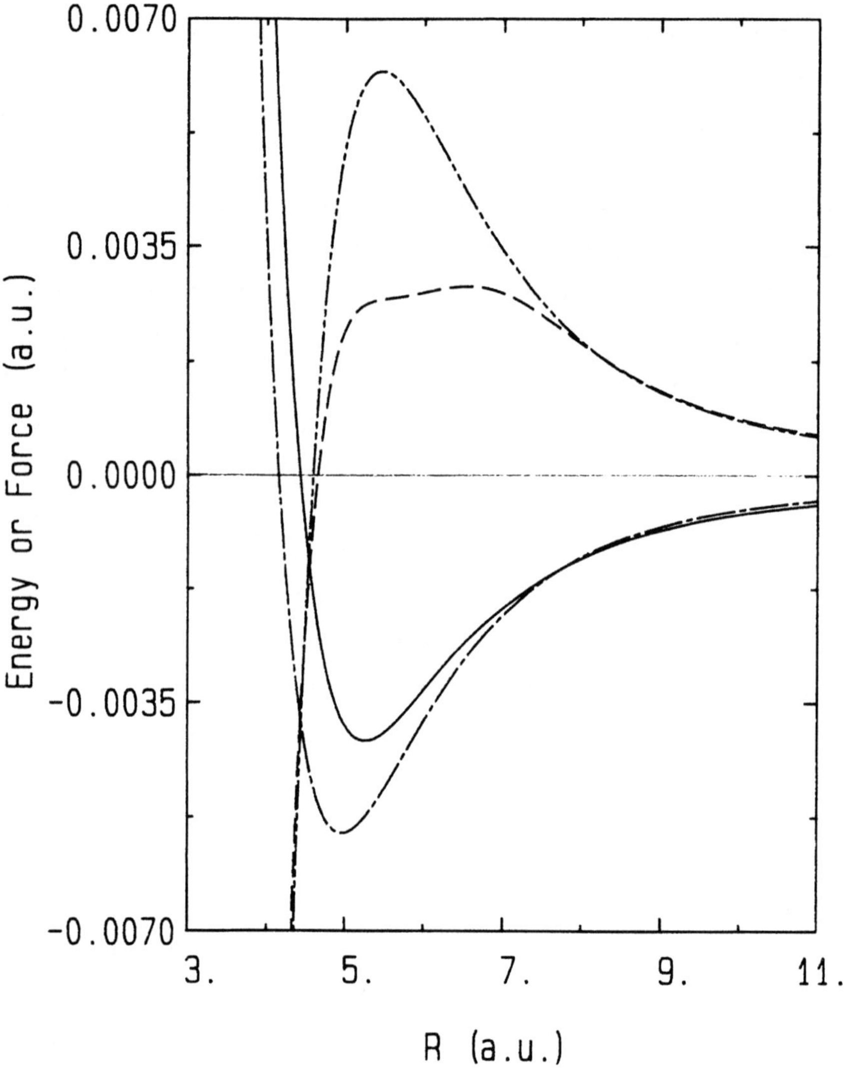

Figure 7a.

Figure 7. V_{int} and F_{int} for HF - HF collisions with $r_1 = r_2 = r_e$ in the same $H_1 - F_1 \ldots H_2 - F_2$ orientation as used for Figure 6. In both part a and part b, the solid line is the RB V_{int}, and the long-short dashed line is the PBS V_{int}. a) dashed curve, $F_{int}^{(1)}$ from RB V_{int}; long-short-short dashed curve, $F_{int}^{(1)}$ from PBS V_{int}. b) dashed curve, $F_{int}^{(2)}$ from RB V_{int}; long-short-short dashed curve, $F_{int}^{(2)}$ from PBS V_{int}.

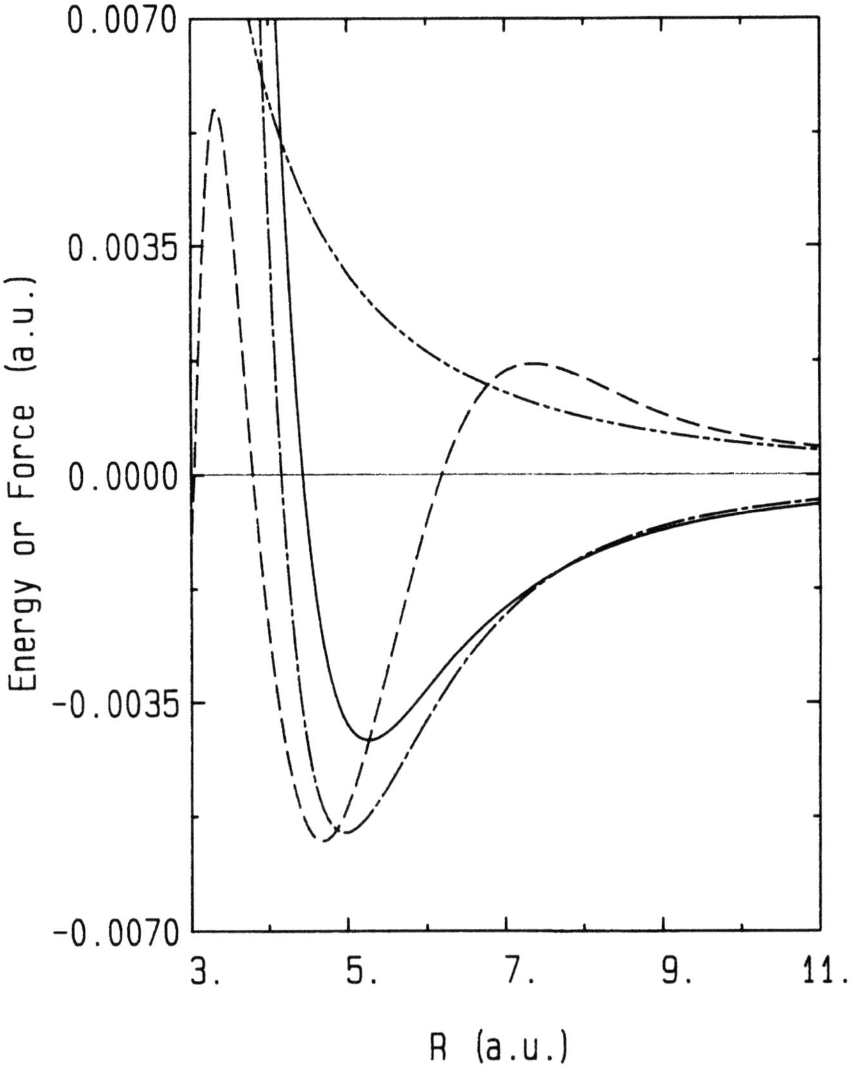

Figure 7b.

would appear to totally rule out the PBS value, but the difference between 18.5 and 13.1 kcal/mol may be a consequence of the inclusion of dispersion contributions in the BR surface.

4. CONCLUDING REMARKS

In this chapter we have reviewed some recent ab initio calculations of interaction potentials and whole potential energy surfaces and their use for studying molecular collisions. We have pointed out that in many cases some aspects of the dynamical results can be related to rather specific features of the potential energy surfaces. This is important because it allows one to concentrate on the convergence of these specific surface features rather than on convergence of the whole potential energy surface. One does still, however, require a global fit to the surface to perform full dynamics calculations, and we have also briefly discussed some new functional forms that may be useful for such fits. An important question discussed here, but still far from settled, is the reliability of various levels (methods + basis sets) of electronic structure calculations for various potential energy surface features.

5. ACKNOWLEDGMENTS

The authors are grateful to Professor Isaiah Shavitt for helpful discussions. The work at the University of Minnesota, except for variational transition state theory, is supported in part by the National Science Foundation under grant no. CHE8317944 and by a computing time grant from the Minnesota Supercomputer Institute. The work at the University of Minnesota on variational transition state theory is supported in part by the U.S. Department of Energy, Office of Basic Energy Sciences, under contract no. 79ER10425. The work at Chemical Dynamics Corporation is supported in part by the U.S. Army Research Office under contract no. DAAG-29-84-C-0011.

References:

1. B.C. Garrett and D.G. Truhlar, Proc. Natl. Acad. Sci. U.S.A., 76, 4755 (1979).
2. B.C. Garrett and D.G. Truhlar, J. Chem. Phys. 72, 3460 (1980).
3. B.C. Garrett, D.G. Truhlar, R.S. Grev. and A.W. Magnuson, J. Phys. Chem. 84, 1730 (1980).
4. N.C. Blais, D.G. Truhlar and B.C. Garrett, J. Phys. Chem. 85, 1094 (1981).
5. N.C. Blais and D.G. Truhlar: 1981, in "Potential Energy Surfaces and Dynamics Calculations," Ed. D.G. Truhlar, Plenum, New York, p. 431.
6. B.C. Garrett, D.G. Truhlar and R.S. Grev, ibid, p. 587.
7. B.C. Garrett and D.G. Truhlar, J. Phys. Chem. 86, 1136 (1982).
8. D.G. Truhlar, A.D. Isaacson, R.T. Skodje and B.C. Garrett, J. Phys. Chem. 86, 2252 (1982).
9. D.K. Bondi, D.C. Clary, J.N.L. Connor, B.C. Garrett and D.G. Truhlar, J. Chem. Phys. 76, 4986 (1982).
10. N.C. Blais, D.G. Truhlar and B.C. Garrett, J. Chem. Phys. 76, 2770 (1982).
11. N.C. Blais and D.G. Truhlar, Astrophys. J. 258, L79 (1982).
12. R.T. Skodje, D.G. Truhlar and B.C. Garrett, J. Chem. Phys. 77, 5955 (1982).
13. N.C. Blais and D.G. Truhlar, J. Chem. Phys. 78, 2388 (1983).
14. N.C. Blais, D.G. Truhlar and B.C. Garrett, 78, 2363 (1983).
15. D.G. Truhlar, R.S. Grev and B.C. Garrett, J. Phys. Chem. 87, 3415 (1983).
16. N.C. Blais and D.G. Truhlar, Chem. Phys. Lett. 102, 120 (1983).
17. B.C. Garrett and D.G. Truhlar, J. Chem. Phys. 81, 309 (1984).
18. B.C. Garrett and D.G. Truhlar, J. Phys. Chem., in press.
19. N.C. Blais, D.G. Truhlar and B.C. Garrett, J. Chem. Phys. 82, 2300 (1985).
20. A.D. Isaacson and D.G. Truhlar, J. Chem. Phys. 76, 1380 (1982).
21. D.G. Truhlar and A.D. Isaacson, J. Chem. Phys. 77, 3516 (1982).
22. A.D. Isaacson, M.T. Sund, S.N. Rai and D.G. Truhlar, J. Chem. Phys. 82, 1338 (1985).
23. B.C. Garrett, D.G. Truhlar, A.F. Wagner, and T.H. Dunning, Jr., J. Chem. Phys. 78, 4400 (1983).
24. S.N. Rai and D.G. Truhlar, J. Chem. Phys. 79, 6046 (1983).
25. D.G. Truhlar, K. Runge, and B.C. Garrett: in proceedings of Twentieth International Symposium on Combustion (August 12-17, 1984, Ann Arbor, Michigan), Combustion Institute, Pittsburgh, in press.
26. F.B. Brown, R. Steckler, D.W. Schwenke, D.G. Truhlar and B.C. Garrett, J. Chem. Phys. 82, 188 (1985).
27. R. Steckler, D.G. Truhlar and B.C. Garrett, J. Chem. Phys., in press.
28. R. Steckler, F.B. Brown and D.G. Truhlar, unpublished.
29. S. Green and D.G. Truhlar, Astrophys. J. 231, L101 (1979).

30. D.L. Thompson, N.C. Blais and D.G. Truhlar, J. Chem. Phys. 78, 1335 (1983).
31. D.G. Truhlar, Bull. Amer. Phys. Soc. 24, 70 (1979).
32. D.W. Schwenke and D.G. Truhlar, J. Chem. Phys. 81, 5586 (1984).
33. F.B. Brown, D.W. Schwenke and D.G. Truhlar, Theoret. Chim. Acta, in press.
34. D.W. Schwenke and D.G. Truhlar: in Supercomputer Applications (proceedings of the Supercomputer Applications Symposium, October 31-November 1, 1984, W. Lafayette, Indiana), Ed. R.W. Numrich, Plenum, New York, in press.
35. F.B. Brown and D.G. Truhlar, Chem. Phys. Lett. 113, 441 (1985).
36. D.W. Schwenke and D.G. Truhlar, Chem. Phys. Lett. 98, 217 (1983).
37. D.W. Schwenke and D.G. Truhlar, J. Chem. Phys. 82, 2418 (1985).
38. J.S. Binkley and M.J. Redmon, unpublished.
39. C.C.J. Roothaan, Rev. Mod. Phys. 32, 179 (1960).
40. I. Shavitt: 1977, in "Methods of Electronic Structure Theory," Vol. 1, Ed. H.F. Schaefer, III, Plenum, New York, p. 189.
41. B.O. Roos, P.R. Taylor and P.E.M. Siegbahn, Chem. Phys. 48, 157 (1980).
42. A. Veillard, Theor. Chim. Acta 4, 22 (1966).
43. G. Das and A.C. Wahl, J. Chem. Phys. 44, 3050 (1966).
44. J. Hinze and C.C.J. Roothaan, Theor. Phys. Suppl. 40, 37 (1967).
45. R.M. Pitzer, J. Chem. Phys. 58, 3111 (1973).
46. M. Dupuis, J. Rys and H.F. King, J. Chem. Phys. 65, 111 (1976).
47. R. Shepard, I. Shavitt and J. Simons, J. Chem. Phys. 76, 543 (1982).
48. H. Lischka, R. Shepard, F.B. Brown and I. Shavitt, Int. J. Quantum Chem. Symp. 15, 91 (1981).
49. W.L. Hase: 1981, in "Potential Energy Surfaces and Dynamics Calculations," Ed. D.G. Truhlar, Plenum, New York, p. 1; K. Morokuma and S. Kato, ibid., p. 243; G.C. Schatz, ibid., p. 287; T.H. Dunning, Jr., S.P. Walch and A.F. Wagner, ibid., p. 329; D.A. Micha, ibid., p. 685; J.C. Tully, ibid., p. 805.
50. K. Fukui: 1983, in "New Horizons of Quantum Chemistry," Eds. P.-O. Löwdin and B. Pullman, Reidel, Dordrecht, p. 111; K. Morokuma, S. Kato, K. Kitaura, S. Obara, K. Ohta and M. Hanamura, ibid., p. 221; M. Simonetta, ibid., p. 295.
51. D.G. Truhlar, Int. J. Quantum Chem. Symp. 17, 77 (1983).
52. T.H. Dunning and L.B. Harding, in "The Theory of Chemical Reaction Dynamics," Ed. M. Baer, CRC Press, Boca Raton, FL, in press.
53. R.J. Heidner and J.F. Bott, J. Chem. Phys. 63, 1810 (1975).
54. J.F. Bott, J. Chem. Phys. 65, 1976 (1976).
55. F.E. Bartoszek, D.M. Manos and J.C. Polanyi, J. Chem. Phys. 69, 933 (1978).
56. C.F. Bender, B.J. Garrison and H.F. Schaefer, III, J. Chem. Phys. 62, 1188 (1975).
57. P. Botschwina and W. Meyer, Chem. Phys. 20, 43 (1977).
58. W.R. Wadt and N.W. Winter, J. Chem. Phys. 67, 3068 (1977).

59. A.F. Voter and W.A. Goddard, J. Chem. Phys. 75, 3638 (1981).
60. T.H. Dunning, J. Phys. Chem. 88, 2469 (1984).
61. P.J. Kuntz, E.M. Nemeth, J.C. Polanyi, S.D. Rosner and C.E. Young, J. Chem. Phys. 44, 1168 (1966).
62. J.T. Muckerman, Theor. Chem. Adv. Perspectives 6A, 1 (1981).
63. B.C. Garrett, D.G. Truhlar, R.S. Grev, G.C. Schatz and R.B. Walker, J. Phys. Chem. 85, 3806 (1981).
64. D.M. Neumark, A.M. Wodtke, G.N. Robinson, C.C. Hayden and Y.T. Lee: 1984, in "Resonances in Electron-Molecule Scattering, van der Waals Complexes, and Reactive Chemical Dynamics," Ed. D.G. Truhlar, American Chemical Society, Washington, p. 479.
65. D.M. Neumark, 1984, Ph.D. Thesis, University of California, Berkeley, unpublished.
66. Y.T. Lee, lecture presented at Fifth American Conference on Theoretical Chemistry, Grand Teton National Park, Wyoming, June 19, 1984.
67. A. Bunge, J. Chem. Phys. 53, 20 (1970).
68. S.R. Langhoff and E.R. Davidson, Int. J. Quantum Chem. 8, 61 (1974).
69. B. Liu, J. Chem. Phys. 80, 581 (1984).
70. (a) B. Liu, J. Chem. Phys. 58, 1925 (1973). (b) P. Siegbahn and B. Liu, J. Chem. Phys. 68, 2457 (1978). (c) D.G. Truhlar and C.J. Horowitz, J. Chem. Phys. 68, 2466 (1978); 71, 1514(E) (1979).
71. A.E. DeVries and F.S. Klein, J. Chem. Phys. 41, 3428 (1964).
72. G.O. Wood, J. Chem. Phys. 56, 1723 (1972).
73. J.D. McDonald and D.R. Herschbach, J. Chem. Phys. 62, 4740 (1975).
74. H. Endo and G.P. Glass, Chem. Phys. Lett. 44, 180 (1976).
75. F.S. Klein and I. Veltman, J. Chem. Soc. Faraday Trans. II 74, 17 (1978).
76. R.E. Weston, J. Phys. Chem. 83, 61 (1979).
77. W. Bauer, L.Y. Rusin and J.P. Toennies, J. Chem. Phys. 68, 4490 (1978).
78. W.H. Beck, R. Götting, J.P. Toennies and K. Winkelmann, J. Chem. Phys. 72, 2896 (1980).
79. J.C. Miller and R.J. Gordon, J. Chem. Phys. 78, 3713 (1983).
80. C.A. Wight, F. Magnotta and S.R. Leone, J. Chem. Phys. 81, 3951 (1984).
81. D.L. Thompson, H.H. Suzukawa and L.M. Raff, J. Chem. Phys. 62, 4727 (1975).
82. T.H. Dunning, J. Chem. Phys. 66, 2752 (1977).
83. C.A. Parr and D.G. Truhlar, J. Phys. Chem. 75, 1844 (1971).
84. E.L. Mehler and W. Meyer, Chem. Phys. Lett. 38, 144 (1976).
85. J.H. Parker and G.C. Pimentel, J. Chem. Phys. 51, 91 (1969).
86. A. Ben-Shaul, Y. Haas, K.L. Kompa and R.D. Levine: 1981, in "Lasers and Chemical Change," Springer, Berlin.
87. R.D.H. Brown and A. Maitland, Prog. React. Kinet. 11, 1 (1981).
88. Y.T. Lee, Ber. Bunsenges. Physik. Chem. 86, 378 (1982).
89. R.E. Wyatt, J.F. McNutt and M.J. Redmon, Ber. Bunsenges. Physik. Chem. 86, 437 (1982).

90. K.L. Kompa and G.C. Pimentel, J. Chem. Phys. 47, 857 (1967); K.L. Kompa, J.H. Parker and G.C. Pimentel, J. Chem. Phys. 51, 91 (1969); R.D. Coombe and G.C. Pimentel, J. Chem. Phys. 59, 251 (1973).
91. J.C. Polanyi and D.C. Tardy, J. Chem. Phys. 51, 5717 (1969); K.G. Anlauf, P.E. Charters, D.S. Horne, R.G. MacDonald, D.H. Maylotte, J.C. Polanyi, W.J. Skrlac, D.C. Tardy and K.B. Woodall, J. Chem. Phys. 53, 4091 (1970); J.C. Polanyi and K.B. Woodall, J. Chem. Phys. 57, 1574 (1972).
92. T.P. Schaefer, P.E. Siska, J.M. Parson, F.P. Tully, Y.C. Wong and Y.T. Lee, J. Chem. Phys. 53, 3385 (1970).
93. N. Jonathan, C.M. Melliar-Smith and D.H. Slater, Mol. Phys. 20, 93 (1971); N. Jonathan, C.M. Melliar-Smith, D. Timlin and D.H. Slater, Appl. Opt. 10, 1821 (1971); N. Jonathan, C.M. Melliar-Smith, S. Okuda, D.H. Slater and D. Timlin, Mol. Phys. 22, 561 (1971).
94. W.H. Green and M.C. Lin, J. Chem. Phys. 54, 3222 (1971).
95. H.W. Chang and D.W. Setser, J. Chem. Phys. 58, 2298 (1973).
96. M.J. Berry, J. Chem. Phys. 59, 6229 (1973).
97. E. Wurzberg and P.L. Houston, J. Chem. Phys. 72, 4811 (1980).
98. R.F. Heidner, III, J.F. Bott, C.E. Gardner and J.E. Melzer, J. Chem. Phys. 72, 4815 (1980).
99. D.H. Neumark, A.M. Wodtke, G.N. Robinson, C.C. Hayden and Y.T. Lee, Phys. Rev. Lett. 53, 226 (1984).
100. C.F. Bender, P.K. Pearson, S.V. O'Neal and H.F. Schaefer, III, J. Chem. Phys. 56, 4626 (1972).
101. C.F. Bender, P.K. Pearson, S.V. O'Neal and H.F. Schaefer, III, Science 176, 1412 (1972).
102. S.R. Ungemach, H.F. Schaefer, III and B. Liu, Discuss. Faraday Soc. 62, 330 (1976).
103. P.D. Mercer and H.O. Pritchard, J. Chem. Phys. 63, 1468 (1959).
104. D.G. Truhlar and B.C. Garrett, Acc. Chem. Res. 13, 440 (1980).
105. B.C. Garrett and D.G. Truhlar, J. Phys. Chem. 83, 1079 (1979); 84, 682(E) (1980); 87, 4553(E) (1983).
106. B.C. Garrett, D.G. Truhlar, R.S. Grev, A.W. Magnuson and J.N.L. Connor, J. Chem. Phys. 73, 1721 (1980).
107. R.T. Skodje, D.G. Truhlar and B.C. Garrett, J. Phys. Chem. 85, 3019 (1982).
108. B.C. Garrett and D.G. Truhlar, J. Chem. Phys. 81, 309 (1984).
109. R. Steckler, D.G. Truhlar, B.C. Garrett, N.C. Blais and R.B. Walker, J. Chem. Phys. 81, 5700 (1984).
110. R.B. Walker and E.F. Hayes, J. Phys. Chem. 87, 1255 (1983).
111. D.G. Truhlar, B.C. Garrett and N.C. Blais, J. Chem. Phys. 80, 232 (1984).
112. D.L. Baulch, R.A. Cox, P.J. Crutzen, R.F. Hampson, Jr., J.A. Kerr, J. Troe and R.T. Watson, J. Phys. Chem. Ref. Data 11, 327 (1982).
113. R.J. Duchovic and W.L. Hase, Chem. Phys. Lett. 110, 474 (1984).
114. R.J. Duchovic, W.L. Hase, H.B. Schlegel, M.J. Frisch and K.

Raghavachari, Chem. Phys. Lett. 89, 120 (1982).
115. R.J. Duchovic, W.L. Hase and H.B. Schlegel, J. Phys. Chem. 88, 1339 (1984).
116. G.B. Skinner, D. Rogers and K.B. Patel, Intern. J. Chem. Kinetics 13, 481 (1981).
117. F.B. Brown, I. Shavitt and R. Shepard, Chem. Phys. Lett. 105, 363 (1984).
118. Y.P. Varshni, Rev. Mod. Phys. 29, 664 (1957).
119. D. Steele, E.R. Lippincott and J.T. Vanderslice, Rev. Mod. Phys. 34, 239 (1962).
120. L. Halonen and M.S. Child, Mol. Phys. 46, 239 (1982).
121. L. Halonen and M.S. Child, J. Chem. Phys. 79, 4355 (1983).
122. S. Peyerimhoff, M. Lewerenz and M. Quack, Chem. Phys. Lett. 109, 563 (1984).
123. G.A. Baker and J.L. Gammell, The Padé Approximant in Theoretical Physics (Academic Press, New York, 1970).
124. W. Meyer, P.C. Hariharan and W. Kutzelnigg, J. Chem. Phys. 73, 1880 (1980).
125. M.H. Alexander and E.V. Berard, J. Chem. Phys. 60, 3950 (1974).
126. M.D. Gordon and D. Secrest, J. Chem. Phys. 52, 120 (1970).
127. J.W. Duff and D.G. Truhlar, J. Chem. Phys. 63, 4418 (1975).
128. G. Hall, K. Liu, M.J. McAuliffe, C.F. Giese and W.R. Gentry, J. Chem. Phys. 78, 5260 (1983); J. Chem. Phys. 81, 5577 (1984).
129. L.R. Kahn, P. Baybutt and D.G. Truhlar, J. Chem. Phys. 65, 3826 (1976).
130. T.P. Tsien and R.T. Pack, Chem. Phys. Lett. 6, 54 (1970).
131. M.A. Brandt and D.G. Truhlar, Chem. Phys. Lett. 23, 48 (1973).
132. W.R. Wadt and P.J. Hay, J. Chem. Phys. 82, 284 (1985).
133. R.E. Smalley, L. Warton and D.H. Levy, J. Chem. Phys. 68, 671 (1978).
134. See e.g. D.C. Clary, Chem. Phys. 65, 247 (1982); Mol. Phys. 51, 1299 (1984).
135. M.H. Alexander and A.E. DePristo, J. Chem. Phys. 65, 5009 (1976).
136. L.L. Poulsen, G.D. Billing and J.I. Steinfeld, J. Chem. Phys. 68, 5121 (1978).
137. W.L. Jorgensen and M.E. Cournoyer, J. Amer. Chem. Soc. 100, 4942 (1978).
138. W.L. Jorgensen, J. Chem. Phys. 70, 5888 (1980).
139. R.L. Redington, J. Chem. Phys. 75, 4417 (1981).
140. F.A. Gianturco, V.T. Lamanna and F. Battaglia, Int. J. Quantum Chem. 19, 217 (1981).
141. A.E. Barton and B.J. Howard, Discuss. Faraday Soc. 73, 45 (1982).
142. D.R. Yarkony, S.V. O'Neil, H.F. Schaefer, III, G.P. Baskin and C.F. Bender, J. Chem. Phys. 60, 855 (1974).
143. M.J. Redmon and G.C. Schatz, Chem. Phys. 54, 365 (1981).
144. D.W. Schwenke, D. Thirumalai, D.G. Truhlar and M.E. Coltrin, J. Chem. Phys. 78, 3078 (1983).

THE SPECTRUM, STRUCTURE AND SINGLET-TRIPLET SPLITTING IN METHYLENE CH_2

Philip R. Bunker
Herzberg Institute of Astrophysics
National Research Council of Canada
Ottawa, Ontario, Canada K1A OR6

ABSTRACT. This talk summarizes work done in the last three years, principally at the National Research Council in Ottawa, Canada and at the National Bureau of Standards in Boulder, U.S.A., on the mid-infrared and far-infrared spectrum of the methylene radical. Using the technique of laser magnetic resonance spectroscopy much infrared spectroscopic data has been obtained for $^{12}CH_2$, $^{13}CH_2$ and CD_2. With the help of theoretical calculations using the nonrigid bender hamiltonian, the equilibrium structure and potential energy surface of the electronic ground state - the $\tilde{X}^3B_1$ state - has been determined. The separation in energy between the ground state and the low-lying $\tilde{a}^1A_1$ electronic state has also been obtained. Ab initio results have been of importance in guiding the experimental work.

1. INTRODUCTION

This is a talk about the methylene molecule (CH_2). As a framework for understanding this molecule let's look at Figure 1. This shows approximate pictures of the three outermost occupied molecular orbitals of methylene, each of which involves one of the carbon 2p orbitals. CH_2 has eight electrons: two in C(1s), two in C(2s) and four more to put into these three orbitals. The $1b_2$ orbital wants to straighten the molecule - its energy rises as we bend the molecule. The $3a_1$ orbital bends the molecule - its energy lowers quickly as we bend the molecule. The $1b_1$ orbital does not care if

R. J. Bartlett (ed.), Comparison of Ab Initio Quantum Chemistry with Experiment for Small Molecules, 141–170.

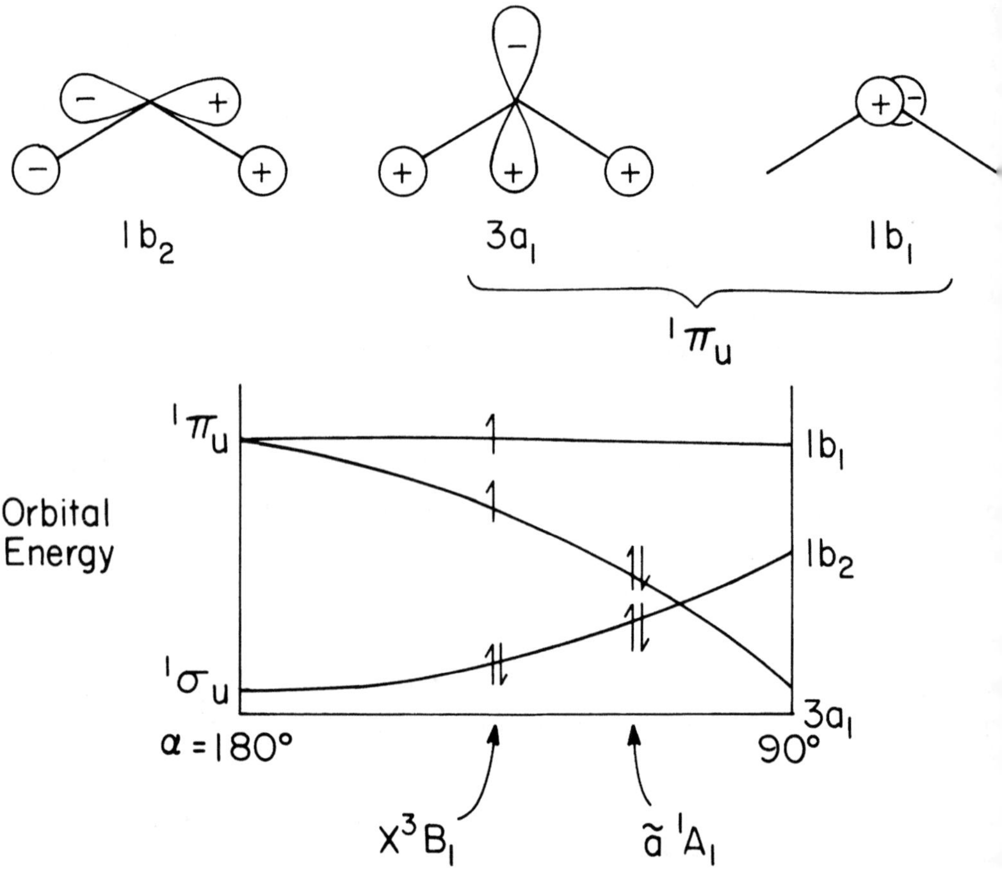

Figure 1. A simple picture of the outermost occupied molecular orbitals of CH_2 and the Walsh diagram of their energies as a function of the bond angle α.

the molecule is linear or bent - its energy does not change. The ground state of CH_2 turns out to have the configuration $...(1b_2)^2(3a_1)^1(1b_1)^1$ with two unpaired electrons, i.e. it is an electronic triplet state. The one electron in the $3a_1$ orbital wants to bend the molecule just about as hard as the two electrons in the $1b_2$ orbital want to straighten it. As a result the electronic energy of the molecule does not change very much as the molecule is bent. The first excited electronic state has the configuration $...(1b_2)^2(3a_1)^2$ and it is a well-bent electronic state. The energy difference between these lowest two electronic states is small. These are the molecular orbitals and the Walsh diagram for their energies as the bending angle is changed. The upshot of this diagram is that the ground state of CH_2 (the $\tilde{X}^3B_1$ state) is a triplet state that is not strongly bent and the first excited state (the $\tilde{a}^1A_1$ state) is a singlet state that is well bent.

These ideas about the triplet and singlet states are put more quantitatively in Figure 2. A cross section through the potential surfaces is drawn here and we see that the barrier to straightening in the triplet state is of the order of the bending energy; a molecule in such a state is said to be QUASILINEAR (although it is officially bent it has many of the attributes of a linear molecule). The singlet state is not much higher in energy than the triplet state but the two electrons in the $3a_1$ orbital really bend the molecule and there is a high barrier to straightening.

In our work we have determined rather accurately the shape of the ground state potential energy curve and, therefore, the equilibrium structure and bending energies. We have also observed transitions and perturbations between levels of the singlet and triplet states, and as a result we have accurately determined the singlet-triplet separation.

Methylene is the simplest polyatomic molecule in which the ground state is a triplet state; its ground state is quasilinear and it has a very low lying singlet state. All this goes to make

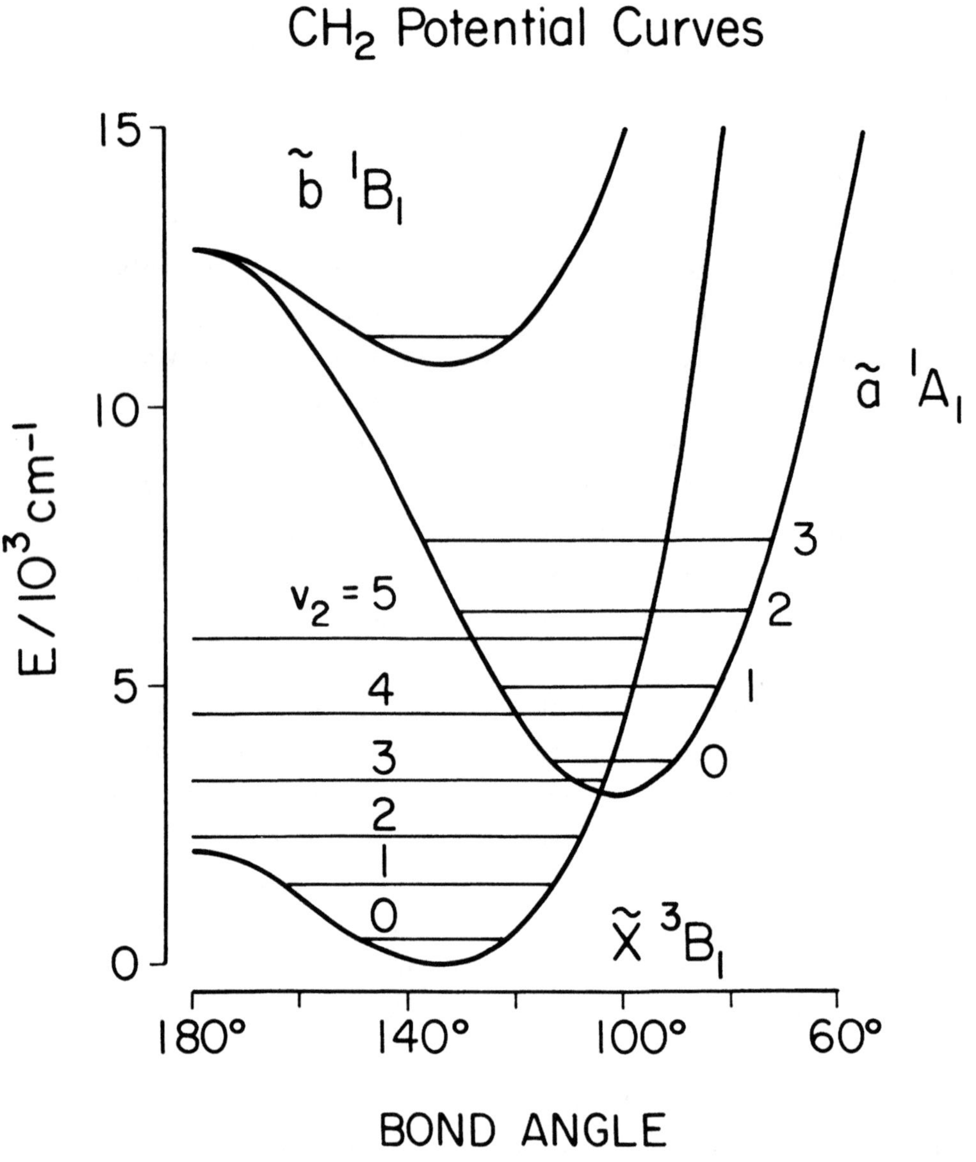

Figure 2. The potential curves of the three lowest electronic states of CH_2 and some bending energy levels.

methylene a very interesting molecule to study. Its position as the prototype of the carbenes and its frequent participation as an intermediate in chemical reactions makes it a very important molecule to study. It is also a free radical and, therefore, very reactive. At pressures of about 1 torr it lasts about 10^{-6} s and this makes it difficult to study. This combination of being interesting, important and difficult to study makes it an ideal research topic and as a consequence it has been the subject of much research. I'll now discuss some of the history of its study.

In 1942 Herzberg was already thinking about how to get a spectrum of CH_2 and he finally succeeded in 1959 with Jack Shoosmith in getting its spectrum. Table 1 shows some history of the study of the triplet state of methylene. Herzberg's spectrum was obtained by flash photolysis of diazomethane and most of the spectrum is diffuse and not amenable to detailed analysis in order to determine the triple state structure; chemical evidence and isotopic substitution proved that the spectrum was of CH_2. The spectrum involves absorption from the triplet state to high-lying Rydberg states. At about the same time Foster and Boys, in one of the early machine CI calculations, came up with an equilibrium structure for the triplet state that was bent. Herzberg interpreted the rather diffuse 1414Å spectral system as showing that the ground state of CH_2 was a linear triplet state. A semiempirical calculation by Jordan and Longuet-Higgins supported this conclusion. For about a decade it was accepted that the ground state of CH_2 was a linear triplet state. However, in the early 70's ESR studies by Wasserman and colleagues and Bernheim and colleagues showed that in a matrix the triplet state of CH_2 was bent. Herzberg and Johns then looked again at the 1414Å system and concluded that indeed this spectrum could be interpreted in terms of a bent equilibrium structure for the ground triplet state. An *ab initio* calculation by Bender and colleagues finally bent CH_2 for most people. The situation remained like this until our work which began in 1981. Up until our work there was no high-resolution gas phase spectroscopic study of triplet CH_2.

Table 1. Previous history of triplet methylene studies.

1959:	HERZBERG and SHOOSMITH	(flash photolysis)
	The 1414Å band of CH_2; largely diffuse	
1960:	FOSTER and BOYS	(*ab initio* CI)
	CH_2 ground state is bent triplet	
1961:	HERZBERG	(analysis of 1414Å)
	CH_2 ground state is linear triplet	
1962:	JORDAN and LONGUET-HIGGINS	(semi-empirical)
	CH_2 ground state is linear triplet	
1970:	WASSERMAN, BERNHEIM et al.	(matrix esr)
	CH_2 ground state is bent triplet	
1971:	HERZBERG and JOHNS	(reanalysis of 1414Å)
	CH_2 ground state could be bent triplet	
1972:	BENDER et al.	(*ab initio* CI)
	CH_2 ground state is bent triplet	

Table 2. Participants

BUNKER, JENSEN, MCKELLAR, SEARS	NRC OTTAWA
HOY	UWO LONDON
EVENSON, JENNINGS, SAYKALLY	NBS BOULDER
LOVAS, SUENRAM	NBS WASHINGTON
HIROTA, YAMADA	IMS OKAZAKI
BROWN	CD SOUTHAMPTON
LANGHOFF	NASA CALIFORNIA

Our work has involved a collaboration between 14 people in seven different laboratories and four different countries. Table 2 lists all the people involved.

2. THE $\tilde{X}^3B_1$ STATE OF CH_2

Our work in Ottawa began in February 1981 when I looked at an *ab initio* paper by Harding and Goddard in which, among other things, some *ab initio* points on the triplet state bending potential curve were calculated. The equilibrium CH bond length in the triplet state was also calculated. Given the shape of the bending potential curve and the bond length, we can use the RIGID BENDER hamiltonian and computer program to calculate the rotation-bending energies E_{rb}. This hamiltonian and computer program has been developed over the last ten years at NRC.

We digress here to discuss the problem of calculating the rotation-vibration energies of quasilinear molecules such as CH_2 in its ground triplet state. For a normal, rather rigid molecule such as, say, benzene, the vibrations are of small amplitude and to a good first approximation we can take the moments of inertia as being those of the molecule at equilibrium. The effects of vibration and rotation on the moments of inertia are small. This means that the rotational energy level pattern will be essentially the same in each vibrational state and that this structure can be calculated rather well if we know the equilibrium structure of the molecule. To a good first approximation we can separate the problem of calculating the rotational energies from the problem of calculating the vibrational energies. This is not the case for a quasilinear molecule. A quasilinear molecule undergoes large amplitude bending motion and the moments of inertia change very much during this motion. This means that the rotational energy level pattern will change very much from one bending vibrational state to another. Also, because of the extreme bending flexibility, centrifugal distortion effects on the structure of the molecule will be large and cannot be neglected (as they can be to a good first approximation for benzene, say). To calculate the rotation and bending energy levels in a quasilinear molecule we must treat the motions together, and in particular allow for the dependence of the moments of inertia on the bending angle. This is done in the rigid bender hamiltonian in which the rotation bending energy levels of a triatomic molecule are calculated from an assumed bending potential function and bond lengths.

The calculation of the rotation-bending energy levels of CH_2 in the triplet state, using the rigid bender with Harding and Goddard's potential surface and bond length, predicted that the bending vibration frequency would be 990 cm^{-1}, much lower than all previous estimates and right in the region where a CO_2 laser operates. Given this result, Sears and McKellar did not need much encouragement to search for the spectrum using CO_2 laser magnetic resonance, particularly so since a far infrared LMR spectrum of CH_2 had been obtained by Evenson and coworkers. This spectrum of Evenson and coworkers was a spectrum that had so far resisted being assigned; it had been proven to be due to CH_2 on the basis of chemical evidence and on the appearance of triplet nuclear hyperfine structure.

Thus in the Spring of 1981, Sears began the search for the bending fundamental band of triplet CH_2 using the CO_2 laser magnetic resonance apparatus at NRC in Ottawa. The important fact about LMR spectroscopy is the enormous sensitivity of the technique in detecting low concentrations. Typically, LMR can detect 10^8 molecules per cc which is about a million times more sensitive than a Fourier Transform Infrared Spectrometer. This is crucial here since CH_2 can only be produced in rather low concentrations owing to its high reactivity.

A block diagram of the NRC apparatus, built by Johns and McKellar, is shown in Figure 3. An LMR spectrum is obtained by passing a fixed laser frequency through the sample under study and monitoring the output power as a magnetic field applied to the

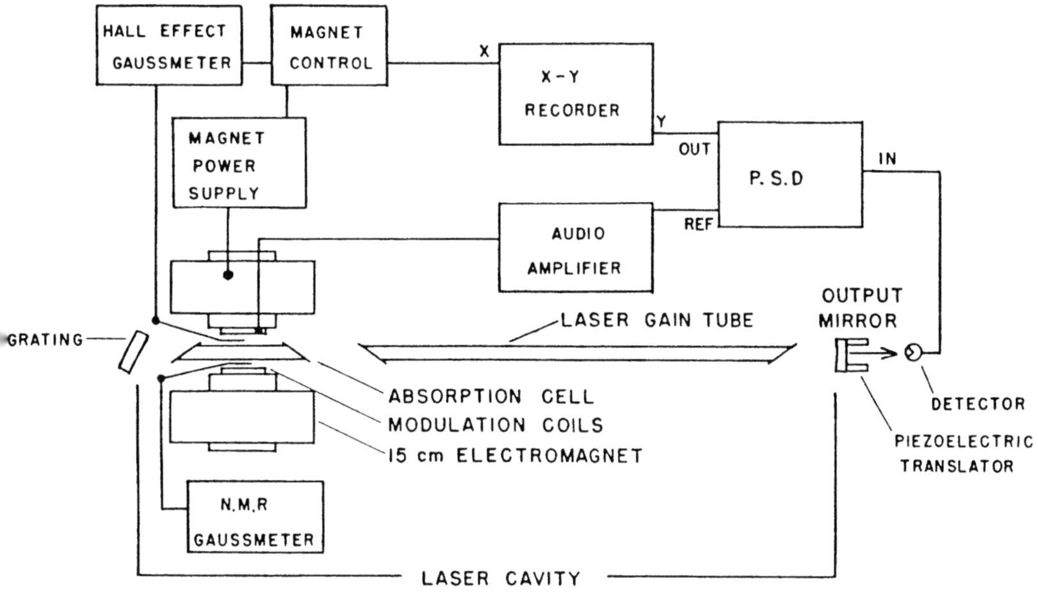

Figure 3. A block diagram of the LMR apparatus constructed at NRC by Johns and McKellar.

sample is changed. If an absorption line of the sample is Zeeman tuned into coincidence with the laser, a drop in output power occurs. Such resonances are recorded as a function of magnetic field. Clearly, only molecules whose energy levels suffer a shift in a magnetic field are amenable to study by this technique; this limits the technique to the study of molecules with unpaired electron spin. Searching for an LMR spectrum can be tedius and the assignment of the spectrum can be very difficult (note that Evenson's Far IR LMR spectrum of CH_2 had resisted assignment as of the Spring of '81).

Using the NRC LMR apparatus and a chemical reaction cell as shown in Figure 4 (gas A was CF_4 and gas B was CH_4) Sears finally obtained some resonances due to CH_2 triplet on May 8, 1981. The chemical reaction producing CH_2 is that between CH_4 and $F^{\bullet}$ (the F atoms are produced by passing CF_4 through an rf discharge). Figure 5 shows one of these early LMR spectra in derivative form.

Assigning LMR spectra is often a very difficult business and it was certainly difficult in the case of our early spectra from triplet CH_2. One proceeds essentially by trial and error, initially guessing the rotational assignment and then adjusting the electron spin fine structure of the levels involved to see if there is any way of reproducing the observed spectrum. McKellar is a master of this and for the CH_2 spectra he developed a fast way of generating guessed spectra using a desk top computer and plotter. Fortunately, the guessing was somewhat restricted by the results of the ab initio plus rigid bend calculation; only five low N lines were calculated to fall within 50 cm^{-1} of the CO_2 laser lines used and all trial-and-error spectra were restricted to these lines. There were also ab initio calculations (and ESR measurements) of the fine structure splitting which was a help. After plotting hundreds of guessed spectra McKellar finally assigned the three rotational transitions that were involved in the resonances seen by Sears (within the bending fundamental band). This was the end of the first stage of the work and in August 1981, we submitted a short

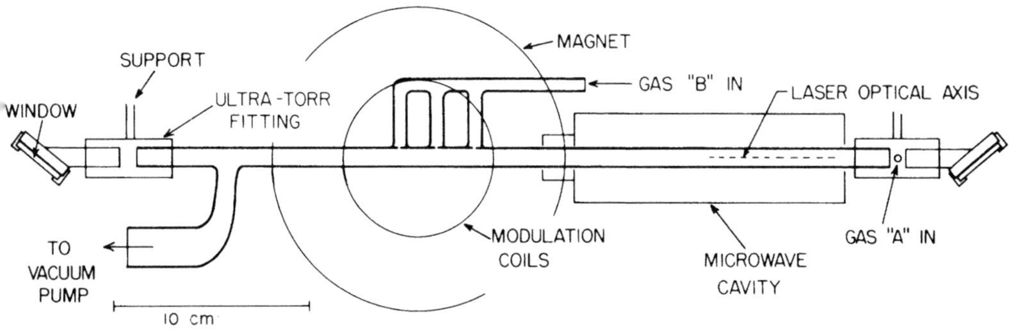

Figure 4. A schematic picture of the chemical reaction cell used to produce CH_2. Gas A was CF_4 and gas B was CH_4. The microwave discharge produces F atoms from CF_4 and the F atoms react with CH_4 to produce CH_2.

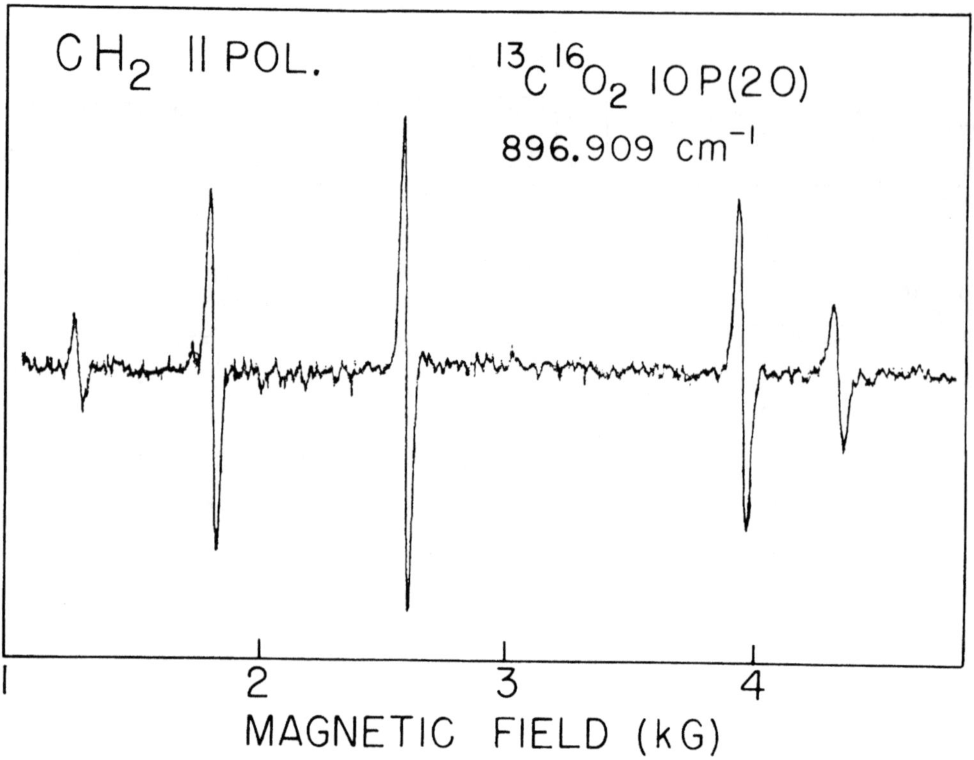

Figure 5. One of the first LMR spectra obtained for CH_2.

note of our findings to the Journal of Chemical Physics.

At this point our collaboration with Evenson at the NBS laboratory in Boulder, U.S.A., began. He had more unassigned and unpublished far-infrared LMR spectra of triplet CH_2 and with our experience of assigning three lines in the ν_2 band it did not take long to assign one of Evenson's spectra. At this point, then, we had three ν_2 band transitions (from NRC) and one pure rotation transition (from NBS) whose frequencies were accurately known. It was a simple matter to adjust the <u>ab initio</u> bending potential and the bond length so that when used in the rigid bender the observed four transition frequencies were produced exactly. The predictions of this calculation were then considered (and found) to be rather reliable indicators of where to look for further spectra, i.e. which laser lines to use in the search for further CH_2 triplet LMR spectra.

This further searching was pursued both in the far infrared (in Boulder) and in the ν_2 band (in Ottawa). A total of 13 pure rotational transitions were finally identified in the far-infrared LMR spectra as shown on the energy level diagram in Figure 6. From these results, the position of a microwave transition (the 4_{04} - 3_{13} transition) was predicted and subsequently seen at the NBS laboratory in Washington, D.C. In Ottawa, nine transitions were seen in the ν_2 band all involving $\Delta K=-1$; these transitions are shown in Table 3. The spectrum of gas phase triplet CH_2 was at last well characterized.

There was one puzzle with the data we obtained in Ottawa: Why did we not see any $\Delta K=+1$ transitions in the ν_2 band? We knew very precisely where to look. After a great deal of effort and using a new, more sensitive LMR spectrometer, we were finally able to see one such transition $1_{11} \leftarrow 2_{02}$; this is shown in Figure 7 and it is indeed much weaker than the $\Delta K=-1$ transitions. After some thought, it became apparent that the weakness of the $\Delta K=+1$ transitions in the ν_2 band of a quasilinear molecule is a general result as we can appreciate by looking at Figure 8.

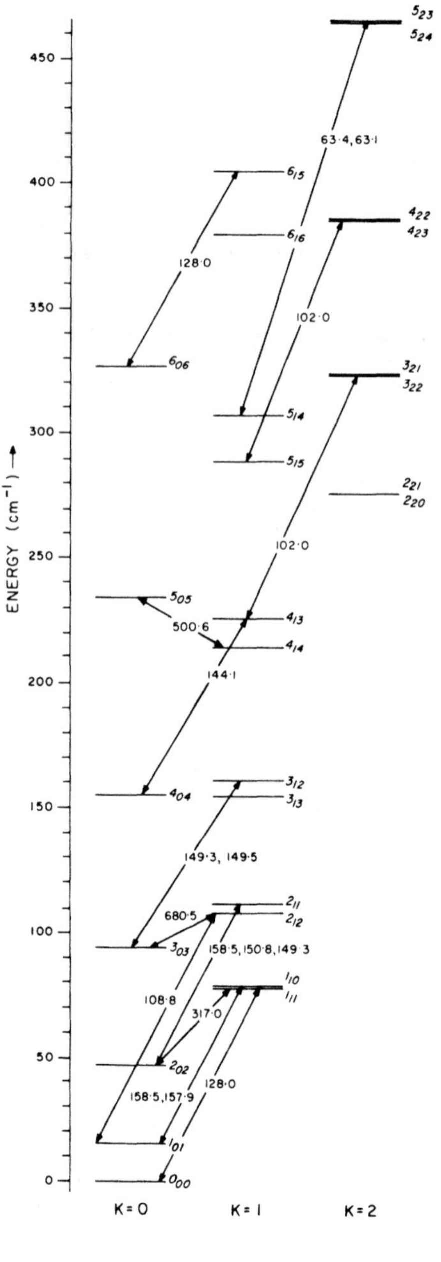

Figure 6. Rotational energy levels for triplet CH_2. The arrows and associated numbers indicate the observed transitions and the wavelengths (in μm) of the laser lines used for the observations.

Table 3. Observed vibration-rotation transitions in $\tilde{X}^3B_1$ CH_2.

Transition	Laser Line(s) Used	
$N'_{K_aK_c}(v'_2=1)\leftarrow N''_{K_aK_c}(v''_2=0)$	Identification	Frequency (cm^{-1})
$4_{14}-4_{23}$	(a) 10P(38)	830.3705
$0_{00}-1_{11}$	(a) 10P(34)	884.1843
$2_{11}-2_{20}$	(a) 10P(28)	889.7561
$4_{04}-4_{13}$	(a) 10P(26)	891.5739
$2_{02}-2_{11}$	(a) 10P(20)	896.9094
	(a) 10P(18)	898.6488
$1_{01}-1_{10}$	(a) 10P(18)	898.6488
$3_{13}-2_{20}$	(a) 10R(22)	929.9930
	(b) 10P(34)	931.0014
	(a) 10R(24)	931.3092
$2_{02}-1_{11}$	(b) 10P(34)	931.0014
	(a) 10R(24)	931.3092
$3_{03}-2_{12}$	(c) 10R(30)	947.2924

(a) $^{13}C^{16}O_2$

(b) $^{12}C^{16}O_2$

(c) $^{13}C^{18}O_2$

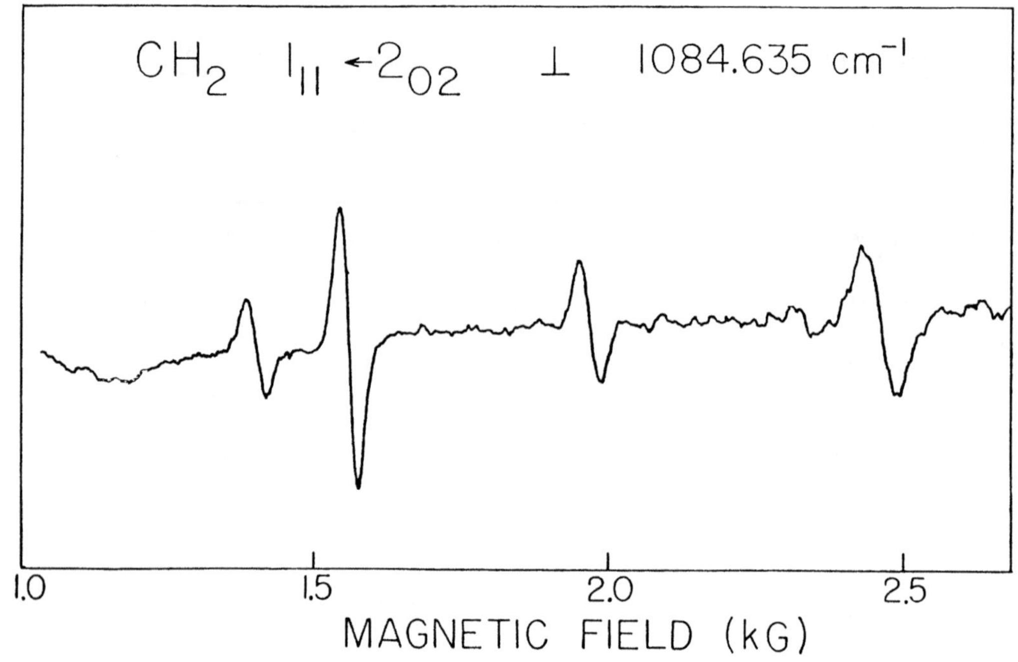

Figure 7. LMR spectrum of the $1_{11} \leftarrow 2_{02}$ transition in the υ_2 band of CH_2; the only $\Delta K_a=+1$ transition seen in the υ_2 band.

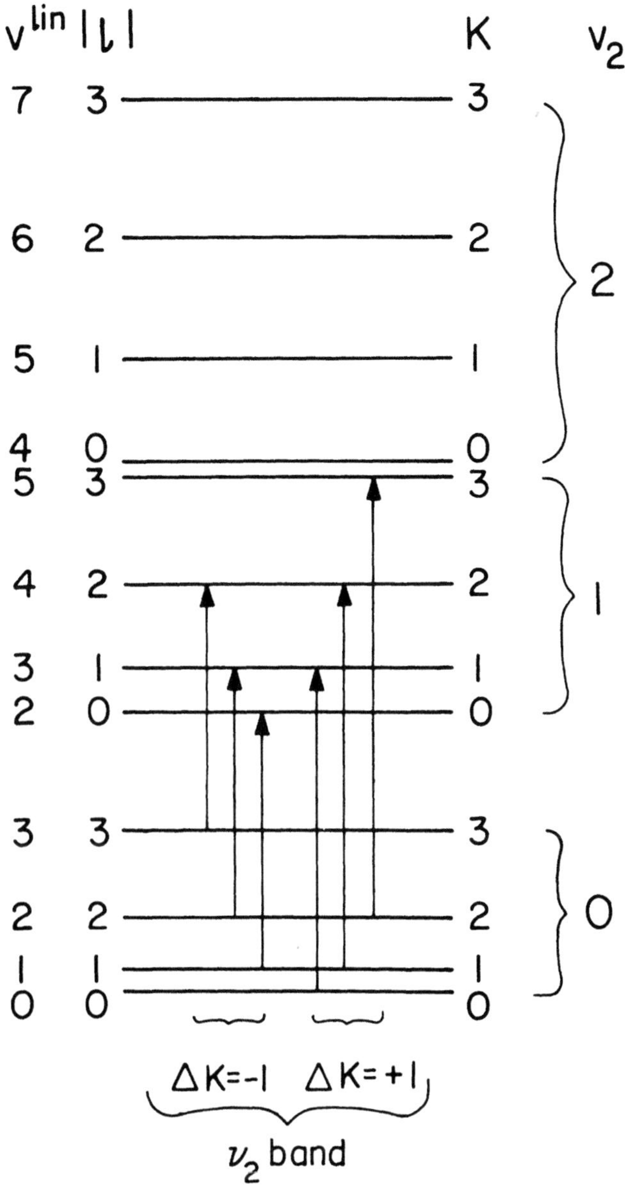

Figure 8. The J=3 levels of triplet CH_2 for v_2=0,1 and 2 with asymmetry splitting and fine structure suppressed. On the right the levels are labelled with the quantum numbers v_2 and K appropriate for a bent molecule, and on the left left by v^{lin} and ℓ appropriate for a linear molecule.

Figure 8 shows the J=3 energy levels of the v_2=0,1 and 2 levels with the asymmetry splittings and fine structure suppressed. This figure is drawn to scale and one can immediately appreciate how the rotational energy levels pattern changes with change in the bending quantum number v_2. In Figure 8, the rotation-bending energy levels are labelled on the right by the quantum number v_2 and K, appropriate for a bent molecule, and on the left by the quantum numbers v^{lin} and ℓ appropriate for a linear molecule. These labels are related by $v^{lin}=2v_2+K$, and $|\ell|=K$. We see that, in a $\Delta v_2=1$ band, the $\Delta K=-1$ transitions involve $\Delta v^{lin}=1$ whereas the $\Delta K=+1$ transitions involve $\Delta v^{lin}=3$. In the linear harmonic oscillator limit $\Delta v^{lin}=3$ transitions are forbidden. Since we can view a quasilinear molecule as a linear molecule with bending anharmonicity, where this anharmonicity diminishes as the height of the barrier to linearity diminishes, we see that the ratio of the intensities of $\Delta K=+1$ to -1 transitions in a $\Delta v_2=1$ band goes down as the barrier goes down, and goes down as v_2'' increases. The weakness of $\Delta K=+1$ transitions means that in a very cold source (such as an inert gas matrix or interstellar space), when nearly all molecules are in K=0 levels, it will be more difficult to detect the ν_2 band of CH_2.

At this stage of the work we had obtained 14 pure rotation transitions and 10 transitions in the ν_2 band. The standard spectroscopic approach to all this data is to fit it to an effective hamiltonian and to obtain a lot of molecular parameters. The result of doing this is shown in Table 4. For a quasilinear molecule many of the parameters determined in fits like this are of rather limited usefulness. In particular, it is not possible to use them to make accurate predictions of the rotation-vibration energy levels of other isotopes, or to make accurate predictions of higher rotation and bending energy levels. However, the fine structure parameters D and E, and the spin rotation parameters, are well determined, useful, and in rather good agreement with _ab initio_ calculations. The nuclear hyperfine splitting parameters were

Table 4. Molecular parameters[a] (in cm^{-1}) of CH_2 ($\tilde{X}^3B_1$).

	(000) state[b]	(010) state
ν_0	---	963.09866(41)
A	73.05775(11)	184.12263(82)
B	8.415172(76)	8.36216(15)
C	7.219272(45)	7.09284(20)
Δ_K	1.991049(47)	23.969[c]
Δ_{NK}	-0.019660(27)	-0.07889(14)
Δ_N	0.0003013(34)	0.000162(18)
δ_N	$0.1012(12)\times10^{-3}$	0.1012×10^{-3}[d]
Φ_K	0.0	1.52[c]
Φ_{KN}	-0.0019417(21)	0.0
Φ_{NK}	$0.1281(86)\times10^{-4}$	$-4.671(57)\times10^{-4}$
Φ_N	$0.251(59)\times10^{-6}$	$0.78(58)\times10^{-6}$
ϕ_N	$0.195(30)\times10^{-6}$	0.0
D	0.77842(14)	0.79560(48)
E	0.039906(38)	0.03540(53)
ε_{aa}	0.000446(78)	0.00352(54)
ε_{bb}	-0.005148(18)	-0.00470(23)
ε_{cc}	-0.004106(27)	-0.00458(21)

[a] Parameters not given were fixed at zero. The numbers in parentheses are one standard deviation from the least-squares fit in units of the last quoted digit. The precise values and standard deviations of the rotational and centrifugal distortion parameters are not very significant, particularly for the (010) state. In the Zeeman hamiltonian, g factors were fixed at appropriate values.

[b] (000) state parameters from Sears, Bunker, McKellar, Evenson, Jennings and Brown.

[c] Δ_K and Φ_K were fixed at values determined from the semirigid bender model.

[d] δ_N was fixed at its ground (000) state value.

also determined for the ground vibrational state of triplet CH_2.

A more useful approach to the fitting of the rotation-bending energy levels of a quasilinear molecule, from the point of view of being able to make predictions, involves using the NONRIGID BENDER hamiltonian. As schematically indicated in Table 5, this involves adding the effects of the centrifugal stretching of the bonds to the rigid bender hamiltonian. This involves incorporating the effects of the stretch-stretch and stretch-bend force constants into the model; the values of these force constants are then determined by fitting to the data. A nonrigid bender fit to the $^{12}CH_2$ data was made and from the results of the fit we were able to predict the rotation-bending energy levels of $^{13}CH_2$ and CD_2. Subsequent LMR and diode spectra of these isotopes were obtained and the predictions were both useful and reliable. A complete fit of the eigenvalues of the nonrigid bender hamiltonian to all the available 61 rotation-vibration data for the three isotopes was made and some of the more important results of the fit are given in Table 6. The equilibrium structure of triplet CH_2 was determined to have a CH bond length of 1.0748Å and a bond angle of 133.84°. The height of the barrier to linearity was determined to be 1940 cm^{-1}. Also, from this fit it was possible to determine how the optimum CH bond length changes with bond angle, and the increase in optimum bond length from 1.060Å at α=180° to 1.088Å at α=109.5° can be compared with the equilibrium bond lengths of 1.061Å and 1.086Å in C_2H_2 and CH_4 respectively. From these fits it was also possible to predict the frequencies of the stretching fundamentals. Subsequent _ab initio_ calculations by Langhoff and myself showed, however, that both stretching fundamentals will be extremely weak.

2. THE SINGLET-TRIPLET ENERGY SEPARATION

The final part of this talk involves the singlet state $\tilde{a}\,^1A_1$, the observation of singlet-triplet transitions, the analysis of singlet-triplet perturbations, and the determination of the singlet-triplet splitting.

Table 5. Nonrigid bender.

Rigid bender plus centrifugal distortion of bonds							
R_e	$f_{\alpha\alpha}$	ρ_e	+	f_{rr}	$f_{r\alpha}$	f_{rrr}	f_{rrrr}
$f_{\alpha\alpha\alpha}$	$f_{\alpha\alpha\alpha\alpha}$			$f_{rr\alpha}$	$f_{r\alpha\alpha}$		$f_{rr'}$

Table 6. Some triplet state parameters.

α/deg	r(opt)/Å	
109.5	1.088	
133.8	1.075	equilibrium
180.0	1.060	saddle point

Height of barrier is 1940 cm^{-1}

υ_1 predicted at 2950 cm^{-1}

υ_3 predicted at 3080 cm^{-1}

but $\mu(\upsilon_3)$ and $\mu(\upsilon_1)$ are both only

0.003 Debye (*ab initio*)

This work really begins with a talk given at Columbus in 1979 by Evenson and coworkers (see Table 7). In that work, an LMR spectrum of CH_2 was found in which there was stimulated emission (i.e. gain); as the magnetic field was swept there was an increase in laser output at a particular magnetic field. This spectrum was retaken in November, 1981, and we began puzzling as to what caused this population inversion to be set up in CH_2. A good possibility is that the singlet state is involved, i.e. this could be a singlet-singlet transition where the lower level is mixed with a triplet state level and depleted in population (the mixing gives the transition magnetic character so that it shifts in a magnetic field -- a pure singlet-singlet transition would not be tuneable by LMR).

In 1968 Duxbury suggested possible singlet-triplet perturbations in the spectrum of singlet methylene.

Table 7. An important methylene observation.

Paper TF5. At Columbus Ohio 1979. Far Infrared LMR Spectra of Methlyene. Evenson, Saykally and Hougen.

...Using the 171.8 μm laser line of $^{13}CH_3OH$ we have observed a very weak sequence of hfs triplets occuring in stimulated emissions ...The origin of this population inversion presents an interesting question.

The transition occurs at 171.8 μm (58.2 cm^{-1}). The Zeeman pattern shows that the transition is a J=7 to J=7 Q-branch transition and the triplet nuclear hyperfine structure means that the transition can only involve levels for which the sum of K_a and K_c is an odd number. From the analysis of the electronic spectrum involving the singlet state of CH_2 by Herzberg and Johns we can determine the term values of the ground vibrational state of the $\tilde{a}^1A_1$ state and look for a transition fulfilling the above requirements. There are two possibilities: 7_{25}-7_{16} (at about 62 cm^{-1}) and 7_{43}-7_{34} (at about 59 cm^{-1}).

Further experiments were performed using other laser lines near 60 cm^{-1}. In all we saw LMR resonances on eight laser lines in this region and were able to unambiguously assign the transitions as shown in Figure 9. The numbers in circles in this figure indicate with how many laser lines a particular transition was observed. From these observations the singlet state levels involved are well-identified, but we can only determine the value of N (i.e. J-S), and that K_a and K_c must both be odd for the perturbing triplet level; we do not know the vibrational assignment of this triplet state level.

The possible vibrational assignments of this triplet state level can be appreciated by looking at the potential curves and energy levels shown in Figure 10. Although we did not know the precise disposition of the curves, it was clear that the vibrational level of the triplet state most likely was (1,0,0), (0,0,1), or $(0,v_2,0)$ with v_2=2 or 3. At this stage, help in making the assignment was sought from an _ab initio_ calculation of the spin-orbit mixing matrix element between the triplet and singlet state. The results of the calculation are given in Table 8 in which we see that the assignment of the perturbing triplet level as $(0,3,0)6_{15}$ produces a theoretical value for the spin-orbit matrix element of 3.0 cm^{-1}, n good agreement with the experimental value of 3.201 cm^{-1}. The strong overlap of the triplet state (0,3,0) level with the singlet state (0,0,0) level can be appreciated by looking at Figure 11. The wavefunctions are drawn in this figure and the three points at which the _ab initio_ electronic spin-orbit matrix element was calculated are indicated.

To achieve an assignment of the triplet level involved here based purely on experiments we did further LMR experiments in order to find more singlet-triplet perturbations. We found three more singlet state levels that were perturbed by triplet state levels and we could determine the N values of the triplet state levels involved. Using the nonrigid bender to calculate the positions of

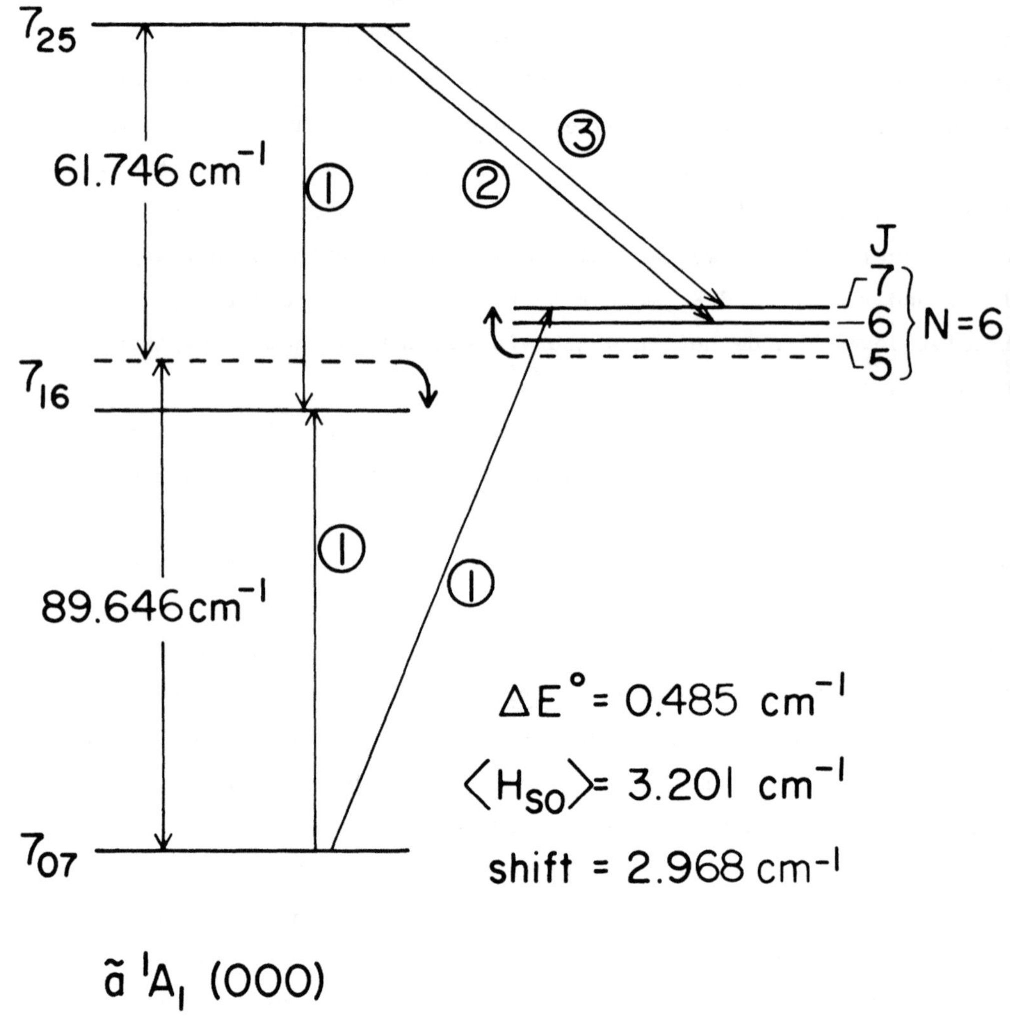

Figure 9. Diagram showing the perturbation between the J=7, N=6 triplet state level and the 7_{16} singlet state level; the dashed lines indicate the positions of the levels before perturbation. The numbers in circles give the number of laser lines used to observe each transition.

Table 8. Values of *ab initio* coupling matrix elements $W_{SO} = \langle H_{SO} \rangle$ for the $\tilde{a}^1A_1$ (000) 7_{16} level with various $\tilde{X}^3B_1$ $(v_1v_2v_3)$ $6_{K_aK_c}$ levels.

v_1	v_2	v_3	K_a	K_c	$W_{SO}(cm^{-1})$
1	0	0	any	any	<0.3[a]
0	0	1	any	any	<0.3[a]
0	any	0	5	1	<0.4[a]
0	any	0	3	3	<2.4[a]
0	0	0	1	5	-1.3
0	1	0	1	5	2.1
0	2	0	1	5	-2.8
0	3	0	1	5	3.0
0	4	0	1	5	-2.9
0	5	0	1	5	2.3
0	6	0	1	5	-1.7

[a] Absolute value of W_{SO} is less than this value.

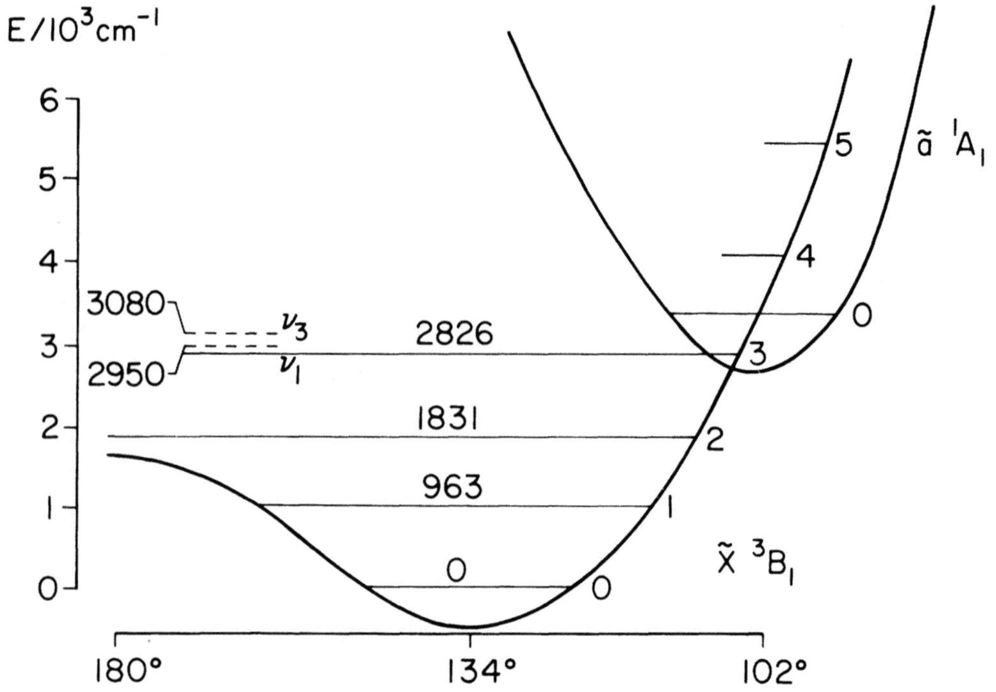

Figure 10. Approximate disposition of singlet and triplet potential curves and vibrational levels.

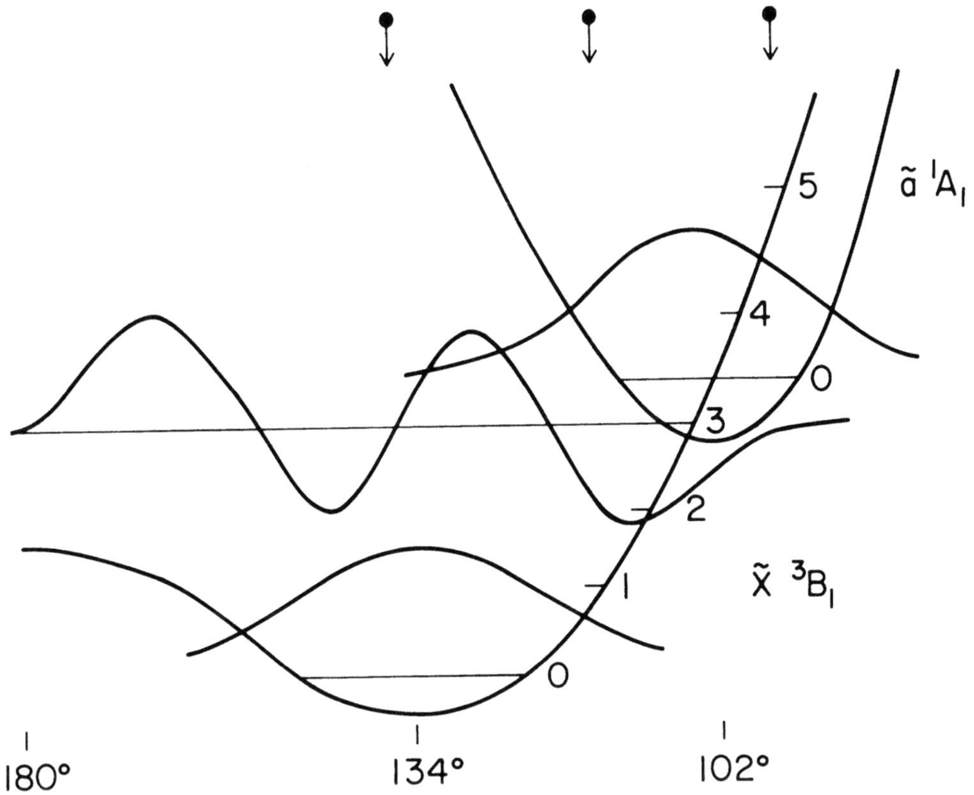

Figure 11. The bending potential curves of the singlet and triplet state determined by semirigid bender fits. Wavefunctions for $v_2=0$ of the singlet state and $v_2=0$ and 3 of the triplet state are shown. The arrows at the top of the figure indicate the three angles for which the <u>ab initio</u> calculations of H_{SO} were carried out.

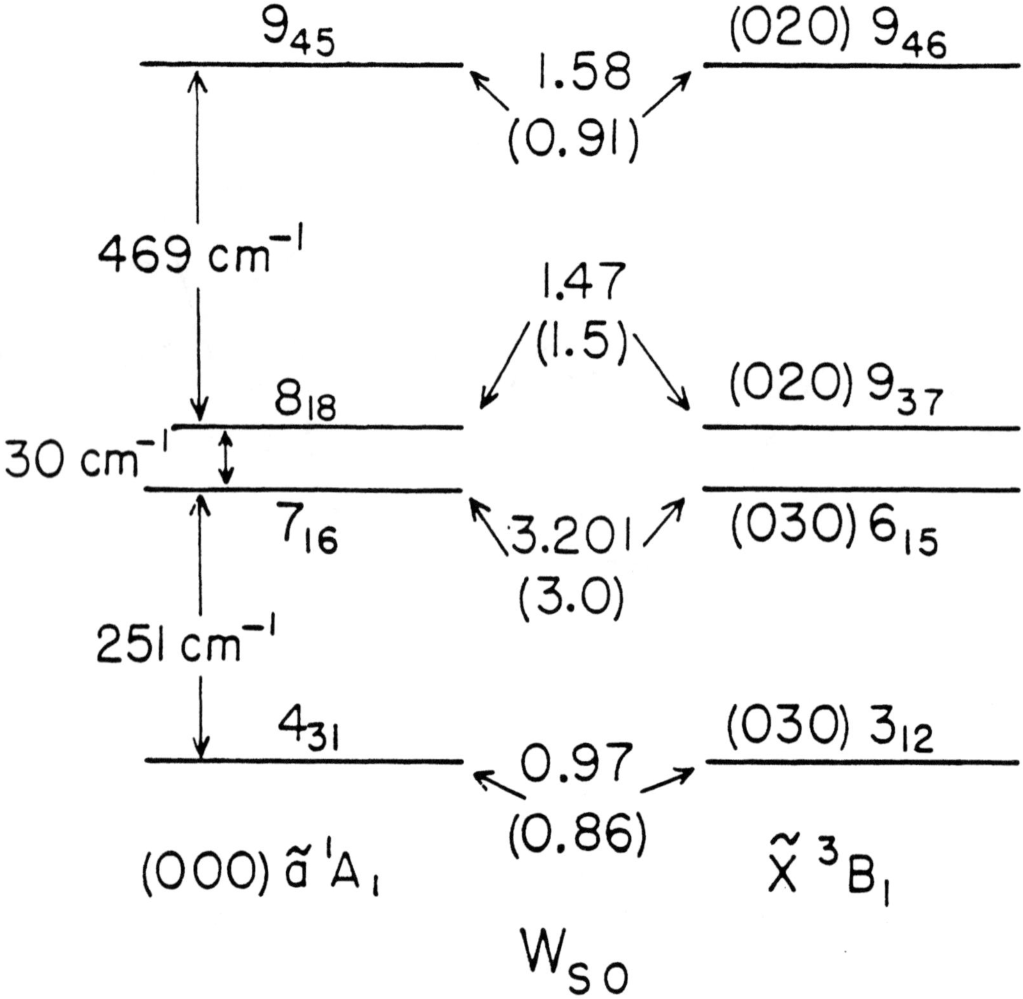

Figure 12. Perturbed energy levels in the singlet and triplet states. The numbers in the center of the figure are the values (in cm^{-1}) of the spin-orbit interaction matrix elements; the *ab initio* theoretical values are in parentheses.

all triplet state levels having v_2 less than 7 we could only find one set of four energy levels with N values, symmetry, and separations that matched the observations. This unique assignment is shown in Figure 12. In this figure, the spin-orbit matrix elements obtained from experiment are compared with the *ab initio* values (the latter are in parentheses). The 9_{45} level of the singlet is apparently also subject to a second perturbation which we have not yet identified in detail.

The energy ladders of the triplet and the singlet states of CH_2 are thus accurately located relative to each other by the perturbation links identified in Figure 12. Knowing the energies of the levels indicated in Figure 12 above the 0_{00} levels of the ground states of the singlet and triplet (the singlet from the work of Herzberg and Johns, and the triplet from the energies taken from the nonrigid bender) we can determine the singlet-triplet energy separation. In this way we determine that $T_0(\tilde{a}\,^1A_1)=3165\pm20\ cm^{-1}$. This is the most precise determination of this quantity to date.

References:

<u>Historical</u>

1. G. Herzberg and J. Shoosmith, Nature, London, 183, 1801 (1959).
2. J.M. Foster and S.F. Boys, Rev. Mod. Phys. 32, 305 (1960).
3. G. Herzberg, Proc. Roy. Soc., London, A262, 291 (1961).
4. P.C.H. Jordan and H.C. Longuet-Higgins, Mol. Phys. 5, 121 (1962).
5. G. Herzberg and J.W.C. Johns, Proc. Roy. Soc., London, A295, 107 (1966); G. Duxbury, J. Mol. Spectrosc. 25, 1 (1968). G. Duxbury, J. Mol. Spectrosc. 25, 1 (1968)
6. R.A. Bernheim, H.W. Bernard, P.S. Wang, L.S. Wood and P.S. Skell, J. Chem. Phys. 53, 1280 (1970).
7. E. Wasserman, W.A. Yager and V.J. Kuck, Chem. Phys. Lett. 7, 409 (1970).
8. E. Wasserman, V.J. Kuck, R.S. Hutton and W.A. Yager, J. Am. Chem. Soc., 92, 7491 (1970).
9. G. Herzberg and J.W.C. Johns, J. Chem. Phys. 54, 2276 (1971).
10. D.R. McLaughlin, C.F. Bender and H.F. Schaeffer III, Theoret. Chim. Acta (Berl) 25, 352 (1972); C.F. Bender, H.F. Schaefer III, D.R. France Schetti, L.C. Allen, J. of Am. Chem. Soc. 94, 688 (1972).

C.F. Bender, H.F. Schaefer III, D.R. France Schetti and L.C. Allen, J. of Am. Chem. Soc. 94, 688 (1972).
11. L.B. Harding and W.A. Goddard, Chem. Phys. Lett. 55, 217 (1978).
12. J.A. Mucha, K.M. Evenson, D.A. Jennings, G.B. Ellison and C.J. Howard, Chem. Phys. Lett. 66, 244 (1979).

Our Work:

1. T.J. Sears, P.R. Bunker and A.R.W. McKellar, J. Chem. Phys. 75, 4731 (1981).
2. T.J. Sears, P.R. Bunker, A.R.W. McKellar, K.M. Evenson, D.A. Jennings and J.M. Brown, J. Chem. Phys. 77, 5348 (1982). Figure 6 is from this paper.
3. T.J. Sears, P.R. Bunker and A.R.W. McKellar, J. Chem. Phys. 77, 5363 (1982). Figures 5 and 7 are from this paper.
4. P. Jensen, P.R. Bunker and A.R. Hoy, J. Chem. Phys. 77, 5370 (1982).
5. A.R.W. McKellar and T.J. Sears, Can. J. Phys. 61, 480 (1983).
6. P.R. Bunker, T.J. Sears, A.R.W. McKellar, K.M. Evenson and F.J. Lovas, J. Chem. Phys. 79, 1211 (1983).
7. A.R.W. McKellar, C. Yamada and E. Hirota, J. Chem. Phys. 79, 1220 (1983).
8. P.R. Bunker and P. Jensen, J. Chem. Phys. 79, 1224 (1983).
9. P. Jensen and P.R. Bunker, J. Mol. Spectrosc. 99, 348 (1983).
10. F.J. Lovas, R.D. Suenram and K.M. Evenson, Ap. J. 267, L131 (1983).
11. P.R. Bunker and S.R. Langhoff, J. Mol. Spectrosc. 102, 204 (1983). Figure 8 is from this paper.
12. P. Jensen, Comp. Phys. Reports 1, 1 (1983).
13. A.R.W. McKellar, P.R. Bunker, T.J. Sears, K.M. Evenson, R.J. Saykally and S.R. Langhoff, J. Chem. Phys. 79, 5251 (1983). Figures 9, 11 and 12 are from this paper.
14. T.J. Sears and P.R. Bunker, J. Chem. Phys. 79, 5265 (1983).
15. T.J. Sears, A.R.W. McKellar, P.R. Bunker, K.M. Evenson and J.M. Brown, Ap. J., 276, 399 (1984).
16. K.M. Evenson, T.J. Sears and A.R.W. McKellar, J. Opt. Soc. Am. B 1, 15 (1984).

AB INITIO SCF AND CI STUDIES ON THE GROUND STATE OF THE WATER MOLECULE. III. VIBRATIONAL ANALYSIS OF POTENTIAL ENERGY AND PROPERTY SURFACES

Walter C. Ermler, Department of Chemistry and Chemical Engineering, Stevens Institute of Technology, Hoboken, New Jersey 07030

Bruce J. Rosenberg, Harris Corporation, Government Support System Division, Syosset, New York 11791

Isaiah Shavitt, Department of Chemistry, The Ohio State University, Columbus, Ohio 43210

ABSTRACT. Ab initio SCF and CI potential energy and property surfaces for the ground electronic state of the water molecule have been used in vibrational analyses of water molecules including the D and T isotopic variants. The formalism is based on a perturbation theory approach and the use of a proper non-linear transformation of internal coordinates of each surface point to normal coordinates. Vibrational excitation energies, zero-point corrections to one-electron properties, and transition intensities are reported and compared with experimental data and other ab initio results. The general usefulness of the method and its dependence on such variables as the quality of the electronic wavefunctions, subsequent analytical representations of the potential energy, and the particular application of the perturbation theory are discussed.

1. INTRODUCTION

This is the third and final article in a series dealing with the use of self-consistent field (SCF) and configuration interaction (CI) calculations in the study of the X^1A_1 state of the water molecule. Paper I [1] involved a comparison of the quality of

R. J. Bartlett (ed.), Comparison of Ab Initio Quantum Chemistry with Experiment for Small Molecules, 171–216.

basis sets in terms of energies and one-electron properties for a single geometry. This established a 39-term basis set of Slater-type functions (STO's) to be used for the computation of SCF and CI energies and properties at 36 nuclear configurations near the equilibrium geometry. The results were used to obtain potential energy and property surfaces, represented as expansions in internal displacement coordinates, as presented in Paper II [2].

The potential energy surfaces of Paper II are used here in vibrational analyses of isotopic species of the water molecule following the perturbation theory formalism of Kern and co-workers [3] and incorporating the further refinement [4] of a proper non-linear transformation from internal to normal coordinates. The results for the SCF and CI surfaces are compared with each other, with those for other ab initio surfaces, and with corresponding results for surfaces derived from experimental data. The effects of electron correlation on the shapes of the surfaces and on the results of the vibrational analysis are examined. These comparisons also help in assessing the dependability of the method for vibrational analysis in the context of the residual errors in the computed surfaces.

The results of the vibrational energy analysis are given in Section 2 and expectation values of various one-electron properties are presented in Section 3. Vibrational transition intensities computed from these results and using the transition matrix elements over corresponding dipole moment operators are given in Section 4.

2. VIBRATIONAL ENERGIES

As was shown in Ref. 3c, the purely vibrational Hamiltonian can be expressed through second-order of perturbation theory in the form

$$H = H_0 + H_1 + H_2, \tag{1}$$

where, in atomic units,

$$H_0 = 1/2 \sum \omega_i p_i^2 + 1/2 \sum \omega_i q_i^2 = T_0 + V_0, \tag{2a}$$

$$H_1 = \sum k_{ijk} q_i q_j q_k = V_1, \tag{2b}$$

and

$$H_2 = \sum k_{ijk\ell} q_i q_j q_k q_\ell + 1/8 \sum_\alpha (I_\alpha)^{-1} (4\Pi_\alpha^2 - 1)$$

$$= V_2 + V_2'. \tag{2c}$$

The sums are over the N nondegenerate modes of vibration, with q_i the reduced normal coordinates and p_i their conjugate momenta. The coefficients ω_i, k_{ijk}, and $k_{ijk\ell}$ are the harmonic frequencies and cubic and quartic force constants when the potential energy is expressed as an expansion in reduced normal coordinates, viz.,

$$V = V_0 + V_1 + V_2. \tag{3}$$

The second term in Eqn. (2c) corresponds to the correction due to the Coriolis interaction , with I_α being the equilibrium principal moments of inertia and Π_α the vibrational angular momenta, and α=A,B,C.

Application of standard methods of perturbation theory to Eqn. (1) leads to a solution for the first-order vibrational wavefunction as an expansion in terms of zeroth-order (i.e. harmonic oscillator) wavefunctions. The total vibrational energy relative to the bottom of the potential well may subsequently be expressed as

$$E_v = G + \sum_i \omega_i (v_i + 1/2) + \sum_{i<j} X_{ij} (v_i + 1/2)(v_j + 1/2), \tag{4}$$

where ω_i are the harmonic frequencies and G and the X_{ij} are constants obtained as algebraic expressions in terms of the potential constants (ω_i, k_{ijk}, and $k_{ijk\ell}$) and the I_α and certain Coriolis coefficients [3,7].

The 36 energy points for the SCF, SD (CI including all single and double replacements), and SDQ (the SD values as modified by the application of an approximate correction for the effects of quadruple replacements) calculations of Paper II have been used to compute the constants in Eqn. (4), using both a linear [3] and a proper non-linear [4] transformation from internal to normal coordinates. Table 1 contains the cubic and quartic force constants (Eqn. (2), V_1 and V_2) for H_2O, HDO, and D_2O and the corresponding values derived [6] from spectroscopic data. (The harmonic frequencies, which are the first terms in the expansion of the potential energy, also appear as the leading terms in Eqn. (4), so they have been included in the tabulation of the results for the vibrational energy expansion constants in Table 2.) The designation "(ℓ)" (for linear) in some column headings in Table 1 indicates constants derived by simple three or four index transformations of the anharmonic force constants expressed relative to internal coordinates (Table IV of Paper II), where transformation matrices relating internal and normal coordinates are the eigenvectors of the GF matrix [3c]. The designation "(n)" (for non-linear) indicates the expansion coefficients obtained from a least-squares fit of the 36 energies (Table I of Paper II) after the internal coordinates (R_1, R_2, and θ) for each geometry were transformed directly to reduced normal coordinates using the Eckart conditions and the harmonic force constants as described in Ref. 4.

The values of the constants G, ω_i, and X_{ij} of Eqn. (4), as derived from data in the corresponding columns of Table 1, are given in Table 2. Also included are results for the HTO, DTO, and T_2O molecules. The values of these constants as obtained [6] by direct fitting of the experimentally observed frequencies to Eqn. (4) are also given for comparison. Finally, as an example of how the raw theoretically obtained data (the energies at each geometry as given in Table I of Paper II) are transformed into values that may be compared directly with raw experimental data (the observed band origins), Table 3 presents vibrational energies relative to the zero-point energy for the states having $v_1+v_2+v_3 \leq 3$ for H_2O, HDO, and D_2O. Thus, the results reported in Table I of Paper II and Tables 1-3 here provide comparisons of theoretically and

Table 1. Cubic and Quartic Force Constants[a]

Constant[b]	SCF(ℓ)	SCF(n)	SD(ℓ)	SD(n)	SDQ(ℓ)	SDQ(n)	Experimentally derived[c]
A. H_2O							
k_{111}	-304.2	-302.4	-303.9	-305.5	-304.4	-304.1	-301.95
k_{112}	15.7	8.2	19.9	11.8	21.5	13.3	18.65
k_{122}	-15.4	58.8	16.8	53.8	-17.6	50.6	23.90
k_{222}	-62.3	-52.1	-55.6	-44.5	-53.7	-42.4	-60.10
k_{133}	-307.7	-307.3	-305.4	-305.6	-306.4	-306.6	-294.43
k_{233}	32.2	41.5	36.7	43.4	40.1	46.5	58.72
k_{1111}	27.5	26.3	30.6	30.9	30.1	30.3	31.38
k_{1112}	-2.3	-2.3	-3.1	-2.6	-3.1	-2.5	-29.66
k_{1122}	0.4	-11.4	0.6	-12.0	0.7	-12.4	-7.21
k_{1222}	1.9	5.6	1.6	6.0	1.4	5.9	-1.76
k_{2222}	-5.0	-4.5	-4.0	-2.8	-5.9	-4.7	-5.12
k_{1133}	27.9	26.6	30.7	30.9	35.1	35.2	30.02
k_{1233}	-2.7	-5.9	-3.5	-4.7	-8.9	-10.1	-15.60
k_{2233}	-0.5	-13.6	-0.4	-14.9	1.1	-13.7	-12.88
k_{3333}	28.5	26.6	31.2	30.9	36.2	36.0	31.79
B. HDO							
k_{111}	-268.6	-265.8	-267.0	-267.1	-267.7	-268.0	-258.45
k_{112}	13.8	14.2	17.0	15.8	18.8	17.9	25.01
k_{122}	-7.6	8.6	-8.6	8.9	-9.1	7.5	-8.87
k_{222}	-51.4	-41.9	-45.9	-35.7	-44.4	-34.2	-48.74
k_{113}	13.8	14.0	13.5	14.0	13.8	14.2	11.08
k_{123}	-6.1	-14.2	-6.1	-14.0	-6.8	-14.5	-15.43
k_{223}	-8.8	81.2	-9.5	73.4	-9.9	70.4	57.01
k_{133}	-20.7	-20.6	-22.0	-21.6	-22.1	-21.7	-24.72
k_{233}	22.0	20.6	25.9	26.1	28.2	28.0	36.52
k_{333}	-432.9	-430.6	-430.3	-430.3	-431.6	-431.2	-418.10

Table 1. Cubic and Quartic Force Constants (cont.)

Constant[b]	SCF(ℓ)	SCF(n)	SD(ℓ)	SD(n)	SDQ(ℓ)	SDQ(n)	Experimentally derived[c]
B. HDO (cont.)							
k_{1111}	29.4	25.5	32.5	33.0	36.5	35.7	31.61
k_{1112}	-1.9	4.2	-2.6	-3.1	-5.6	-14.3	-16.00
k_{1122}	-0.1	-4.7	0.0	-4.2	0.5	-3.1	-1.67
k_{1222}	1.0	-3.9	0.9	1.4	0.8	5.5	-3.33
k_{2222}	-3.8	-0.2	-3.0	1.9	-4.5	0.8	-0.06
k_{1113}	-1.6	-1.5	-1.8	-1.8	-3.0	-3.0	-1.57
k_{1123}	0.2	0.5	0.2	0.7	1.8	2.9	-2.81
k_{1223}	0.3	0.1	0.4	0.0	-0.2	-0.6	1.22
k_{2223}	1.1	12.4	0.9	5.4	0.8	-0.1	2.45
k_{1133}	0.3	0.3	0.3	0.2	-0.5	-0.6	0.88
k_{1233}	-0.1	0.9	-0.1	0.8	0.9	2.5	-5.18
k_{2233}	-0.1	-18.3	0.0	-20.0	0.7	-21.1	-18.08
k_{1333}	2.5	2.8	3.0	2.6	2.3	2.4	3.23
k_{2333}	-3.3	-1.7	-4.3	-5.8	-9.8	2.1	-24.57
k_{3333}	55.6	55.2	61.4	62.0	69.5	71.8	62.03
C. D_2O							
k_{111}	-187.2	-185.2	-187.3	-186.2	-187.7	-186.7	-184.92
k_{112}	7.0	-1.3	9.7	1.1	10.6	2.1	7.11
k_{122}	-7.7	37.9	-8.6	34.7	-9.1	32.7	16.65
k_{222}	-39.5	-30.7	-35.3	-26.0	-34.2	-24.8	-37.02
k_{133}	-192.3	-192.3	-190.9	-191.2	-191.5	-191.9	-184.28
k_{233}	16.8	26.3	19.5	27.4	21.6	29.2	37.36

Table 1. Cubic and Quartic Force Constants (cont.)

Constant[b]	SCF(ℓ)	SCF(n)	SD(ℓ)	SD(n)	SDQ(ℓ)	SDQ(n)	Experimentally derived[c]
C. D_2O (cont.)							
k_{1111}	14.5	13.8	16.1	16.2	15.9	15.8	15.84
k_{1112}	-1.0	-0.2	-1.4	-0.3	-1.4	-0.2	-14.71
k_{1122}	0.1	-6.3	0.2	-6.6	0.2	-6.8	-4.24
k_{1222}	1.1	2.8	1.0	2.9	0.9	2.8	-1.04
k_{2222}	-2.6	-2.2	-2.1	-1.3	-3.1	-2.3	-2.83
k_{1133}	14.8	14.3	16.4	16.5	18.9	19.0	16.42
k_{1233}	-1.2	-3.0	-1.5	-2.4	-4.4	-5.2	-8.24
k_{2233}	-0.3	-7.3	-0.3	-7.9	0.4	-7.4	-7.00
k_{3333}	15.3	14.1	16.8	16.4	19.4	19.1	16.93

a All values in cm^{-1} relative to Eqns. (2) and (3).
b The subscripts 1,2,3 refer, respectively, to the symmetric stretch, bend and asymmetric stretch for H_2O and D_2O and to the OD stretch, bend, and OH stretch for HDO.
c Ref. 6.

Table 2. Vibrational Energy Expansion Constants[a]

Constant[b]	SCF(ℓ)	SCF(n)	SD(ℓ)	SD(n)	SDQ(ℓ)	SDQ(n)	Experimentally derived[c]
A. H_2O							
ω_1	4133	4132	3932	3931	3856	3856	3832.17
ω_2	1771	1772	1702	1703	1688	1688	1648.47
ω_3	4237	4238	4036	4036	3959	3959	3942.53
X_{11}	-43.3	-43.7	-43.2	-41.8	-46.2	-45.0	-42.6
X_{22}	-16.5	-23.3	-13.8	-18.6	-16.3	-20.5	-16.8
X_{33}	-46.8	-51.1	-46.8	-48.7	-42.4	-44.3	-47.6
X_{12}	-1.5	-3.3	-0.3	-9.3	-0.2	-13.5	-15.9
X_{13}	-171.2	-175.8	-168.3	-166.1	-151.4	-149.6	-165.8
X_{23}	23.6	-5.3	21.7	-16.7	30.2	-11.7	-20.3
G	1.6	-3.9	4.8	1.1	6.0	2.6	--
B. HDO							
ω_1	3040	3040	2894	2894	2838	2839	2824.32
ω_2	1552	1553	1492	1492	1479	1479	1440.21
ω_3	4186	4184	3986	3986	3910	3909	3889.84
X_{11}	-45.5	-49.6	-44.6	-43.8	-41.1	-42.5	-43.4
X_{22}	-11.4	-18.4	-9.3	-12.5	-11.3	-13.7	-11.8
X_{33}	-86.3	-85.1	-84.8	-83.9	-77.5	-73.7	-82.9
X_{12}	-1.0	-19.2	-0.9	-14.6	1.9	-7.3	-8.6
X_{13}	-5.3	-7.0	-7.6	-9.5	-13.2	-15.0	-13.1
X_{23}	11.2	6.4	10.6	-10.4	14.4	-19.2	-20.1
G	-12.9	-17.7	-9.9	-12.0	-6.6	-8.4	--

Table 2. Vibrational Energy Expansion Constants (cont.)

Constant[b]	SCF(ℓ)	SCF(n)	SD(ℓ)	SD(n)	SDQ(ℓ)	SDQ(n)	Experimentally derived[c]
C. D_2O							
ω_1	2979	2978	2834	2833	2779	2779	2763.80
ω_2	1296	1297	1246	1246	1235	1235	1206.39
ω_3	3106	3107	2958	2958	2901	2901	2888.78
X_{11}	-22.5	-22.6	-22.6	-21.6	-24.2	-23.3	-22.6
X_{22}	-8.7	-12.7	-7.2	-10.2	-8.6	-11.2	-9.2
X_{33}	-25.5	-28.8	-25.4	-27.5	-23.1	-25.1	-26.1
X_{12}	-1.4	-1.2	-0.7	-4.2	-0.6	-6.2	-7.6
X_{13}	-90.1	-91.1	-88.6	-86.4	-78.4	-76.3	-87.2
X_{23}	12.2	2.6	11.3	-8.6	15.3	-6.5	-10.6
G	0.9	2.6	2.5	0.1	3.2	0.8	--
D. HTO							
ω_1		2551.1		2427.5		2381.4	
ω_2		1473.5		1415.4		1402.8	
ω_3		4181.9		3984.3		3907.2	
X_{11}		-35.0		-31.2		-30.8	
X_{22}		-21.8		-12.9		-13.2	
X_{33}		-83.5		-83.1		-70.4	
X_{12}		-20.3		-18.0		-18.7	
X_{13}		2.9		0.8		2.7	
G		-19.5		-13.7		-9.8	

Table 2. Vibrational Energy Expansion Constants (cont.)

Constant[b]	SCF(ℓ)	SCF(n)	SD(ℓ)	SD(n)	SDQ(ℓ)	SDQ(n)	Experimentally derived[c]
E. DTO							
ω_1		2542.5		2419.4		2373.0	
ω_2		1198.4		1151.6		1141.5	
ω_3		3050.6		2904.3		2848.2	
X_{11}		-32.1		-29.6		-25.2	
X_{22}		-11.7		-9.4		-10.1	
X_{33}		-45.4		-44.4		-41.1	
X_{12}		-0.3		-1.0		4.7	
X_{13}		-8.2		-9.6		-14.2	
X_{23}		-1.1		-5.8		-13.6	
G		-11.2		-8.4		-6.1	
F. T_2O							
ω_1		2479.3		2358.9		2313.4	
ω_2		1090.8		1048.3		1039.2	
ω_3		2624.6		2497.3		2449.6	
X_{11}		-15.5		-15.0		-16.2	
X_{22}		-9.1		-7.3		-8.0	
X_{33}		-21.3		-20.5		-18.3	
X_{12}		-0.5		-2.5		-3.7	
X_{13}		-62.2		-59.5		-51.6	
X_{23}		-2.0		-6.1		-5.0	
G		-2.1		-0.3		0.2	

a All values are in cm^{-1} relative to Eqn. (4).
b See footnote b of Table 1.
c Ref. [6].

Table 3. Vibrational Transition Energies[a]

$v_1v_2v_3$[b]	SCF(ℓ)	SCF(n)	SD(ℓ)	SD(n)	SDQ(ℓ)	SDQ(n)	Observed[c]
A. H_2O							
000	5008	4991	4777	4761	4702	4683	4634
100	3960	3955	3761	3760	3688	3684	3657
010	1749	1721	1686	1652	1670	1634	1595
001	4069	4045	3869	3847	3814	3790	3756
110	5708	5672	5446	5403	5357	5305	5235
101	7858	7824	7462	7441	7350	7325	7250
011	5842	5760	5577	5483	5514	5413	5331
111	9629	9536	9169	9067	9050	8934	8807
200	7834	7822	7435	7436	7283	7278	7201
020	3465	3395	3343	3267	3307	3228	3151
002	8045	7988	7645	7597	7543	7491	7445
012	9841	9698	9374	9216	9273	9102	9000
021	7582	7429	7256	7081	7181	6994	6871
102	11663	11591	11069	11025	10928	10877	
120	7422	7343	7104	7009	6994	6885	6775
201	11561	11515	10968	10951	10794	10769	10613
210	9580	9536	9120	9069	8952	8886	8762
300	11621	11602	11023	11028	10785	10782	
030	5148	5022	4973	4845	4911	4780	4667
003	11927	11828	11327	11250	11187	11105	11032

Table 3. Vibrational Transition Energies (cont.)

$v_1v_2v_3$[b]	SCF(ℓ)	SCF(n)	SD(ℓ)	SD(n)	SDQ(ℓ)	SDQ(n)	Observed[c]
B. HDO							
000	4342	4328	4142	4130	4075	4062	4032
100	2946	2928	2800	2794	2750	2742	2782
010	1534	1510	1478	1455	1465	1439	1403
001	4017	4014	3818	3808	3755	3744	3707
110	4480	4419	4278	4234	4217	4173	4100
101	6958	6935	6610	6592	6492	6472	6452
011	5563	5531	5307	5252	5234	5164	5090
111	8503	8432	8099	8022	7973	7884	
200	5801	5756	5511	5501	5418	5400	
020	3046	2984	2938	2885	2907	2850	2724
002	7862	7858	7466	7448	7355	7341	
012	9419	9381	8966	8882	8848	8742	
021	7086	7010	6778	6672	6691	6556	6415
102	10797	10772	10251	10223	10079	10054	
120	5990	5873	5737	5649	5661	5576	
201	9808	9756	9314	9289	9147	9114	
210	7334	7228	6988	6926	6887	6824	
300	8566	8486	8133	8120	8004	7973	
030	4536	4420	4380	4289	4326	4234	4146
003	11534	11532	10945	10921	10800	10791	

Table 3. Vibrational Transition Energies (cont.)

$v_1v_2v_3$[b]	SCF(ℓ)	SCF(n)	SD(ℓ)	SD(n)	SDQ(ℓ)	SDQ(n)	Observed[c]
C. D_2O							
000	3657	3648	3488	3479	3431	3421	3389
100	2888	2887	2744	2745	2691	2691	2672
010	1284	1269	1237	1219	1225	1207	
001	3016	3002	2868	2855	2823	2809	2788
110	4171	4155	3980	3960	3916	3891	
101	5814	5798	5523	5514	5436	5423	5374
011	4313	4269	4116	4066	4064	4009	3956
111	7109	7063	6771	6720	6676	6617	6533
200	5731	5728	5443	5447	5334	5335	5292
020	2551	2513	2459	2418	2433	2390	
002	5981	5947	5865	5655	5600	5568	
012	7290	7211	6944	6858	6856	6762	
021	5591	5510	5349	5256	5287	5187	5105
102	8689	8651	8252	8227	8134	8107	
120	5436	5397	5202	5155	5123	5069	
201	8567	8548	8134	8129	8000	7992	7900
210	7012	6995	6678	6657	6558	6530	
300	8529	8524	8097	8105	7928	7933	
030	3800	3732	3666	3597	3624	3552	
003	8895	8834	8452	8401	8331	8277	

[a] Calculated using the spectroscopic constants in Table 2. All values are in cm^{-1}.
[b] See footnote b of Table 1.
[c] Ref. [6].

experimentally derived data at four logical levels of development.

The same second-order perturbation theory treatment permits the definition of vibration-rotation coupling constants relative to the three principal moments of inertia (or Eckart) axes in terms of the cubic force constants and the Coriolis coupling coefficients [7]. Values of these coefficients are given in Table 4 together with experimentally derived results.

Table 5 contains values of spectroscopic constants for H_2O derived from the potential energy surfaces of Paper II together with values due to other high quality ab initio surfaces [8-10] and values derived from spectroscopic data [6]. Vibrational transition energies of H_2O derived using ab initio potential energy surfaces and the perturbation theory based method of Refs. 3 and 4 and variational procedures [11-13] are compared in Table 6.

Botschwina [14] has used SCF and CEPA potential energy surfaces and the variational procedure of Handy and co-workers [12,13] to determine vibrational transitions for water including several ^{17}O, D and T isotopic substituents. He has chosen to adjust certain harmonic internal coordinate force constants to reproduce the experimental fundamental frequencies. This procedure leads to computed excited vibrational frequencies for levels up to $v_1+v_2+v_3 \leq 2$ that agree with observed values to better than 5 cm^{-1}. These results cannot be directly compared to those reported here. In a generalization of the second-order perturbation theory procedure, Harding and Ermler [15] have shown that an adjustment of the harmonic force constants to reproduce harmonic frequencies of formaldehyde leads to frequencies that are in agreement with observed values to within 20 cm^{-1} for low-lying levels. These empirical approaches indicate that the rather large error made in second derivatives of ab initio potential energy surfaces is predominantly responsible for the errors in computed vibrational excitation energies, the anharmonicities being quite well represented, even at the SCF level of calculation, by power series in internal displacement coordinates.

Table 4. Rotational Constants and Vibration-Rotation Coupling Constants[a]

Constant	SCF	SD	SDQ	Experimentally Derived[b]
A. H_2O				
A_e	29.495	27.951	27.466	27.379
B_e	14.828	14.653	14.582	14.584
C_e	9.867	9.613	9.525	9.526
α_1^A	0.593	0.665	0.697	0.750
α_2^A	-2.900	-2.632	-2.538	-2.941
α_3^A	1.090	1.132	1.175	1.253
α_1^B	0.213	0.214	0.216	0.238
α_2^B	-0.144	-0.165	-0.168	-0.160
α_3^B	0.103	0.0990	0.0966	0.078
α_1^C	0.161	0.171	0.176	0.202
α_2^C	0.145	0.146	0.149	0.139
α_3^C	0.129	0.137	0.142	0.144
B. HDO				
A_e	24.733	23.557	33.185	22.691
B_e	9.343	9.186	9.127	9.129
C_e	6.781	6.609	6.549	6.551
α_1^A	0.264	0.285	0.298	0.253
α_2^A	-1.933	-1.755	-1.693	-1.798
α_3^A	0.930	0.982	1.019	1.087
α_1^B	0.162	0.171	0.176	0.199
α_2^B	-0.117	-0.127	-0.129	-0.147
α_3^B	0.0190	0.0146	0.0125	0.0125
α_1^C	0.0928	0.0980	0.1010	0.1098
α_2^C	0.0813	0.0822	0.0836	0.0710
α_3^C	0.0701	0.0749	0.0774	0.0881

Table 4. Rotational Constants and Vibration-Rotation Coupling Constants (cont.)

Constant	SCF	SD	SDQ	Experimentally Derived[b]
C. D_2O				
A_e	16.408	15.549	15.279	15.222
B_e	7.419	7.332	7.297	7.290
C_e	5.109	4.982	4.938	4.935
α_1^A	0.177	0.211	0.226	0.246
α_2^A	-1.154	-1.047	-1.010	-1.161
α_3^A	0.487	0.502	0.518	0.593
α_1^B	0.0906	0.0909	0.0916	0.0958
α_2^B	-0.0587	-0.0663	-0.0676	-0.0823
α_3^B	0.0321	0.0305	0.0298	0.0418
α_1^C	0.0639	0.0663	0.0679	0.0768
α_2^C	0.0545	0.0551	0.0560	0.0495
α_3^C	0.0450	0.0488	0.0508	0.0538

[a] All results are in cm^{-1}. They were obtained using the nonlinear transformation to normal coordinates described in Ref. [15] and the formulas of Ref. [7].

[b] Ref. [6].

Table 5. Spectroscopic Constants of H_2O (cm^{-1})

Constant	SD[a]	SDQ[a]	SDQ-MBPT(4)[b]	SD[c]	SDQ[c]	MCSCF[d]	MR-CI[d]	Experimentally derived[e]
ω_1	3931	3855	3866	3954	3869	3783	3818	3832
ω_2	1703	1687	1688	1685	1670	1730	1666	1648
ω_3	4036	3958	3976	4062	3980	3884	3933	3942
X_{11}	-41.8	-45.0	-42.6	-60.6	-62.9	-66.9	-62.3	-42.6
X_{22}	-18.6	-20.5	-17.7	-18.6	-18.5	-18.9	-18.1	-16.8
X_{33}	-48.7	-44.3	-48.8	-65.7	-66.9	-67.4	-63.2	-47.6
X_{12}	-9.3	-13.5	-12.8	-14.7	-14.9	-10.9	-15.1	-15.9
X_{13}	-166.1	-149.6	-166.0	-227.1	-232.6	-245.8	-227.8	-165.8
X_{23}	-16.7	-11.7	-20.5	45.9	46.2	51.2	49.2	-20.2
A_e	27.951	27.466	27.519	28.367	27.875	22.297	27.220	27.379
B_e	14.653	14.582	14.591	14.655	14.588	14.539	14.608	14.584
C_e	9.613	9.525	9.535	9.663	9.576	9.486	9.506	9.526
$\alpha_1 A$	0.665	0.697	0.696	0.856	0.905	0.920	0.897	0.750
$\alpha_2 A$	-2.632	-2.538	-2.558	-2.837	-2.742	-2.418	-2.581	-2.941
$\alpha_3 A$	1.132	1.175	1.147	1.389	1.395	1.348	1.304	1.253
$\alpha_1 B$	0.214	0.216	0.216	0.254	0.261	0.286	0.268	0.238
$\alpha_2 B$	-0.165	-0.168	-0.166	-0.150	-0.153	-0.168	-0.162	-0.160
$\alpha_3 B$	0.099	0.097	0.120	0.105	0.117	0.152	0.133	0.078
$\alpha_1 C$	0.171	0.176	0.176	0.210	0.219	0.233	0.223	0.202
$\alpha_2 C$	0.146	0.149	0.148	0.141	0.143	0.155	0.146	0.139
$\alpha_3 C$	0.137	0.142	0.141	0.169	0.175	0.181	0.177	0.144

a This work.
b Potential energy surface from Ref. [8].
c Ref. [9].
d Ref. [10].
e Ref. [6].

The shortcomings of the linear approximation for the definition of the normal coordinates are evident in the anharmonic force constants shown in Table 1. The constants most affected are those involving coupling between the two stretching modes and the bending mode (i.e. k_{112}, k_{122}, k_{1122}, k_{1233}, and k_{2233}). The correct nonlinear transformation results in better agreement with the experimentally derived values, the constant k_{1112} being an exception for H_2O and D_2O. The SCF, SD and SDQ values of the coupling force constants are generally very close in magnitude to each other.

The spectroscopic constants in Table 2 are generally improved when going from SCF to SD to SDQ. This is most apparent in the harmonic frequencies, where the error is reduced from 8% to 3% to 1%, respectively. The anharmonicity constants are of about the same quality for the different surfaces with the exception of X_{12}, which shows a marked improvement for the surfaces that include electron correlation. It is clearly disastrous to use the linear approximation for computing X_{12} and X_{23}, the latter having the wrong sign in this approximation. The problems with the linear approximation, a direct extension of the GF-matrix procedure, are discussed in detail in Ref. 4.

The transition frequencies given in Table 3 are the only quantities that may be directly compared with spectral data. The SDQ (non-linear) results are in best agreement with the observed values, with an average error of about 1%. The SD values would improve to nearly this accuracy if the harmonic force constants were constrained to reproduce the experimentally derived values [14,15]. The SCF results would show a similar, although not as marked, improvement. This is apparent from the results shown in Table 1, where it is seen that the anharmonicity constants are quite comparable for the SCF, SD, and SDQ surfaces. Carrying out the vibrational analysis using force constants obtained by performing all fittings with respect to the experimental equilibrium geometry, as opposed to the geometry corresponding to the minimum of the ab initio potential energy surface, has also been shown to lead to frequencies that are in better agreement with experiment (P. Pulay, this conference). We have chosen not to introduce these empirical corrections in our analyses, maintaining a clear distinction between truly ab initio and experimentally obtained results.

The rotational constants and vibration-rotation coupling constants in Table 4 are generally in good agreement with those derived from the spectra. Of course, the rotational constants A_e, B_e, and C_e are directly related to the equilibrium geometry. The trends in computed versus experimental geometries for the various electronic wavefunctions have been discussed in detail in Paper II and in Refs. 8, 9 and 10. The differences seen in the rotational constants can be attributed to the quality of the electronic wavefunctions, with the changes from H_2O to HDO to D_2O being straightforward isotope shifts. The vibration-rotation coupling constants are of nearly the same quality for the SD and SDQ surfaces, each showing a clear improvement over those due to the SCF surface. As mentioned above, these constants depend only on the cubic force constants of Eqn. (2), on the matrix that defines the transformation from mass-weighted cartesian displacement to reduced normal coordinates, and on the Coriolis coupling coefficients [7]. The favorable agreement with experimentally derived values is a clear indication that the ab initio potential energy surfaces provide a good approximation to the cubic anharmonicity, and that the non-linear transformation procedure [4] is reliable.

Comparisons with other results derived from ab initio potential energy surfaces are shown in Tables 5 and 6. The effect of the extension of the STO basis results to high-order many body perturbation theory by Bartlett et al. [8] has been tested by analyzing their fourth-order surface, SDQ-MBPT(4), using the same procedures as used for the SCF, SD, and SDQ surfaces. Table V shows that the SDQ-MBPT(4) surface leads to small improvements in the harmonic frequencies over the SDQ-CI values, and to anharmonicity constants that are nearly of the same quality as the SD and SDQ results, with the exception of X_{13} and X_{23} which show a clear improvement. The fact that vibration-rotation coupling constants agree quite well among those derived from the SD, SDQ, and MBPT(4) surfaces again suggests that the cubic force constants are of good quality and that the differences in the X_{ij}'s are due mainly to the quartic force constants.

The spectroscopic constants reported for the SD and SDQ surfaces of Hennig et al. [9] are seen to parallel those computed from the surfaces of Paper II, with the exception that the X_{ij}'s are in somewhat poorer agreement with the experimentally derived results. This is particularly true for X_{11}, X_{33}, X_{13}, and X_{23}. Since the latter constants show behavior similar to that observed for those computed using the linear approximation for the transformation from internal displacement coordinates to normal coordinates, it is surmised that the method these authors used for obtaining the anharmonic force constants of Eqn. (2) is different from the nonlinear transformation employed here. This also appears to be the case for the MCSCF and MRD-CI surfaces of Kraemer et al [10]. The quality of the basis set of Gaussian-type functions (GTO) of both Refs. 9 and 10 was quite good, but the incremental change in geometry used to generate the surface points was larger than that of Paper II. This may have affected the values of the third and fourth derivatives with respect to the internal displacement coordinates evaluated at the minimum geometry, because of the neglected effects of higher derivatives. This would also have affected the cubic and quartic anharmonicity constants of Eqn. (2). Again, since the agreement is noticably better for the vibration-rotation coupling constants, it is suspected that the quartic force constants are the source of the discrepencies.

The comparison of actual transition frequencies shown in Table 6 is of interest for two reasons. First, it is the level at which the ab initio and experimental spectral data can be directly compared. Second, it permits us to gauge the quality of the second-order perturbation theory approach in comparison with the accurate variational methods [11,12,13]. Fortunately, a direct comparison between the two methods is possible, becaue Handy and co-workers [12,13] and Carney et al. [11] have analyzed the surfaces of Paper II as well as those of Ref. 3. The SCF surfaces referred to in the first two columns of Table 6 are seen to be of comparable quality, both showing about 8% average errors. (The surface of Ref. 3c was derived using a large basis set of GTO's.) The results in the third column are from the variational procedure of Carney and co-workers, and show average errors that are nearly the same as those from second-order perturbation theory. The variational method of Handy and co-workers applied to the same expansion of Ref. 3c is seen to lead to fundamental frequencies that are close, but not identical, to those of Carney et al. These differences are of comparable size to those between the present perturbation theory results and the variational results, at least for low-lying states. Clearly, the variational procedures are preferred for those cases where excitations to the higher levels are to be computed, since the perturbation theory method is restricted to small distortions of the molecule from equilibrium.

The comparison between the perturbational and variational analyses of the SD surface in columns 6 and 7 supports the assessment that the two methods lead to comparable results for the low-lying vibrational levels. (The better agreement for the perturbation theory approach can only be called fortuitous. The variational procedure of Carney et al. requires the use of a large basis of harmonic oscillator functions, the choice of which could explain some of the discrepancies with the present results and with those of Handy et al.) The SD results of Whitehead and Handy [13] shown in the eighth column are in closer agreement with the observed fundamental frequencies. Their results were derived from electronic SD-CI energies computed at geometries specifically chosen as the quadrature points needed for the numerical solution of the nuclear motion Schrödinger equation. They computed up to 96 SD energies in a basis set of STO's somewhat smaller than that used for the present surfaces. It is interesting to note that the resulting fundamental frequencies are lower than those due to both the perturbation theory and the variational analysis of the SD surface of Paper II (columns 6 and 7). This suggests (as was pointed out in Ref. 13) that one problem with the surfaces of Paper II is that the chosen distortions from equilibrium were too small. However, a recent thorough investigation of the optimal choice of distortion grid points by Sexton and Handy [21] leads to the conclusion that the grid used in Paper II was quite appropriate. It is not clear that a direct comparison with the surface of Ref. 13 is possible, since all of the computed points of that surface correspond to asymmetric distortions from equilibrium. The choice of the surface grid of Paper II was made in order to obtain an analytical form appropriate for the perturbation theory analysis, i.e. a force constant expansion corresponding to a Taylor series. It was determined that larger distortions than those chosen (0.03 bohr and 3°) led to large errors in the cubic and quartic force constants. It has been shown [11,21] that an improved fit is obtained when the Simons-Parr-Finlan modification of the Taylor series is used.

We conclude that although variational procedures are superior in accuracy to those based on perturbation theory, there remain certain advantages to using the latter. The perturbation theory approach is seen to yield comparable results to those obtained from the variational methods when based on the same form for the potential energy function (at least for low-lying vibrational states). This indicates that the main problem is the quality of the potential energy surface and its analytical representation, and that reliable results can be expected from perturbation theory for those surfaces that are represented as force constant expansions. Spectroscopic constants and expectation values derived from the perturbation theory procedures, which are given by analytical formulas, are easily computed once the normal coordinate force constants of Eqn. (2) are computed. (In our procedure this is accomplished by a least squares fitting of the surface after all surface points have been non-linearly transformed to their respective normal coordinates [4].) The variational procedures require choices of basis sets [11] or of numerical quadratures [12,13], and can be costly in terms of computer time. The extension of the perturbation theory approach to the computation of expectation values of properties other than the energy and of transition moments is also straightforward, the results similarly expressed as analytical forms. Analogous conclusions have been reached by Hennig et al. [9] in their analyses of ab initio potential energy surfaces of water.

3. VIBRATIONAL EFFECTS ON ONE-ELECTRON PROPERTIES

The dependence of various molecular properties on vibrational motion may be expressed as expectation values over the vibrational wavefunction and cast into the form [3c]

$$\langle P\rangle_v = P_o + G + \sum_i A_i(v_i+1/2) + \sum_{i\leq j} B_{ij}(v_i+1/2)(v_j+1/2). \tag{5}$$

The constants G, A_i, and B_{ij} are algebraic expressions involving the potential constants of Eqn. (2) and the constants that define the expansion of the property P as a function of the reduced normal coordinates,

$$P = P_o + \sum\alpha_i q_i + \sum\beta_{ij}q_iq_j + \sum\gamma_{ijk}q_iq_jq_k + \cdots . \tag{6}$$

The SCF and SD wavefunctions for each of the 36 geometries were used to compute a set of one-electron property surfaces as described in Paper II. Using the normal coordinate representation of each of the 36 geometries as described above, property surfaces were fitted to Eqn. (6) and values of the constants in Eqn. (5) were computed. Table 7 contains values of the expansion constants of Eqn. (6) for the H_2O, HDO, and D_2O dipole moments, as defined in the space-fixed coordinate systems that satisfy the Eckart con-

Table 7. Dipole Moment Expansion Coefficients

Coef.[a]	H_2O				HDO			
	μ_y(SCF)[b]	μ_y(SD)	μ_z(SCF)[b]	μ_z(SD)	μ_y(SCF)	μ_y(SD)	μ_z(SCF)	μ_z(SD)
$10\alpha_1$			0.633	0.520 (0.216)	-0.629	-0.542 (-0.75)	0.435	0.357 (0.194)
$10\alpha_2$			-2.208	-2.068 (-1.61)	-0.593	-0.583 (-0.70)	-2.062	-1.932 (-1.25)
$10\alpha_3$	1.148	0.989 (0.950)			0.834	0.716 (0.897)	0.441	0.366 (0.014)
$10^2\beta_{11}$			-0.42	-0.150	0.164	0.216	0.125	0.075
$10^2\beta_{12}$			0.286	0.379	-1.484	-1.436	0.039	0.040
$10^2\beta_{22}$			-0.789	-0.995	1.115	1.014	-0.733	-0.922
$10^2\beta_{13}$	-0.168	-0.250			-0.037	-0.037	-0.207	-0.250
$10^2\beta_{23}$	2.862	2.812			2.051	2.032	0.472	0.509
$10^2\beta_{33}$			0.400	0.400	-0.188	-0.276	0.207	0.158
$10^3\gamma_{111}$			0.34	0.41	0.23	0.29	0.28	0.33
$10^3\gamma_{112}$			-0.07	-0.12	0.20	0.28	0.09	0.07
$10^3\gamma_{122}$			0.28	0.24	-0.42	-0.36	-0.29	-0.30
$10^3\gamma_{222}$			-0.98	-0.51	2.44	2.56	-0.35	-0.04
$10^3\gamma_{113}$	-0.35	-0.42			0.06	0.06	0.02	0.01
$10^3\gamma_{123}$	-0.51	-0.65			0.11	0.11	-0.05	-0.06
$10^3\gamma_{223}$	0.40	0.26			0.00	-0.13	0.66	0.72
$10^3\gamma_{133}$			-0.32	-0.35	-0.11	-0.13	-0.01	-0.03
$10^3\gamma_{233}$			-0.10	-0.06	-0.69	-0.82	-0.17	-0.22
$10^3\gamma_{333}$	0.03	-0.03			-0.36	-0.45	-0.44	-0.52

Table 7. Dipole Moment Expansion Coefficients (cont.)

Coef.[a]	D_2O μ_y(SCF)	μ_y(SD)	μ_z(SCF)	μ_z(SD)
$10\alpha_1$			0.615	0.514 (0.24)
$10\alpha_2$			-1.878	-1.760 (-1.37)
$10\alpha_3$	0.927	0.790 (0.868)		
$10^2\beta_{11}$			-0.048	-0.132
$10^2\beta_{12}$			0.231	0.304
$10^2\beta_{22}$			-0.569	-0.718
$10^2\beta_{13}$	-0.201	-0.258		
$10^2\beta_{23}$	2.142	2.100		
$10^2\beta_{33}$			0.356	0.351
$10^3\gamma_{111}$			-0.20	-0.24
$10^3\gamma_{112}$			-0.06	-0.09
$10^3\gamma_{122}$			0.20	0.16
$10^3\gamma_{222}$			-0.61	-0.31
$10^3\gamma_{113}$	-0.19	-0.23		
$10^3\gamma_{123}$	-0.35	-0.43		
$10^3\gamma_{223}$	0.30	0.22		
$10^3\gamma_{133}$			-0.21	-0.24
$10^3\gamma_{233}$			-0.01	0.02
$10^3\gamma_{333}$	-0.07	-0.10		

[a] Expansion coefficients are relative to Eqn. (6) and give dipole moments in Debyes. P_o(SCF)=1.9484 D, P_o(SD) = 1.9070 D. See also footnote b of Table 1. Quantities in parentheses were derived from Stark effect measurements (Ref. 16). Values for HDO and D_2O were obtained from isotope shift formulas using the H_2O data. The HDO values were rotated to the Eckart axes frame (Refs. 4 and 11).

[b] The subscripts y and z refer to the small and large components along the axes frame satisfying the Eckart conditions (see Ref. 4).

ditions [4]. These dipole moment functions are of the proper form for comparison to values of experimentally derived dipole moment derivative constants [16], since the α_i, β_{ij}, and γ_{ijk} correspond, respectively, to the first, second, and third derivatives with respect to the reduced normal coordinates.

The constants in Eqn. (5) for the dipole moments of H_2O, HDO, and D_2O are listed in Table 8, and the theoretical values for the vibrational states $v_1+v_2+v_3 \leqslant 3$ may be obtained from Table 9 by adding the respective entries to the values of P_0 at the mimina of the SCF and SD potential wells. An analysis of the Stark effect in microwave transitions for H_2O in the $(v_1v_2v_3)=(010)$ state has led [17] to a value of -0.028 ± 0.005 D for the change in the dipole moment relative to its ground state value of 1.8546 D. The corresponding value obtained for the SD surface from Table 9 by subtracting the entries in the third and first rows is -0.0283 D. This good agreement raises again the question as to the accuracy of the value of 0.0216 D for the dipole moment derivative with respect to the symmetric stretching normal coordinate [16]. Values calculated from SCF dipole moment surfaces have been reported by Smith et al. [18] (0.0584) and by Luh and Lie [19] (0.0571). These compare favorably with the SCF and SD values of 0.0633 and 0.0520 given in Table 7.

Expansions of the form of Eqn. (6) were derived for other one-electron properties using the SCF and SD wavefunctions. These were used to compute the constants G, A_i, and B_{ij} of Eqn. (5) for each property. Using these results, zero-point vibrational corrections to the properties reported in Paper II have been computed. Table 10 contains results for properties related to second moments, Table 11 for those related to electric field gradients, Table 12 for diamagnetic shielding constants, and Table 13 for electrostatic forces on the nuclei and the correlation energy. Expectation values of internal displacement coordinates for vibrational states $v_1+v_2+v_3 \leqslant 2$ are given in Table 14.

The zero-point vibrational corrections to the quadrupole moments are seen from Table X to be rather small, on the order of

Table 8. Vibrational quantum number expansion constants for the dipole moment[a]

Constant	μ_z (H_2O)		μ_z (HDO)		μ_z (D_2O)		μ_y (HDO)	
	SCF	SD	SCF	SD	SCF	SD	SCF	SD
$10A_1$	1.65	1.48	1.78	1.64	1.04	0.93	-1.42	-1.17
$10A_2$	-3.01	-2.83	-2.70	-2.54	-2.14	-2.01	0.20	0.25
$10A_3$	3.35	3.19	2.48	2.44	2.68	2.55	2.50	2.23
10^4B_{11}	-1.82	-2.28	-1.90	-2.32	-0.94	-1.17	1.35	1.77
10^4B_{22}	-3.22	-1.91	-3.92	-3.16	-1.78	-1.06	3.85	4.41
10^4B_{33}	-1.59	-1.94	-3.22	-3.86	-1.04	-1.26	-1.99	-2.42
10^4B_{12}	4.86	4.37	1.59	0.26	2.79	2.47	0.27	2.22
10^4B_{13}	-6.85	-8.13	-1.49	-0.42	-3.79	-4.48	-0.51	0.45
10^4B_{23}	4.18	3.23	10.52	11.19	2.49	1.88	-3.59	-6.50
10^4G_o	-0.74	-0.56	-2.24	-2.19	-0.49	-0.33	-0.58	-1.31

a Expansion constants are relative to Eqn. (5) and are given in Debyes.

Table 9. Vibrational corrections to dipole moments

State	$10^2\mu_z$ (H_2O)		$10^2\mu_z$ (HDO)		$10^2\mu_z$ (D_2O)		$10^2\mu_y$ (HDO)	
$v_1v_2v_3$	SCF	SD	SCF	SD	SCF	SD	SCF	SD
000	0.98	0.90	0.76	0.75	0.78	0.72	0.64	0.64
100	2.59	2.32	2.51	2.34	1.80	1.62	-0.75	-0.51
010	-2.05	-1.93	-1.96	-1.80	-1.37	-1.29	0.90	0.93
001	4.29	4.03	3.24	3.17	3.44	3.24	3.08	2.79
110	-0.39	-0.47	-0.19	-0.21	-0.32	-0.37	-0.49	-0.24
101	5.83	5.37	4.98	4.75	4.42	4.09	1.70	1.64
011	1.30	1.23	0.62	0.73	1.31	1.25	3.31	3.02
111	2.89	2.61	2.38	2.32	2.32	2.12	1.92	1.85
200	4.16	3.69	4.22	3.89	2.80	2.49	-2.11	-1.62
020	-5.14	-4.80	-4.75	-4.41	-3.55	-3.32	1.23	1.31
002	7.56	7.12	5.64	5.51	6.07	5.73	5.48	4.89
012	4.62	4.35	3.13	3.18	3.97	3.76	5.67	5.05
021	-1.75	-1.61	-2.07	-1.77	-0.85	-0.77	3.61	3.33
102	9.03	8.38	7.39	7.09	7.01	6.53	4.11	3.75
120	-3.44	-3.30	-2.97	-2.82	-2.48	-2.37	-0.16	0.12
201	7.33	6.66	6.69	6.30	5.38	4.91	0.34	0.53
210	1.23	0.95	1.53	1.34	0.71	0.53	-1.85	-1.37
300	5.69	5.02	5.89	5.39	3.78	3.34	3.44	2.70
030	-8.30	-7.71	-7.63	-7.09	-5.77	-5.37	1.65	1.78
003	10.81	10.17	7.98	7.77	8.68	8.19	7.84	6.94

[a] All values are in Debyes, and were computed using Eqn. (5) and the constants given in Table 5.

Table 10. Zero-point vibrational corrections to quadrupole moments, $\langle r^2 \rangle$, and the average diamagnetic susceptibility[a]

Property	$P_0(H_2O)$	Vib. Corr.	$P_0(HDO)$	Vib. Corr.	$P_0(D_2O)$	Vib. Corr.
θ_{yy}	2.651	0.056	2.703	0.048	2.749	0.040
$(2.6 \pm 0.02)^b$	2.620	0.046	2.672	0.038	2.719	0.033
θ_{zz}	-0.185	0.006	-0.289	0.004	-0.382	0.003
$(-0.13 \pm 0.03)^b$	-0.120	0.016	-0.224	0.014	-0.318	0.010
θ_{yz}	0.000	0.000	-0.116	-0.030	0.000	0.000
	0.000	0.000	-0.114	-0.031	0.000	0.000
$\langle r^2 \rangle$	5.291	-0.043	5.252	-0.048	5.128	-0.029
$(5.1 \pm 0.7)^b$	5.417	-0.048	5.375	-0.052	5.338	-0.032
χ^d_{av}	14.964	0.123	14.855	0.134	14.757	0.083
$(14.6 \pm 2.0)^b$	15.321	0.137	15.203	0.147	15.097	0.092

a All values are computed relative to the center of mass and the Eckart axis system. Quadrupole moments are in Buckinghams, $\langle r^2 \rangle$ in 10^{-6} cm^2 and diamagnetic susceptibilities in 10^{-6} cm^3/mole. Conversion factors are given in Paper I. The first row for each entry corresponds to the SCF and the second row to the SD surface.

b Experimental values for H_2O. Sources are given in Paper I.

Table 11. Zero-point vibrational corrections to ^{17}O and D quadrupole coupling tensors[a]

Property	P_0	Zero-Point Correction H_2O	HDO	D_2O
$(eQq/h)_{xx}(^{17}O)$	-11.162	-0.247	-0.212	-0.182
(-10.17 ± 0.07)[b]	-10.809	-0.230	-0.199	-0.169
$(eQq/h)_{yy}(^{17}O)$	10.374	0.052	0.043	0.038
(8.89 ± 0.03)[b]	9.724	0.011	0.005	0.008
$(eQq/h)_{yz}(^{17}O)$	0.000	---	-0.094	---
	0.000	---	-0.089	---
$\eta(^{17}O)$[d]	0.859	-0.028	-0.025	-0.021
(0.75 ± 0.01)[b]	0.799	-0.033	-0.029	-0.024
$(eQq/h)_{aa}(D)$	-385.52	---	5.6	5.4
(-307.95 ± 0.14)[b]	-362.46	---	6.6	6.2

Table 11. Zero-point vibrational corrections to ^{17}O and D quadrupole coupling tensors (cont.)

Property	P_0	Zero-point Correction		
		H_2O	HDO	D_2O
$(eQq/h)_{bb}(D)$	168.51	---	-3.1	-2.8
$(133.13 \pm 0.14)^b$	157.70	---	-3.6	-2.8
$\alpha(D)^c$	0.761	---	0.156	0.140
$(1.266)^b$	0.984	---	0.110	0.102
$\eta(D)^d$	0.1258	---	0.0032	0.0025
$(0.1350 \pm 0.0007)^b$	0.1298	---	0.0040	0.0032

[a] Coupling constants centered at ^{17}O are in KHz and those at D are in MHz. The first row for each entry corresponds to the SCF and the second row to the SD surface.

[b] Experimental values for $HD^{17}O$. Sources are given in Paper I.

[c] The angle α measures the deviation of the principal axis system from the bond axis (see Fig. 1 of Ref. 3c).

[d] Asymmetry parameter (dimensionless).

Table 12. Zero-Point vibrational corrections to the diamagnetic shielding tensor at O, H and D[a]

Property	P_0	Zero-Point Correction H_2O	HDO	D_2O
$\sigma_{av}^{d}(O)$[b]	416.53 416.41	-0.26 -0.27	-0.23 -0.24	-0.19 -0.20
$\sigma_{xx}^{d}(O)$	417.69 417.67	-0.34 -0.35	-0.29 -0.30	-0.25 -0.25
$\sigma_{yy}^{d}(O)$	415.60 415.47	-0.19 -0.20	-0.16 -0.17	-0.14 -0.14
$\sigma_{zz}^{d}(O)$	415.32 416.16	-0.20 -0.21	-0.17 -0.18	-0.14 -0.15
$\sigma_{yz}^{d}(O)$	0.00 0.00	--- ---	0.019 0.018	--- ---
$\sigma_{av}^{d}(H,D)$[b]	104.07 102.83	-1.14 -1.20	-1.13 (-0.84)[c] -1.19 (-0.88)	-0.83 -0.87
$\sigma_{aa}^{d}(H,D)$	131.52 130.05	-1.69 -1.76	-1.61 (-1.32) -1.67 (-1.38)	-1.23 -1.28

Table 12. Zero-point vibrational corrections to the diamagnetic shielding tensor at O, H and D (cont.)

Property	P_0	Zero-Point Correction		
		H_2O	HDO	D_2O
$\sigma_{bb}^{d}(H,D)$	48.49	-0.29	-0.34 (-0.14)	-0.20
	47.64	-0.30	-0.36 (-0.15)	-0.22
$\sigma_{xx}^{d}(H,D)$	132.20	-1.22	-1.20 (-0.91)	-0.89
	130.79	-1.28	-1.26 (-0.96)	-0.93
$\alpha(H,D)^{d}$	1.695	0.087	0.079 (0.079)	0.063
	1.719	0.021	0.021 (0.022)	0.015

a In parts per million. The first row for each entry corresponds to the SCF and the second row to the SD surface.

b Experimental values for H_2O are 414.6 ppm (O) and 102.4 ppm (D). Sources are given in Paper I.

c Values in parentheses are the zero-point corrections for the property centered at D.

d Angle, in degrees, that measures the deviation of the principal axis system from the bond axis (see Fig. 1 of Ref. 3c).

Table 13. Zero-Point vibrational corrections to electrostatic forces on the nuclei and to the correlation energy[a]

Property	P_0	Zero-Point Correction H_2O	HDO	D_2O
$F_y(O)$	0.000 0.000	--- ---	0.004 0.005	--- ---
$F_z(O)$	0.890 0.653	-0.019 -0.005	-0.018 -0.006	-0.017 -0.007
$F_y(H,D)$	-0.035 0.090	-0.006 -0.009	-0.007 (-0.001)[b] -0.010 (-0.003)	-0.002 -0.004
$F_z(H,D)$	-0.024 0.064	-0.003 -0.006	-0.004 (0.000) -0.007 (-0.002)	-0.001 -0.003
ΔE_{corr}	-0.27526	-0.00111	-0.00097	-0.00081

[a] Forces are in mdyn and energies in hartrees. The first row for each entry corresponds to the SCF and the second row to the SD surface.

[b] Values in parentheses are in the zero-point corrections to the forces centered at D.

2%. The corrections to the expectation value $\langle r^2 \rangle$ are also small, but are in a direction such that the agreement with experiment is improved. The average diamagnetic susceptibilities show about 1% corrections. When compared with the experimental values, the signs of the corrections suggest that the true values are likely to be closer to the high end of the indicated error range. Note that the procedure for obtaining the vibrational expectation values follows that suggested in Ref. 4, in that the second moment electron expectation value for each surface point was translated and rotated to the Eckart coordinate reference frame prior to fitting to a fourth degree Taylor series in terms of the internal displacement coordinates. This transformation improves the values of P_o over those reported in Paper II.

The corrections to the quadrupole coupling constants shown in Table 11 are of a magnitude similar to those obtained earlier using a 75-point SCF surface [3c]. The main differences are in P_o, which are improved by the use of the SD wavefunction. The ^{17}O and D coupling constants are still in error by about 6-18%, even though the vibrational correction is of the correct sign. This indicates that the electric field gradients are extremeley sensitive to the quality of the basis set and to the level of CI [20].

The diamagnetic shielding tensors have small zero-point vibrational corrections, as shown in Table 12. The magnitudes and signs of the corrections are such that better agreement with experimental values is obtained.

The electrostatic forces show quite small corrections due to zero-point vibrational motion (Table 13). In each case the correction causes the value of the force to become closer to zero, although not by a large fraction of P_o. Very small corrections are also obtained for the correlation energy which, as was discussed in Paper II, varies nearly linearly as a function of geometry, with a slope of approximately zero, in the region of small displacements from equilibrium.

Table 14. Expectation values of internal displacement coordinates and root mean square amplitudes of vibration of H_2O[a]

Vib. State	$\langle \Delta R \rangle$	$\langle \Delta\theta \rangle$	$\langle \Delta R^2 \rangle^{1/2}$	$\langle \Delta\theta^2 \rangle^{1/2}$
000	0.013	0.065	0.048	8.71
	0.014	0.102	0.067	8.76
	0.015	0.173	0.067	8.75
	0.014[b]	0.183[b]	0.068[b]	8.72[b]
100	0.027	-0.117	0.092	8.71
	0.030	-0.346	0.094	8.76
	0.031	-0.447	0.095	8.75
010	0.010	1.134	0.065	15.09
	0.011	0.870	0.067	15.17
	0.011	0.758	0.068	15.16
001	0.028	-0.778	0.092	8.71
	0.030	-1.063	0.095	8.76
	0.031	-1.176	0.095	8.75
200	0.042	-0.281	0.113	8.71
	0.045	-0.589	0.115	8.76
	0.047	-0.722	0.117	8.75

Table 14. Expectation values of internal displacement coordinates... (cont.)

Vib. State	$\langle \Delta R \rangle$	$\langle \Delta \theta \rangle$	$\langle \Delta R^2 \rangle^{1/2}$	$\langle \Delta \theta^2 \rangle^{1/2}$
110	0.024	0.969	0.092	15.09
	0.026	0.627	0.095	15.17
	0.028	0.484	0.096	15.16
101	0.042	-0.943	0.113	8.71
	0.046	-1.842	0.067	8.76
	0.047	-1.450	0.117	8.75
020	0.006	2.220	0.066	19.48
	0.008	1.842	0.067	19.58
	0.008	1.690	0.068	19.57
011	0.025	0.308	0.092	15.09
	0.027	-0.064	0.095	15.17
	0.028	-0.244	0.096	15.16
002	0.043	-1.604	0.113	8.71
	0.046	-1.970	0.116	8.76
	0.048	-2.178	0.117	8.75

a Bond lengths are in Ångstroms and angles are in degrees. The first row for each entry is for the SCF, the second row the SD and the third row the SDQ potential energy surface.

b Experimental values. Sources are given in Ref. 3b.

4. VIBRATIONAL TRANSITION DIPOLE MOMENTS AND INTENSITIES

The transition dipole moment of a polyatomic molecule having N nondegenerate modes of vibration can be represented by [5]

$$R_{vw} = (\prod_{i} \Gamma_{v_i w_i})(G + \sum_{i} A_i V_i + \sum_{i \leq j} B_{ij} V_i V_j), \tag{7}$$

where V_i is defined by

$$V_i = (v_i + w_i + 1)/2 \tag{8}$$

and

$$\Gamma_{v_i w_i} = \Gamma_{w_i v_i} = \begin{cases} (w_i!/v_i!)^{1/2}, & w_i > v_i \\ 1 , & w_i = v_i \\ (v_i!/w_i!)^{1/2}, & w_i < v_i \,. \end{cases} \tag{9}$$

Eqn. (7) is a generalization of Eqn. (5), to which it reduces when $w_i = v_i$. The constants G, A_i, and B_{ij} depend on the force constants of Eqns. (2) and the dipole moment expansion constants of Eqn. (6).

The integrated band absorption intensity A for a transition $(v_1 v_2 v_3) \leftrightarrow (w_1 w_2 w_3)$ having a frequency $\tilde{\nu}$ in cm^{-1} and transition moment R_{vw} in Debyes is given by [5]

$$A = [8\pi^3 N_o/3hc) \, \tilde{\nu} |R_{vw}|^2 e^{-hc\tilde{\nu}/kT}] \; Q(T)^{-1} \tag{10}$$

where the units of A are km/mole, N_o is Avogadro's number, and Q(T) is the vibrational partition function.

Transition dipole moments are given in Table 15 for H_2O, HDO, and D_2O for the SCF and SD surfaces. Values computed using the ab initio surfaces and the variational procedures of Carney and co-workers [11] are also shown, as are available experimental results. Integrated absorption intensities at 296 K, computed using the

Table 15. Transition dipole moments (Debyes) for excitations from the ground vibrational state

Upper State	SCF[a]	SCF[b]	SCF[c]	SD[a]	SD[d]	Expt.[e]
A. H_2O						
0 1 0	-0.157	-0.157	-0.161	-0.146	-0.149	0.121
1 0 0	0.044	0.041	0.041	0.035	0.035	0.0149
0 0 1	-0.079	-0.083	-0.078	-0.066	-0.073	0.0708
0 2 0	-0.0081	-0.0080	-0.0073	-0.0093	-0.0092	0.0070
1 1 0	0.00078	0.0021	0.0023	0.0022	0.0024	0.0028
0 1 1	-0.031	-0.016	-0.024	-0.031	-0.024	0.0208
1 0 1	0.011	0.013	0.011	0.011	0.0099	0.0157
2 0 0	-0.0039	-0.0047	-0.0045	-0.0044	-0.0038	0.0042
0 0 2	-0.0011	-0.0020	-0.0046	-0.00076	-0.00011	0.00131
B. HDO						
0 1 0	-0.146	-0.149	---	-0.136	---	0.124
1 0 0	0.031	0.028	---	0.025	---	0.044
0 0 1	-0.057	-0.029	---	0.025	---	0.0569
0 2 0	-0.00056	-0.0069	---	-0.0031	---	0.016
1 1 0	-0.0024	-0.0020	---	-0.0014	---	0.0091
0 1 1	0.021	-0.0099	---	0.0032	---	0.0076
1 0 1	-0.0012	-0.0018	---	-0.0026	---	---
2 0 0	-0.0018	-0.0026	---	-0.0019	---	0.0042
0 0 2	-0.0022	-0.0027	---	-0.0023	---	---

Table 15. Transition dipole moments (Debyes) for excitations from the ground vibrational state (cont).

Upper State	SCF[a]	SCF[b]	SCF[c]	SD[a]	SD[d]	Expt.[e]
C. D_2O						
0 1 0	-0.133	-0.134	---	-0.124	---	---
1 0 0	0.043	0.041	---	0.035	---	---
0 0 1	-0.064	-0.075	---	0.053	---	---
0 2 0	-0.0077	-0.076	---	-0.0083	---	---
1 1 0	0.00041	0.0013	---	0.0015	---	---
0 1 1	-0.023	-0.012	---	0.023	---	---
1 0 1	0.0082	0.010	---	0.0082	---	---
2 0 0	-0.0031	-0.0038	---	-0.0035	---	---
0 0 2	-0.00054	-0.0016	---	-0.00031	---	---

a This work.

b This work. Perturbation theory results using the potential energy and dipole moment functions of Ref. 3c.

c Ref. 11b. Variational results using the potential energy and dipole moment functions of Ref. 3c.

d Ref. 11b. Variational results using the potential energy and dipole moment functions of Paper II.

e Sources are given in Refs. 5b and 11b. The signs are undetermined.

transition moments of Table 15 and the transition energies of Table 3, are given in Table 16.

The transition dipole moments shown in Table 15, computed using the perturbation theory approach of Ref. 5, are in good agreement with the variational results of Carney et al. [11b] as well as with the values determined experimentally. There is a small improvement in the SD results as compared to those due to the SCF surfaces. The fact that the variational results derived from the same potential energy surfaces are nearly identical to the perturbation theory values is a clear indication that the latter approach can be used with confidence for transitions involving reasonably small values of the vibrational quantum numbers. Similar agreement was observed for hot band transitions. Note that the transitions which are forbidden in the harmonic oscillator approximation and have small transition dipole moments, such as $(000)\rightarrow(200)$, are also well represented by the anharmonic potential energy and dipole moment expansions used for the SCF and SD surfaces.

Eqn. (10) shows that the integrated absorption intensities involve both the excitation energy and the transition dipole moment. Errors in the transition energies shown in Table 3 therefore degrade the intensities in a linear fashion. The error in A is compounded by the dependence on the square of the error in the transition dipole moment. The values given in Table 16 are clearly not in as good agreement with experiment as the results given in Tables 3 and 15. However, clear improvement is seen for the SD as compared to the SCF results. This may be attributed principally to the errors in the transition frequencies for the SCF surfaces ($\approx$8%), the dipole moment expansions being of nearly the same quality. It is also true, as was the case for the transition moments, that the intensities derived from the perturbation theory formalism are in good agreement with those due to the variational treatment when the same potential energy and dipole moment surfaces are used.

The results would be improved by employing the SDQ transition

Table 16. Integrated absorption intensities (km/mole) at 298 K for transitions from the ground vibrational state

Upper State	H_2O			HDO			D_2O	
	SCF	SD	Expt[a]	SCF	SD	Expt[a]	SCF	SD
010	106.3	88.5 (91)[b]	71-89	80.8	73.3 (61.8)	45.5	56.3	47.1
100	19.0	11.5 (11.6)	2.93	7.2	14.6 (11.4)	11.6	13.3	8.48
001	63.1	42.3 (43.2)	40-59	32.8	27.6 (23.0)	24.8	30.6	20.3
020	0.56	0.68 (0.47)	0.45	0.0023	0.16 (0.25)	--	0.37	0.42
110	0.013	0.058 (0.11)	0.11	0.065	3.87 (1.28)	--	0.0017	0.021

Table 15. Integrated absorption intensities (km/mole)... (cont.)

Upper State	H_2O			HDO			D_2O	
	SCF	SD	Expt[a]	SCF	SD	Expt[a]	SCF	SD
011	14.3	12.8 (14.3)	5.32-5.68	6.42	6.16 (5.4)	--	5.87	5.27
101	2.20	2.10 (1.95)	2.06-2.19	0.025	0.150 (0.14)	--	0.98	0.92
200	0.30	0.35 (0.38)	0.37	0.047	0.51 (0.47)	--	0.14	0.17
002	0.023	0.0093 (0.0090)	0.74	0.10	1.16 (0.94)	--	0.0044	0.0013

a Sources of experimental values are given in Refs. 5a and 11a.

b Values in parentheses, from Ref. 11a, were obtained based on a variational procedure using an empirical force field and the SD dipole moment functions of Paper II.

energies in conjunction with the SD dipole moment surfaces. We have chosen not to mix the SDQ and SD results, neither in the computation of transition intensities nor in the computation of vibrational corrections to one-electron properties. As discussed above, it is our intention to maintain a clear distinction between ab initio and experimental and semi-empirical results. The formulas for the spectroscopic constants, vibrational expectation values, and transition dipole moments, Eqns. (4), (5) and (7), respectively, all involve the harmonic frequencies and the cubic and quartic force constants defined in terms of normal coordinates. To use energies due to different wavefunctions than those from which the properties were computed removes the formal rigor of our procedure. Preserving internal consistency in our formal approaches serves also to establish the level of applicability of the procedure.

5. SUMMARY

Second-order perturbation theory has been applied in vibrational analyses of potential energy and property surfaces of the water molecule. The method produces detailed information concerning the effects of nuclear motion on a wide variety of one-electron properties in addition to anharmonic spectroscopic constants. Vibration frequencies, transition dipole moments, and intensities compare favorably with those computed using more accurate variational procedures and with available experimentally derived values. SCF and CI wavefunctions for 36 judiciously chosen geometries employed to compute energies and properties were presented as fits to power series in internal displacement coordinates in Paper II. These analytical representations are the sole basis for the straightforward application of perturbation theory to calculate the vibrational properties presented in this work. The results reported here compliment and reinforce other theoretical and experimental studies on the effects of nuclear motion on properties of the water molecule.

Acknowledgement

This work was supported in part by the National Science Foundation under grants CHE-8214689 and 8219408.

References:

1. B.J. Rosenberg and I. Shavitt, J. Chem. Phys. 63, 2162 (1975).
2. B.J. Rosenberg, W.C. Ermler and I. Shavitt, J. Chem. Phys. 65, 4072 (1976).
3. (a) C.W. Kern and R.L. Matcha, J. Chem. Phys. 49, 2081 (1968); (b) W.C. Ermler and C.W. Kern, J. Chem. Phys. 55, 4851 (1971); (c) B.J. Krohn, W.C. Ermler and C.W. Kern, J. Chem. Phys. 60, 22 (1974).
4. W.C. Ermler and B.J. Krohn, J. Chem. Phys. 67, 1360 (1977).
5. (a) B.J. Krohn and C.W. Kern, J. Chem. Phys. 69, 5310 (1978); (b) B.J. Krohn and C.W. Kern, Battelle Memorial Institute - Ohio State University Theoretical Chemistry Group Technical Report No. 98 (1978).
6. A.R. Hoy, I.M. Mills and G. Strey, Mol. Phys. 24, 1265 (1972); A.R. Hoy and P.R. Bunker, J. Mol. Spectrosc. 74, 1 (1979).
7. I.M. Mills, in Molecular Spectroscopy: Modern Research, edited by K.N. Rao and C.W. Mathews (Academic Press, New York, 1972), p. 115.
8. R.J. Bartlett, I. Shavitt and G.D. Purvis, J. Chem. Phys. 71, 281 (1979).
9. P. Hennig, W.P. Kraemer, G.H.F. Diercksen and G. Strey, Theoret. Chim. Acta 47, 233 (1978).
10. W.P. Kraemer, B.O. Roos and P.E.M. Siegbahn, Chem. Phys. 69, 305 (1982).
11. (a) G.D. Carney, L.L. Sprandel and C.W. Kern, Adv. Chem. Phys. 37, 305 (1978); (b) G.D. Carney and C.W. Kern, Int. J. Quantum Chem. Symp. 9, 317 (1975).
12. M.G. Bucknell and N.C. Handy, Mol. Phys. 28, 777 (1974).
13. R.J. Whitehead and N.C. Handy, J. Mol. Spectrosc. 59, 459 (1976).
14. P. Botschwina, Chem. Phys. 40, 33 (1979).
15. L.B. Harding and W.C. Ermler, J. Comput. Chem. 6, 13 (1985).
16. S.A. Clough, Y. Beers, G.P. Klein and L.S. Rothman, J. Chem. Phys. 59, 2254 (1973).
17. H. Kuze, T. Amano and T. Shimizu, J. Chem. Phys. 75, 4869 (1981).
18. J.A. Smith, P. Jorgensen and Y. Öhrn, J. Chem. Phys. 62, 1285 (1975).
19. W.T. Luh and G.C. Lie, J. Quant. Spectrosc. Radiat. Transfer 21, 547 (1979).
20. E.R. Davidson and D. Feller, Chem. Phys. Lett. 104, 54 (1984).
21. G.J. Sexton and N.C. Handy, Mol. Phys. 51, 1321 (1984).

INTERMOLECULAR INTERACTIONS INVOLVING FIRST ROW HYDRIDES: SPECTROSCOPIC STUDIES OF COMPLEXES OF HF, H_2O, NH_3, AND HCN*

K.I. Peterson, G.T. Fraser, D.D. Nelson, Jr.†, and W. Klemperer
Department of Chemistry
Harvard University
Cambridge, MA 02138

ABSTRACT. This article reviews the structural characterization, by means of rotational spectroscopy, of a number of complexes of HF, H_2O, NH_3, and HCN. In addition to the complexes of each of these hydrides with each other, the complexes with CO, CO_2, C_2H_2, and C_2H_4 are also examined. The studies of the complexes of NH_3 are complemented by recently obtained infrared-microwave double resonance results.

1. INTRODUCTION

The hydrogen bond is probably the most extensively studied nonbonding interaction in chemistry. Its importance in biological systems [1], crystals [2], and solution chemistry [3] has been noted numerous times. Obviously, any complete understanding of the hydrogen bond depends on the knowledge of the gas phase structures and bond dissociation energies of hydrogen-bonded dimers and clusters. The simplest molecules which are important to an understanding of hydrogen bonding are the first row hydrides, HF, NH_3, H_2O, as well as HCN. These four hydrides are often thought to be prototype hydrogen bond formers and have therefore been the subject of extensive theoretical and experimental study. They exhibit

*Supported by the National Science Foundation. † National Science Foundation Predoctoral Fellow

R. J. Bartlett (ed.), Comparison of Ab Initio Quantum Chemistry with Experiment for Small Molecules, 217–244.

a wide diversity in properties of which the most important is acidity; HF is the strongest Lewis acid of the group while NH_3 is the weakest. These molecules also differ in their number of lone pair electrons and electric dipole moments (HCN μ=3.0 D, HF μ=1.8 D, H_2O μ=1.9 D, and NH_3 μ=1.5 D). Thus these simple hydrides furnish a suitable testing ground for simple models of hydrogen bonding, and van der Waals interactions in general.

It was originally hoped that an understanding of the van der Waals chemistry of HF would lead to an ability to predict the structures of the weakly bound complexes of NH_3 and H_2O. As will be discussed in this article, this has not been the case. The van der Waals chemistry of each of these complexes is quite unique. In fact, it has been shown that the weak interactions of H_2O and NH_3 are not dominated by hydrogen bonding. These results may be summarized by noting that HF behaves typically as a Lewis acid and NH_3 always behaves as a Lewis base, while H_2O and HCN are each amphoteric in nature. In this paper, we review recent advances in our understanding of the van der Waals and hydrogen bonding of NH_3, H_2O, HF and HCN with emphasis on the more recent results obtained for NH_3 and H_2O.

2. EXPERIMENTAL AND DATA ANALYSIS

The van der Waals complexes which will be discussed were all studied by high resolution rotational spectroscopy. Three different techniques were used: molecular beam electric resonance spectroscopy [4]; Fourier transform microwave spectroscopy [5]; and microwave absorption spectroscopy in a static gas cell. The latter technique has been used successfully to study H_2O-HF [6] since it is a relatively strongly bound complex and therefore occurs in adequate concentration in cooled absorption cells. The first two methods, on the other hand, rely on supersonic expansion sources at low rotational temperatures (10 K). Recently, the molecular beam electric resonance technique has been extended for microwave-infrared double resonance experiments [7]. In these experiments the signal intensity of a microwave transition is monitored as a function of CO_2 laser frequency.

Molecular beam electric resonance is combined with mass spectrometric detection. This frequently avoids ambiguity in determining the molecular origin of the spectrum. It also provides the opportunity to study the fragmentation patterns of the complexes resulting from electron impact ionization. While a known rotational transition is monitored, the signal strength at various mass spectral peaks is recorded. The results are surprising. In some cases, sufficient quantities of the parent ion peak are produced, even though the van der Waals bond is very weak. In fact, for the acetylene-water experiments, most of the spectra were taken while monitoring the parent ion peak. In general, for the complex, AB, the amount A^+ is greater than that of B^+ if A has the lower ionization potential. Therefore, most of the ammonia complexes were successfully monitored on the NH_3^+ peak. For the water complexes, the protonated partner is frequently the major daughter ion. A notable counter example is H_2O-CO_2 for which H_2O^+ was the major daughter ion.

Since the experimental resolution approaches 1 kHz, hyperfine structure is resolved and, thus, quadrupole and spin-spin coupling constants are obtained. Dipole moments are determined from a suitable analysis of the electric field perturbed spectra. In most cases, spectra were also obtained for isotopically substituted complexes.

Because of the generally long van der Waals bond, the a inertial axis lies essentially along that bond and the systems are prolate tops. The moments of inertia about the b and c axes, therefore, are used to determine the distance between the centers of mass of the two submolecules, R_{cm}. Due to the small moments of inertia of the submolecules compared to the overall moment of inertia of the complex, R_{cm} is well determined. Deuterium substitution experiments lead to a rough determination of the orientation of the hydride with respect to its binding partner. In Tables 1-4,

Table 1. Molecular constants of H_2O complexes. Included are center of mass distances, nearest atom distances, dipole moments, and stretching force constants.

PARTNER	R_{cm}(A)	r_N(A)	μ_a(D)	k_s(mdyne/A)	STRUCTURE
H_2O[a]	2.98	2.05	2.64307	---	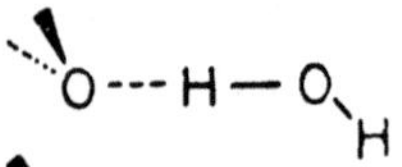
NH_3[b]	3.06	2.02	2.972	---	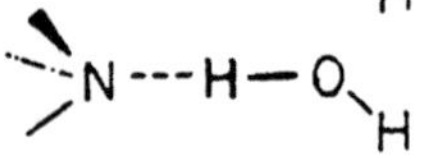
CO[c]	3.953	2.407	---	---	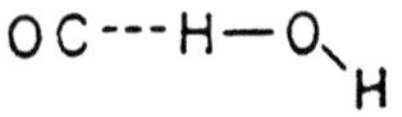
CO_2[d]	2.901	2.836	1.8522	0.064	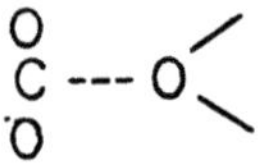
HF[e]	2.680	1.801	4.073	0.15	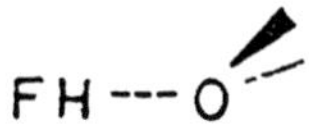
HCN[f]	3.764	2.075	---	---	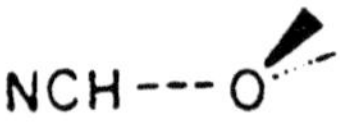
C_2H_2[g]	3.958	2.229	2.0124	0.065	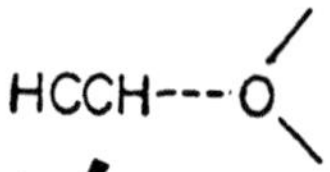
C_2H_4[h]	3.413	2.481	1.0943	---	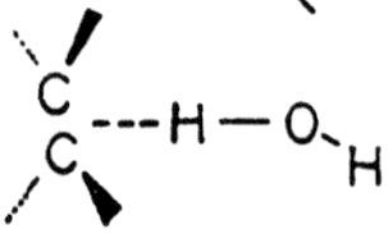
Ar[i]					

[a] T.R. Dyke, K.M. Mack and J.S. Muenter, J. Chem. Phys. 66, 498 (1977).
[b] P. Herbine and T.R. Dyke (to be published). T.R. Dyke, Top. Curr. Chem. 120, 85 (Spring-Verlag, N.Y. 1984).
[c] T.A. Fisher, K.I. Peterson and W. Klemperer (work in progress).
[d] K.I. Peterson and W. Klemperer, J. Chem. Phys. 80, 2439 (1984).
[e] Z. Kisiel, A.C. Legon and D.J. Millen, J. Chem. Phys. 78, 2910 (1983) and J.W. Bevan, Z. Kisiel, A.C. Legon, D.J. Millen and S.C. Rogers, Proc. R. Soc. London Ser. A 372, 441 (1980).
[f] A.J. Fillery-Travis, A.C. Legon and L.C. Willoughby, Chem. Phys. Lett. 98, 369 (1983).
[g] K.I. Peterson and W. Klemperer, J. Chem. Phys. 81, 3842 (1984).
[h] K.I. Peterson and W. Klemperer (manuscript in preparation).
[i] No microwave transitions observed.

Table 2. Molecular constants of NH_3 complexes. Included are center of mass distances, nearest atom distances, dipole moments, induced dipole moments, NH_3 bending angles and stretching force constants.

PARTNER	R_{cm}(A)	R_N(A)	μ_a(D)	μ_{ind}(A)	$\overline{\theta}$(deg)	k_s(mdyne/A)	STRUCTURE
H_2O[a]	3.06	2.02	2.972	--	23	---	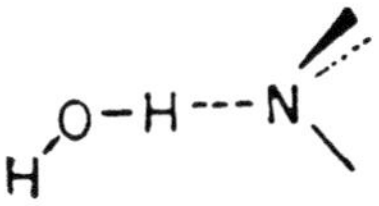
NH_3[b]	3.411	--	0.74	--	--	---	---
CO[c]	--	--	1.248	--	36.8	---	---
CO_2[d]	3.054	2.988	1.7684	0.411	22.71	0.070	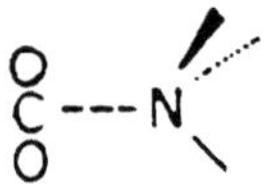
HF[e]	--	1.78	--	1.3	--	---	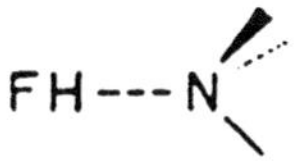
HCN[f]	3.847	2.156	5.2608	0.939	20.40	0.122	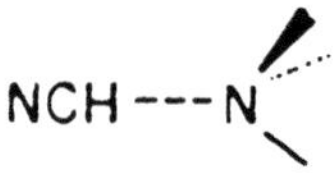
C_2H_2[g]	4.063	2.333	1.9871	0.634	23.2	0.070	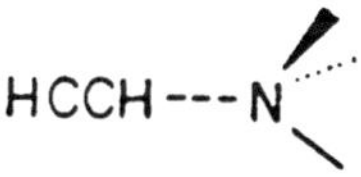
C_2H_4[h]							
Ar[i]							

[a] See H_2O table.
[b] G.T. Fraser, D.D. Nelson, Jr., A. Charo and W. Klemperer, J. Chem. Phys. (to be published 1985).
[c] G.T. Fraser, D.D. Nelson, Jr. and W. Klemperer (work in progress).
[d] G.T. Fraser, K.R. Leopold and W. Klemperer, J. Chem. Phys. 81, 2577 (1984).
[e] B.J. Howard (private communication).
[f] G.T. Fraser, K.R. Leopold, D.D. Nelson, Jr., A. Tung and W. Klemperer, J. Chem. Phys. 80, 3073 (1984).
[g] G.T. Fraser, K.R. Leopold and W. Klemperer, J. Chem. Phys. 80, 1423 (1984).
[h] No microwave transitions observed.
[i] G.T. Fraser, D.D. Nelson, Jr. and W. Klemperer (work in progress).

Table 3. Molecular constants of HF complexes. Included are center of mass distances, nearest atom distances, dipole moments, induced dipole moments, and HF bending angle.

PARTNER	R_{cm}(A)	R_N(A)	μ_a(D)	μ_{ind}(A)	$\bar{\theta}$(deg)	STRUCTURE
H_2O[a]	2.680	1.801	4.073	---	18.3	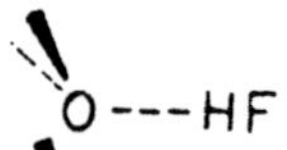
NH_3[b]	---	1.78	---	1.3	---	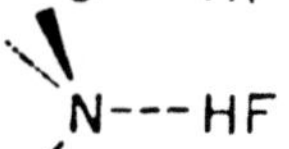
CO[c]	3.646	2.074	2.352	0.55	21.9	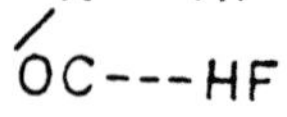
CO_2[d]	3.954	3.907	2.2465	0.60	25.17	
HF[e]	2.758	1.832	2.987	---	23	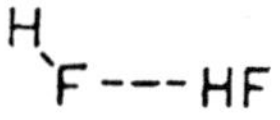
HCN[f]	3.346	1.87	5.612	1.1	23	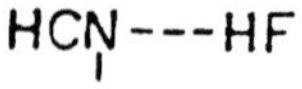
C_2H_2[g]	3.075	2.196	2.3681	0.65	19.8	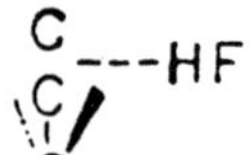
C_2H_4[h]	3.097	2.218	2.384[j]	0.67	19.8	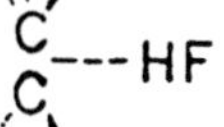
Ar[i]	3.510	2.630	1.3353	---	41.27	

[a] See H_2O table plus A.C. Legon and L.C. Willoughby, Chem. Phys. Lett. 92, 333 (1982).
[b] See NH_3 table.
[c] A. C. Legon, P.D. Soper and W.H. Flygare, J. Chem. Phys. 74, 4944 (1981); and E.J. Campbell, W.G. Read and J.A. Shea, Chem. Phys. Lett. 94, 69 (1983).
[d] F.A. Baiocchi, T.A. Dixon, C.H. Joyner and W. Klemperer, J. Chem. Phys. 72, 6544 (1981).
[e] T.R. Dyke, B.J. Howard and W. Klemperer, J. Chem. Phys. 56, 2442 (1972).
[f] A.C. Legon, D.J. Millen and S.C. Rogers, Proc. R. Soc. London Ser. A 370, 215 (1980); and E.J. Campbell and S.G. Kukolich, Chem. Phys. 76, 225 (1983).
[g] W.G. Read and W.H. Flygare, J. Chem. Phys. 76, 2238 (1982).
[h] J.A. Shea and W.H. Flygare, J. Chem. Phys. 76, 4857 (1982).
[i] T.A. Dixon, C.H. Joyner, F.A. Baiocchi and W. Klemperer, J. Chem. Phys. 74, 6539 (1981).
[j] D.D. Nelson, Jr., G.T. Fraser and W. Klemperer, J. Chem. Phys. (submitted).

Table 4. Molecular constants of HCN complexes. Included are center of mass distances, nearest atom distances, dipole moments, induced dipole moments, and HCN bending angle.

PARTNER	R_{cm}(A)	R_N(A)	μ_a(D)	μ_{ind}(A)	$\overline{\theta}$(deg)	STRUCTURE
H_2O[a]	3.764	2.075	---	---	---	O---HCN
NH_3[b]	3.847	2.16	5.2608	0.939	9.6	N---HCN
CO[c]	4.876	2.606	---	---	14.2	OC---HCN
CO_2[d]	3.592	2.998	3.207	0.361	17.4	OCO---NCH
HF[e]	3.346	1.87	5.612	1.1	17	FH---NCH
HCN[f]	4.450	2.223	6.55	0.77	17.34,11.9	NCH---NCH
C_2H_2[g]	4.215	2.591	---	---	12.3	HCCH---HCN
C_2H_4[h]	4.269	2.645	---	---	12.8	CC---HCN
Ar[i]	4.343	---	2.6254	---	30.8	---

[a] See H_2O table.
[b] See NH_3 table.
[c] E.J. Goodwin and A.C. Legon, Chem. Phys. 87, 81 (1984).
[d] K.R. Leopold, G.T. Fraser, and W. Klemperer, J. Chem. Phys. 80, 1039 (1984).
[e] See HF table.
[f] L.W. Buxton, E.J. Campbell and W.H. Flygare, Chem. Phys. 56, 399 (1981); and E.J. Campbell and S.G. Kukolich, Chem. Phys. 76, 225 (1983).
[g] P.D. Aldrich, S.G. Kukolich and E.J. Campbell, J. Chem. Phys. 78, 3521 (1983).
[h] S.G. Kukolich, W.G. Read, P.D. Aldrich, J. Chem. Phys. 78, 3552 (1983).
[i] K.R. Leopold, G.T. Fraser, F.J. Lin, D.D. Nelson, Jr. and W. Klemperer, J. Chem. Phys. (to be published, Dec. 1984).

R_{cm} and R_N are listed. R_N is calculated from the most probable equilibrium orientation of the subunits as derived from the experimental results.

Although the heavy atom positions can be determined from the rotational constants, the effect of the light hydrogens on these constants is small. This, together with the possibility of tunnelling motions, forbids a reliable determination of the hydrogen positions from the rotational constants alone. Additional data in the form of quadrupole and spin-spin coupling constants and dipole moments are necessary to adequately determine the complete structure.

To the extent that the quadrupole coupling constant of a molecule does not change upon complexation, $\langle\cos^2\theta\rangle$ can be derived from the measured quadrupole coupling constant of the complex. θ is the angle between the appropriate axis of the submolecule and the _a_-axis of the complex. For HF complexes, $\langle\cos^2\theta\rangle$ can also be obtained from the measured spin-spin coupling constant. In Tables 2-4 are listed $\bar{\theta}=\cos^{-1}\langle\cos^2\theta\rangle^{1/2}$. For HCN and NH_3 complexes, the nitrogen quadrupole coupling constant is measured. For HF complexes, the spin-spin hyperfine constant is used except where designated by a 'D' in which case the DF quadrupole coupling constant data was used. In summary, for HF, HCN or NH_3 complexes, $\bar{\theta}$ refers to the angle between the HF, HCN, or NH_3 symmetry axis and the _a_ inertial axis of the complex. $\bar{\theta}$ is not included in Table 1 since it generally has not been determined for the water complexes.

The dipole moment of the complex along the _a_ axis, μ_a, is equal to the sum of the _a_-axis projections of the subunit moments with the _a_-axis projection of the induced dipole moment, μ_{ind}. In the case of HCN, NH_3 and HF complexes, $\bar{\theta}$ gives an estimate of the projection angle so that μ_{ind} can be calculated. This parameter is useful in developing insight into bonding properties of van der Waals interactions. For the water complexes, $\bar{\theta}$ is generally not known so μ_{ind} is not included. As a point of reference, the dipole moments of H_2O, NH_3, HF, CO and HCN are 1.855 [8], 1.472 [9], 1.826

[10], 0.112 [11] and 2.985 D [12], respectively.

The centrifugal distortion constant, Δ_J, is assumed to be predominantly a measure of the interaction of the van der Waals stretching vibration with the overall rotation. This assumption holds if the stretching mode is adequately separated from the bending modes. Such a separation does not hold for Ar-HCN, but this complex appears to be anomolous. No evidence for radial-angular coupling has been observed in other complexes. If the bending and stretching modes can be treated as uncoupled oscillators, Δ_J is a direct measure of the stretching force constant, k_s, for the van der Waals bond. This parameter is listed for the H_2O and NH_3 complexes in Tables 1 and 2, respectively.

3. WATER COMPLEXES

Microwave spectroscopy was first applied to water complexes in a molecular beam electric resonance study of water dimer by Dyke, Mack and Muenter [13]. Although the spectrum obtained is very complicated, transitions which clearly fall into a pattern for a nearly prolate symmetric top rotor enable values for the rotational constant, B + C, to be determined. The dipole moment component along the a-axis is also obtained. These data are consistent with a structure in which one water molecule hydrogen bonds to one of the lone pair orbitals of the other.

In water dimer, evidence for internal rotation or inversion motion arises from the presence of transitions which do not fit a rigid rotor model. In addition, two sets of rigid rotor transitions are observed. This is expected considering the many possibilities for tunnelling splittings to occur. It is noteworthy that, despite all these possibilities, rigid-rotor rotational constants can be obtained which are reliably used for a structural analysis. Unfortunately, the details of the relative orientation of the two water molecules are not definitive. More experiments involving various isotopically substituted species would be helpful. Particularly important is information concerning the barriers to the possible internal rotations.

The water dimer is about as difficult to analyze as one can imagine. It consists of two asymmetric rotor subunits, both capable of either tunneling motions or large amplitude vibrational motions due to the presence of light hydrogens. Many of the other water complexes are somewhat easier to understand since the binding partner of the water often either contributes hyperfine structure or possesses heavy atoms with small amplitude vibrational motions.

All the water complexes studied so far can be divided into two structural types, one in which the water hydrogen-bonds to the most negative part of its partner and another in which the water oxygen-bonds to the positive part. The propensity of the water to hydrogen-bond is in general correlated to its acidity relative to the binding partner. C_2H_2-H_2O [14] is the only complex which contradicts this observation.

The set of molecules studied to which H_2O hydrogen-bonds is H_2O [13], NH_3 [15], CO [16] and C_2H_4 [17]. $(H_2O)_2$ has already been discussed. Dyke [15] has found that in NH_3-H_2O the vibrationally averaged bending angle of the NH_3 is 23°, so that the projection of the NH_3 dipole moment on the a-axis of the complex is 1.35 D. The induced dipole moment is calculated to be 0.37 D. Subtracting these from the complex dipole moment of 2.97 D, one obtains a value of 1.25 D for the projection of the H_2O dipole moment on the a axis. This is reasonable for a linear hydrogen bonded structure, i.e., with atoms, N---H-O, colinear at equilibrium. The details of the NH_3-H_2O study are unpublished, so we will not discuss it further here.

The dipole moment components along the a-axis for both CO-H_2O [16] and C_2H_4-H_2O [17] are low -- 1.05 and 1.09 D, respectively. This means that the C_{2v} axis of H_2O is at a large angle (>53°) with respect to the a-axis of the complex. This is consistent with a structure in which the water is hydrogen-bonding. In the case of C_2H_4-H_2O [17], this implies hydrogen bonding to a π orbital. Although all details of the vibrationally averaged structure are not known, the data point to an equilibrium structure in which one of the water hydrogens is directed into the plane of the ethylene. This structure is analogous to that of C_2H_4-HF and -HCl. In all three cases, the ethylene orientation was determined using the A rotational constant of the complex. The dipole moment of C_2H_4-H_2O implies that the C_{2v} axis of the H_2O is at least 54° from the a-axis of the complex. A reasonable upper estimate, assuming an induced moment of 0.4 D, is 68°. This is interesting because the equilibrium angle must be slightly larger. The angle between the C_{2v} axis and the O-H bond in H_2O is 52.3°. This suggests that the hydrogen is not pointed directly into the center of the π orbital. Although it is tempting to hypothesize that the hydrogen is directed toward one of the carbon atoms, where the electron density is larger, it should be noted that there is no evidence for this sort of behavior in the analogous C_2H_4-HCl complex. In their study of HCl-C_2H_4, Aldrich, Legon, and Flygare [18] found, through an analysis of the chlorine quadrupole coupling constant, that the bending angles of the HCl subunit in and out of the heavy atom plane are the same. This clearly points to hydrogen bonding to the center of the ethylene with virtually no vibrational anisotropy. Therefore, the coarse structures of C_2H_4-HCl and -H_2O appear to be similar, but some differences are indicated in the details of the bonding.

Water hydrogen-bonds to the carbon of carbon monoxide in D_2O- and HDO-CO [16]. If it is assumed that this bond is to the oxygen of CO, an inordinately small hydrogen-bond length is determined. As for the C_2H_4-H_2O, the low dipole moment for D_2O-CO shows that the D_2O C_{2v} axis is greater than 53° from the a-axis of the complex. This could imply a slightly nonlinear hydrogen bond. At this point data have been obtained only on D_2O- and HDO-CO. Further experiments on H_2O-CO will provide more insight into the bonding. It is noteworthy that only the isomer of HDO-CO with the deuterium bound to the CO was observed in the molecular beam. No evidence for the latter isomer was found despite extensive searches that were made. This is presumably due to the lower binding energy of the HOD-CO isomer.

HF [6], HCN [19], and C_2H_2 [14] form a group of molecules all of which hydrogen-bond to water. The details of the bonding, though, are different in each case and are consistent with a van der Waals bond-strength order of HF > HCN > C_2H_2. The weak bond-stretching force constant is largest for the HF-H_2O complex, and the van der Waals bond length is the smallest. C_2H_2-H_2O has the smallest force constant and longest bond length. It is interesting to take a more detailed look at the out-of-plane vibration of the water molecule. The bending motion described by the angle θ, as shown in Fig. 1 for C_2H_2-H_2O, will be perturbed by the presence of a barrier at the planar position, θ=0. Kisiel, Legon, and Millen [6] have observed the rotational spectrum of H_2O-HF in excited van der Waals bending states. They find a potential barrier of 126 cm^{-1}. This is between the energy of the first and second vibrational states of the out-of-plane mode. It appears, then, that the motion of the water molecule is strongly affected by the interaction of its lone-pair orbitals with the HF proton.

The van der Waals bond length in C_2H_2-H_2O is much longer than that in HF-H_2O, and it is thus reasonable to expect that the energy barrier to the out-of-plane bending motion is lower. No excited vibrational states were observed for C_2H_2-H_2O, so one must rely on other data to investigate the height of the barrier at θ=0. This is best accomplished by studying the effect of the deuterium substitution of the water on the dipole moment of the complex. Deuterium substitution lowers the van der Waals bending vibrational frequency of the water, so the ground vibrational state lies deeper within the double minimum potential if it is present. In such a case, the barrier perturbs the vibrational motion more strongly, and this effect can be observed in the dipole moments. For example, the dipole moment of H_2O-HF is higher than that of D_2O-DF where the barrier is known to be high. On the other hand, the dipole moment of C_2H_2-H_2O is smaller than that of C_2H_2-D_2O. This is expected for zero point vibrations around a planar equilibrium configuration.

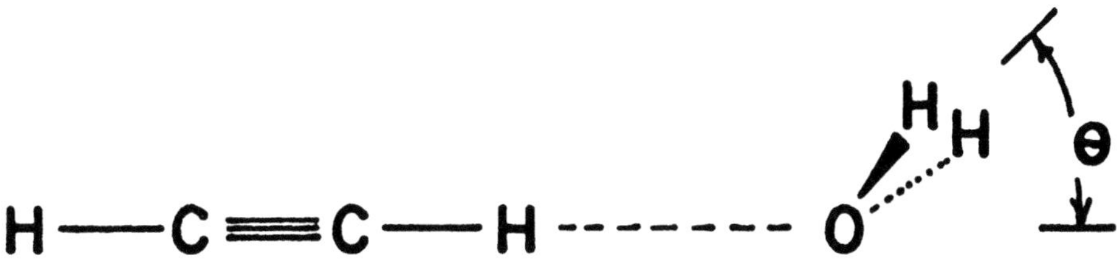

Figure 1. Representation of the out-of-plane vibrational motion in C_2H_2-H_2O. In this complex there is no evidence for a barrier at $\theta=0°$ hindering the motion. In HF-H_2O, a barrier height of 123 cm^{-1} was determined.

In this case, there is no evidence for a barrier at $\theta=0$. This is perhaps due to the charge distribution on the water molecule appearing more spherical to C_2H_2 as a consequence of the longer van der Waals bond length.

H_2O-CO_2 is the only complex in Table 1 which does not involve hydrogen bonding. Instead, the oxygen atom of the water bonds to the carbon of the carbon dioxide molecule forming a planar complex. The rotational spectrum shows clear evidence for the presence of a hindered internal motion. The relevant motion is a torsional motion around the C_{2v} axis of the H_2O relative to the CO_2 subunit. The first two torsional states are observed because the lowest is correlated with rotational states which have a K quantum number which is even, and the first excited state is correlated with those of odd K. It is not possible for the upper state to relax to the lower state even under the conditions of a supersonsic expansion since this is a nuclear spin forbidden process. The rotational spectrum of H_2O-CO_2 including both torsional states cannot be fit using a rigid rotor model while that of D_2O-CO_2 can be. This implies that D_2O-CO_2 is more rigid and is consistent with the lower tunneling frequency which occurs in that complex.

Assuming only the torsional motion is coupled with the general rotation, the spectrum can be analyzed using a method by Quade [20]. Quade treats the problem of hindered rotation with a two-fold potential barrier by assuming the molecule to be rigid except for one degree of freedom for the internal rotation. Rotational constants are obtained which are separated into rigid rotor rotational constants plus correction terms which are strongly dependent on the barrier height. Use of both the D_2O-CO_2 and H_2O-CO_2 spectral data furnishes sufficient information to obtain an upper and lower limit for the barrier. The resulting barrier of 0.9(2) kcal/mole is very large considering that the van der Waals bond strength is only 2-3 kcal/mole. This result suggests that the bonding is dependent on the charge distribution of the lone pair orbitals of the water so that a π-type bond is formed.

Ab initio calculations of van der Waals molecules present a challenge to theoreticians since the bond energies are only a few kcal/mole. Consequently, for the SCF energy difference to be meaningful, extended basis sets are required for each submolecule and for the complex. Electron correlation is a significant fraction of the van der Waals interaction, and the necessity of accounting for this increases the difficulty of the calculations considerably. Despite these problems, reasonable binding energies are predicted. Although the absolute uncertainties in the binding energies of van der Waals complexes are of the same order as (or lower than) those calculated for chemically bound molecules, the relative uncertainties are large for the weakly bound complexes. Thus, these systems provide a critical test for the theoretical models.

An extensive evaluation of all the relevant *ab initio* calculations on complexes containing at least one water molecule is beyond the scope of this report. However, it is useful to point out one work by Frisch, Pople, and Del Bene [21] which addresses some of the problems involved. Frisch et al have used Hartree-Fock wavefunctions in the 6-31G* basis set to optimize the geometry of H_2O-, NH_3-, and HF-C_2H_2. The results they obtain for the hydrogen bond lengths are very close to the experimental values. In the case of the water complex this bond length is 2.26 Å compared to the experimental value of 2.23 Å. Use of a more extensive basis set at this fixed geometry decreased the calculated van der Waals binding energy. On the other hand, inclusion of correlation increased the energy. It is clear from their work that very careful calculations are necessary to acquire any confidence in the resulting energy.

4. COMPLEXES OF NH_3

The complexes of NH_3 with Ar [7], NH_3 [7], HCCH [22], HCN [23], HF [24], CO_2 [25], CO [26], and H_2O [15] have been studied using high resolution microwave spectroscopy. As with the water complexes, the resultant structures show a rich stereochemistry and are, in some cases, somewhat surprising. Most importantly, the results show that NH_3 exhibits no propensity for proton donation.

This uniqueness of NH_3 in not exhibiting simply hydrogen bonding is exemplified by the structure of $(NH_3)_2$. This complex has been studied extensively theoretically [27] with each calculation suggesting or assuming a hydrogen-bonded structure for this complex. The N-H---N arrangement is expected to be nearly linear. It has been shown recently [22], though, that $(NH_3)_2$ does not have this structure. In Figure 2, the J=0-1, K=0 transition of $(NH_3)_2$ is shown. The splitting pattern displayed by this transition is the overlap of two ^{14}N nuclear quadrupole hyperfine patterns for two vibrational states of the complex. The two vibrational states are most likely the A and E internal rotor states for the internal rotation of one of the NH_3 subunits. At least in these states, $(NH_3)_2$, unlike $(HF)_2$, shows no inversion motion exchanging the two subunits. The μ_a dipole moment component measured by the Stark effect of the J=0-1, K=0 transition is 0.74(2) D. This dipole moment component is entirely inconsistent with the dipole moment component expected for the theoretically predicted hydrogen-bonded $(NH_3)_2$. The μ_a dipole moment component of $(ND_3)_2$ (0.55(1) D) is in agreement with the measured μ_a dipole moment component of $(NH_3)_2$. The relative insensitivity of the dimer dipole moment to isotopic substitution, as well as the observation of a pure rotational spectrum for this complex, shows that the measured dipole moments are near in value to the equilibrium dipole moment. It should be noted that the theoretically predicted hydrogen-bonded structure for $(NH_3)_2$ gives a μ_a dipole moment component greater than 2 D.

Interestingly, $(NH_3)_2$ was expected to be the trivial case for which NH_3 would exhibit hydrogen bonding. The lack of a hydrogen bond shows that NH_3 is an extremely poor hydrogen-bond former. We presently do not know the structure of $(NH_3)_2$. The rotational constant $\frac{B + C}{2}$ yields a center-of-mass separation of 3.411 Å between the two NH_3 subunits. This is consistent with the nearest neighbor distance observed in crystalline NH_3. The structure of

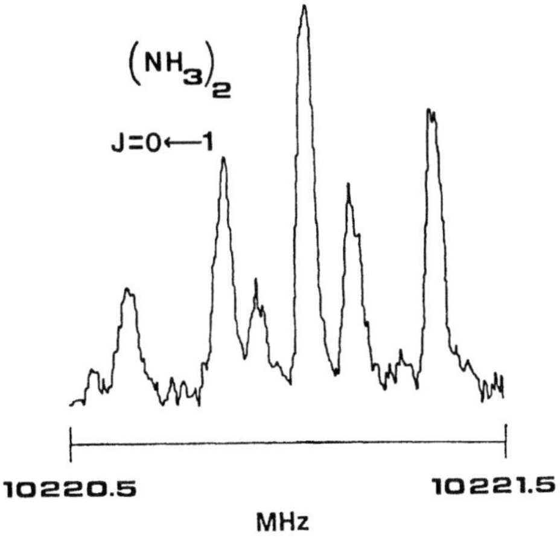

Figure 2. a) High resolution spectrum of the J=0-1 transition of $(NH_3)_2$. The splitting pattern displayed by this transition is due to the overlap of two hyperfine patterns for two vibrational states of the complex.

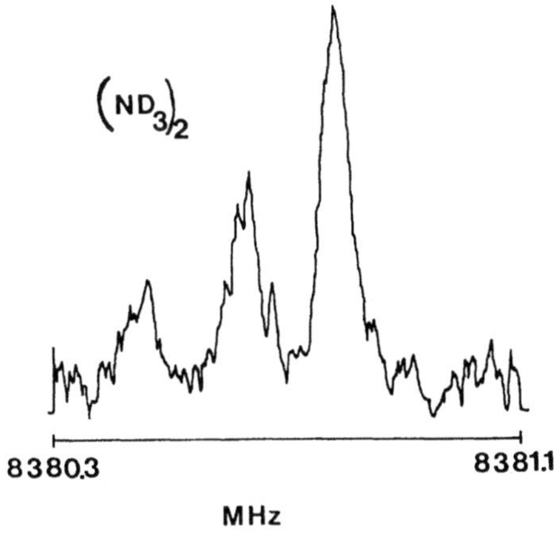

b) High resolution spectrum of the J=0-1 transition of $(ND_3)_2$. Unlike what is observed in $(NH_3)_2$, only one vibrational state is observed. This is most likely the result of the larger moments of inertia of ND_3 compared to that of NH_3, which leads to a quenching of the internal rotation.

crystalline NH_3 is complex and does not show a linear hydrogen bond. The center-of-mass separation observed in $(NH_3)_2$ is surprisingly large compared to those found in $(H_2O)_2$ (2.98 Å), NH_3-H_2O (3.06 Å), $(HF)_2$ (2.758 Å), and HF-H_2O (2.680 Å). This further indicates the uniqueness of the $(NH_3)_2$ interaction. The dissociation energy, D_o, of this complex has been shown to be less than 2.8 kcal/mole by microwave-infrared double resonance studies in a molecular beam electric resonance spectrometer using a line tunable CO_2 laser as the infrared source [7]. The NH_3 umbrella motion υ_2 was excited in the complex. Several CO_2 laser lines between 975 and 985 cm^{-1} have been shown to photodissociate $(NH_3)_2$. The infrared transitions were monitored by recording the signal strength of the J=0-1,K=0 transition as a function of CO_2 laser line.

Another rather intriguing complex of NH_3 is Ar-NH_3 [7]. This complex has an extremely dense spectrum in the region between 13-21 GHz as shown in Figure 3. Since NH_3 has an inversion spectrum at 23 GHz, the clump of transitions at 19-20 GHz is highly suggestive of a Q-branch inversion spectrum for the complex. This requires the μ_a dipole moment component of the complex to be antisymmetric with respect to the inversion. One model presently being examined considers the NH_3 subunit as virtually free, since it is sitting in a nearly isotropic potential due to the argon. The observation of an infrared transition at 938.69 cm^{-1} [7] for Ar-NH_3 supports this model. The frequency of this transition shows that the NH_3 was not affected sufficiently by complexation to perturb the umbrella motion of the NH_3.

The structures of the complexes of NH_3 with HCN [23], HCCH [22], HF [24], CO_2 [25], and H_2O [15] are more typical for the binding of NH_3. In each of these complexes NH_3 behaves as a Lewis base with the bond forming along the C_3 axis of NH_3. For the complexes NCH-NH_3, HCCH-NH_3, and FH-NH_3, this leads to symmetrical-top structures with a hydrogen bond to the NH_3 subunit. The complexes HCN-NH_3 and HF-NH_3 are the gas phase species which condense to form the ionic solids ammonium cyanide and ammonium

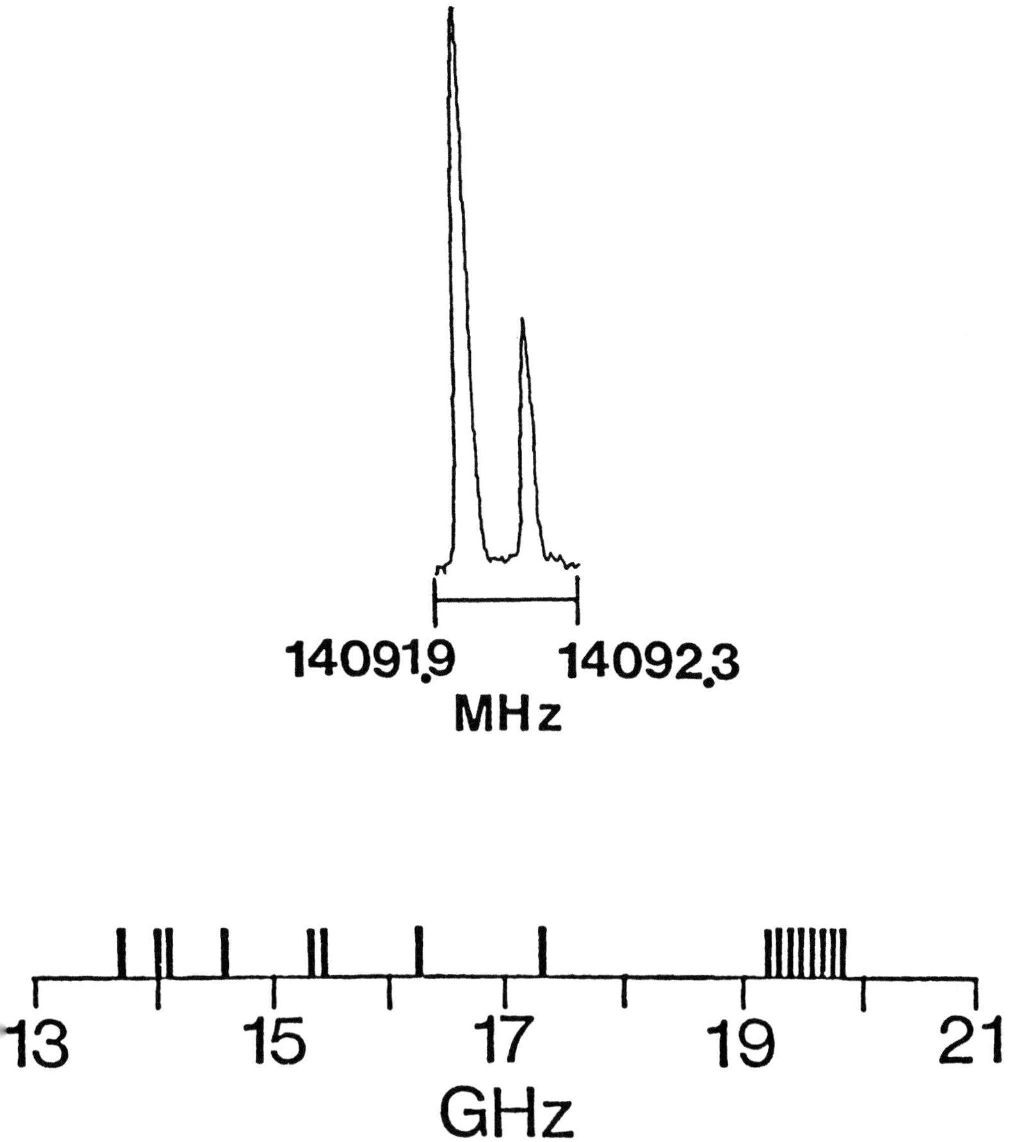

Figure 3. Observed microwave spectrum of Ar-NH_3 from 13 GHz to 21 GHZ. The spectral regions from 65 kHz-91 MHz and from 3.75 GHz-13.0 GHz were also searched, but no resonances were found in these regions. Also shown are two hyperfine components of the only microwave transition which was observed in microwave-infrared double resonance.

fluoride. Ammonimum cyanide shows no ionic character in the gas phase. This is most clearly shown by examination of the ^{14}N nuclear quadrupole coupling constants of the NH_3 in the complex. As NH_3-HX approaches NH_4^+-X^- the electric field gradient (and thus the quadrupole coupling constant) at the nitrogen nucleus approaches zero. This is not observed in HCN-NH_3. Both the large stretching force constants ($k_s^{NH_3-HCN}$ = 0.12 mydn/Å) and large induced electric dipole moments ($\mu_{in}^{NH_3-HF}$ = 1.3 D [24] and $\mu_{in}^{NH_3-HCN}$ = 0.94 D) indicate that the binding strengths of HF-NH_3 and HCN-NH_3 are greater than that observed in the other van der Waals complexes of NH_3 which have been studied.

The complexes NCH-NH_3 and HCCH-NH_3 form part of an interesting class of molecules which have an acetylenic hydrogen forming a bond. The symmetrical structure of HCCH-NH_3 was at first puzzling since it had been previously determined that the complex Ar-HCCH [28] has a T-shaped structure. Both the symmetrical HCCH-NH_3 and T-shaped Ar-HCCH structures, though, can be easily explained by the frequently used HOMO-LUMO bonding picture [22]. Due to symmetry in the T-shaped configuration, the highest unoccupied π_g^* orbitals of HCCH have zero overlap with the lone pair orbital of NH_3. With Ar the π_g^* orbital is of the correct symmetry for overlap with the p_x and p_y orbitals of Ar. The next highest unoccupied orbital of HCCH, the σ* orbital, is centered on the hydrogens and is of the correct symmetry to achieve overlap with the lone pair orbital of NH_3. A quantitative theoretical study of the binding of NH_3-HCCH was undertaken by Frisch, Pople, and Del Bene [21]. Using SCF plus Møller-Plesset (many-body) perturbation theory at the MP4SDQ/6-31G** level, they predicted a symmetrical complex with an equilibrium N---H van der Waals bond length of 2.329 Å. These results appear to be in excellent agreement with the vibrationally averaged bond length of 2.333 Å obtained from the microwave studies. These authors also have calculated the bond dissociation energy, D_o, of NH_3-HCCH to be 3.6 kcal/mole. They indicated, though, that larger basis set calculations suggest a binding energy closer to kcal/mole. As will be discussed below, the observed infrared photodissociation of NH_3-HCCH at 984.38 cm^{-1} places the binding energy, D_o, at less than 2.8 kcal/mole.

NH_3-CO_2 [25] is a T-shaped complex in which the carbon bonds to the nitrogen. Such a complex displays a six-fold barrier to internal rotation of the NH_3 subunit against the CO_2. For stable molecules six fold barrier heights are low (on the order of a few calories per mole), and the spectra tend to be that of a nearly free internal rotor. This is also observed in NH_3-CO_2. Interestingly, the weak bond-stretching force constant for NH_3-CO_2 is identical to that of NH_3-HCCH (0.070 mdyn/Å), suggesting that the binding strength for these two systems might be very similar. The average bending amplitudes of the NH_3 subunits are also nearly the same. This similarity, though, is most likely the result of the bending amplitude's inverse fourth root dependence on the bending force constant. The measured asymmetry of the quadrupole coupling constant in NH_3-CO_2 allows the determination of the difference in amplitude of the in-plane and out-of-plane bend of the NH_3 subunit. This difference is 1.0(4)° with the amplitude of the in-plane bend being greater. The measured quadrupole coupling constants of the ground and the first excited internal rotor states show that the average bending amplitude of the NH_3 subunit is insensitive to the internal rotation of the NH_3 (bending amplitude changes by less than 0.2°). Thus the presence of 6 cm^{-1} of internal rotation energy does not lead to a gyroscopic reduction of the bending amplitude.

In Table 2 we summarize some of the molecular properties of several weakly bound complexes of NH_3. Listed are stretching force constants, k_s; van der Waals bond lengths, R_N; electric dipole moment components, μ_a; induced dipole moments, μ_{ind}; average bending angles of the NH_3 subunit measured from the a inertial axis, $\overline{\theta}$; and the distance between the centers of mass of the two bonding partners, R_{cm}.

We have recently measured infrared spectra of both NH_3-CO_2 and

NH_3-HCCH in the region of the υ_2 umbrella motion of NH_3 [7] (the υ_2 band origin for noninverting free NH_3 is at 950.3 cm^{-1}). This was done by injecting the radiation from a line tunable CO_2 laser into the molecular beam electric resonance spectrometer. Due to the small rotational constants of van der Waals complexes and their 10 K rotational temperature in the molecular beam [29] the infrared bands of such complexes are dense and extend over 5 cm^{-1}. With a line tunable laser of 1 cm^{-1} spacings and 25 to 50 MHz linewidth, many complexes of NH_3 have an infrared spectrum which overlaps at least one CO_2 laser line. In Figure 4 we show a simulated parallel band for NH_3-HCCH. For NH_3-HCCH infrared spectral searches were made by line tuning the CO_2 laser from 983.25 to 988.65 cm^{-1}. A resonance was detected at 984.38 cm^{-1}. This resonance photodissociates the complex and thus places the dissociated energy, D_o, at less than 2.8 kcal/mole. Infrared-microwave double resonance studies indentified the initial state in the resonance as J=4, K=1. Thus the 984.38 cm^{-1} infrared transition is either the P(4), Q(4), or R(4) transition of the υ_2 NH_3 parallel band of this complex.

Because the transition is observed to be photodissociative, an estimate of the upper state lifetime is of interest. Since at least five CO_2 laser lines should overlap the infrared band, the observation of only one CO_2 laser coincidence with one rotational state of the complex suggests a narrow linewidth for the infrared transitions. A simple, probabalistic model suggests a linewidth of 150 MHz which corresponds to an excited state lifetime of 1 ns. The assignment of the infrared transition allows the determination of the band origin as 984.4(9) cm^{-1} for υ_2 NH_3-HCCH. Thus complexation of NH_3 with HCCH causes a blue shift of the non-inverting NH_3 υ_2 frequency of 34 cm^{-1}.

Similarly, we have also obtained the NH_3 υ_2 infrared spectrum of NH_3-CO_2. The spectrum should be characteristic of a parallel band of an asymmetrical top (k = -0.72, $\frac{B + C}{2}$ = 3756.178(3) MHz) with an effectively free internal rotor. As in NH_3-HCCH, the infrared transitions are photodissociative, and they thus place the

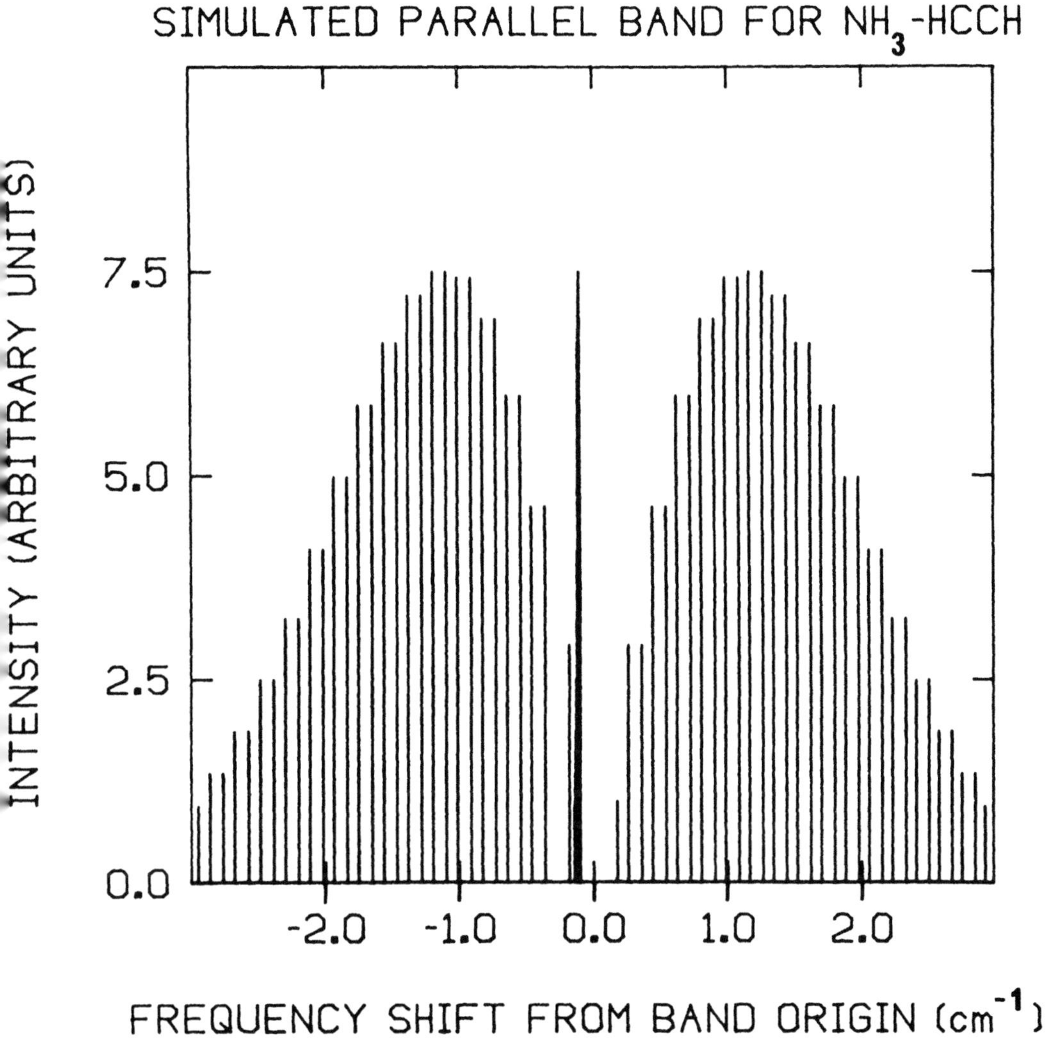

Figure 4. Infrared spectrum of NH_3-HCCH assuming $\Delta A = -3$ GHz and $\Delta B = -5$ MHz. The intensities are derived from Boltzman factors at 10 K. The relative intensities of the K=0 and K=1 states are assumed equal.

bond dissociation energy, D_o, at less than 2.8 Kcal/mole.

Infrared-microwave double resonance studies are summarized in Figure 5. The J=0-1, K=0 and J=1-2, K=0 double resonance studies place R(1) near 987.62 cm^{-1} and P(2) near 986.57 cm^{-1}. In contrast to NH_3-HCCH, each laser line overlaps several rotational states of the complex. This is the result of both a more complicated infrared band as well as a shorter upper state lifetime. The band origin is 987.1(2) cm^{-1} and the infrared transition linewidth is 0.45(20) cm^{-1}. This corresponds to an upper state lifetime of 8 to 20 ps. In Table 5 we summarize infrared band origins, linewidths, and lifetimes for NH_3-HCCH, NH_3-CO_2, NH_3-N_2O, and NH_3-OCS. The binding energy for both NH_3-N_2O and NH_3-OCS is less than 2.8 kcal/mole. No transition linewidths or lifetimes were determined for these two complexes.

Table 5. Infrared band origins, linewidths and lifetimes.

	υ_o	Γ	τ
NH_3-HCCH	984.4(9) cm^{-1}	150 MHz	1 ns
NH_3-CO_2	987.1(2) cm^{-1}	0.45(20) cm^{-1}	8-20 ps
NH_3-OCS	981.5(15) cm^{-1}	---	---
NH_3-N_2O	980.(2) cm^{-1}	---	---

5. DISCUSSION

It was originally thought that complexes of NH_3 and H_2O with Lewis bases would structurally mimic the corresponding complexes of HF and HCN. As shown in Tables 1-4, this is not generally true. In fact, the van der Waals chemistries of both NH_3 and H_2O are surprisingly subtle.

The complex stereochemistry of NH_3 occurs predominantly in its interactions with Lewis bases. When NH_3 is compelled to bind to a Lewis base, the resulting structure is difficult to predict with either simple chemical intuition or even extensive <u>ab initio</u> calcu-

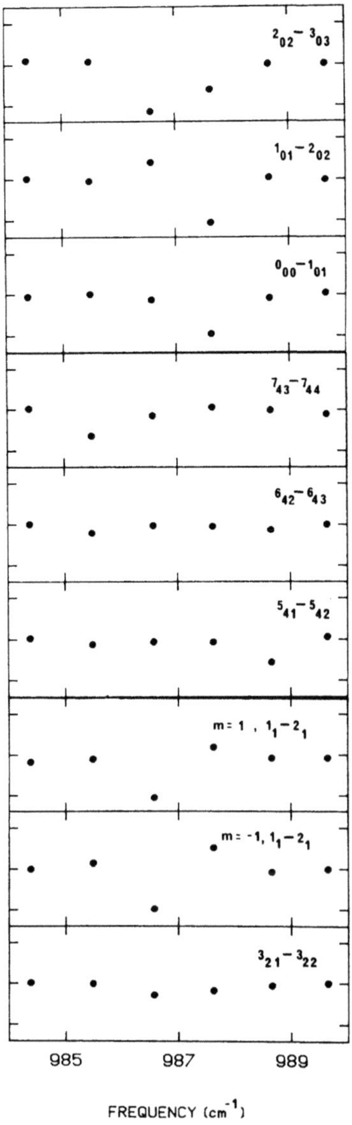

Figure 5. Microwave-infrared double resonance study of NH_3-CO_2. The microwave transitions which were monitored are indicated. The three tick marks along the y-axis in each figure indicate 100%, 0%, and -100% change in the microwave signal strength. For example, at 985.49 cm^{-1} the 6_{42} - 6_{43} transition is detuned by 21(6)%. All the transitions are in the ground internal rotor state except where otherwise noted.

lations. Both HOMO-LUMO arguments and simple Lewis Acid-Base theory predict proton donation by the NH_3 to form a hydrogen-bonded complex. However, attempts to synthesize such a complex have been unsuccessful to date. The NH_3 dimer was expected to be the classic example of NH_3 proton donation as predicted by ab initio calculations. This structure is clearly contradicted by experimental evidence and demonstrates the difficulty of understanding NH_3 van der Waals chemistry. NH_3 systems often display dynamics which are just as complex and unpredictable as their structures. This is a direct consequence of the large rotational constants and of the nuclear spin functions of free NH_3 which prevent the intermolecular potential from completely quenching the NH_3 rotations. The facile NH_3 inversion can add further complexity. Such large amplitude motions are seen in Ar-NH_3, NH_3-CO_2, NH_3-CO, and $(NH_3)_2$.

H_2O, unlike NH_3, is quite amphoteric in nature. In this regard it more closely resembles HCN than either NH_3 or HF in terms of the structures of its van der Waals complexes. In fact, HCN and H_2O almost always agree in their choice of whether or not to hydrogen bond with a particular binding partner. The dynamics of H_2O complexes, however, are more difficuilt to predict. H_2O complexes, like those of NH_3 do exhibit complicated internal motions. These motions include internal rotation (H_2O-CO_2), inversion (H_2O-HF), and possibly exchange tunnelling (H_2O dimer).

In Tables 1-4, we compare various complexes of H_2O, NH_3, HF, and HCN. Examination of these tables shows the large amount of data that has been obtained on these systems. Some interesting observations about H_2O and NH_3 complexes should be noted. Those complexes in which both NH_3 and H_2O behave as Lewis bases (i.e. with CO_2, HCCH, HCN, and HF) display interesting trends. Binding strength, as inferred from the weak bond force constant and bond length, increases in the order HF, HCN, HCCH, CO_2 for both NH_3 and H_2O complexes. Interestingly, NH_3-CO_2 and NH_3-HCCH have identical weak bond-stretching force constants. With the same Lewis acid binding partner, the force constants and bond lengths are similar;

though, for the H_2O complexes they are both somewhat smaller. This suggests that the bonding strength of the water complexes should be similar to that of the corresponding NH_3 complexes. Hence, it is likely that the dissociation energies, D_o, of H_2O-CO_2 and H_2O-HCCH are less than 2.8 kcal/mole.

In general, ab initio calculations have been unable to quantitatively account for these surprisingly low bond-dissociation energies. Of course, calculating these energies is one of the most difficult challenges for ab initio theory. Theoretical predictions of ground state structures, on the other hand, have been more successful. Due to the large dimensionality of the problem, theoretical predictions have been aided in part by the use of chemical intuition to choose the region of the potential surface where the absolute energy minimum is likely to lie. As the complexity of the binding partners increase, chemical intuition is of less use. This leads to an effective increase in the dimensionality of the problem. Predicting the structure of these complex van der Waals molecules will continue to be a challenge for ab initio theories. The ability to place upper bounds to the dissociation energies of weakly bound complexes by spectroscopic means provides a further, most important connection between experiment and theory. It is likely that in the near future a variety of spectroscopic studies will provide relatively precise dissociation energies for these species.

References:

1. F. Franks, "Water, A Comprehensive Treatise," Vol. 4-6, (Plenum, New York, 1972).
2. R. Taylor and O. Kennard, Acc. Chem. Res. 17, 320 (1984).
3. G.C. Pimentel and A.D. McClellan, "The Hydrogen Bond," (Freeman, San Francisco, 1960).
4. T.C. English and J.C. Zorn, "Methods of Experimental Physics," Vol. 3 (2nd Edition), D. Williams, ed. (Academic Press, Inc., New York, 1972).
5. T.J. Balle, E.J. Campbell, M.R. Keenam and W.H. Flygare, J. Chem. Phys. 71, 2723 (1979).
6. Z. Kisiel, A.C. Legon and D.J. Millen, Proc. R. Soc. London A 381, 419 (1982).

7. G.T. Fraser, D.D. Nelson, Jr., A. Charo and W. Klemperer, J. Chem. Phys., to be published.
8. S.A. Clough, Y. Beers, G.P. Klein and L.S. Rothman, J. Chem. Phys. 59, 2254 (1973).
9. M.D. Marshall and J.S. Muenter, J. Mol. Spect. 85, 322 (1981).
10. J.S. Muenter and W. Klemperer, J. Chem. Phys. 52, 6033 (1970).
11. J.S. Muenter, J. Mol. Spec. 55, 490 (1975).
12. W.L. Ebenstein and J.S. Muenter, J. Chem. Phys. 80, 3989 (1984).
13. T.R. Dyke, K.M. Mack and J.S. Muenter, J. Chem. Phys. 66, 498 (1977).
14. K.I. Peterson and W. Klemperer, J. Chem. Phys. 81, 3842 (1984).
15. T.R. Dyke, Top. Current Chem. 120, 85 (Spring-Verlag, N.Y., 1984).
16. T.A. Fisher, K.I. Peterson and W. Klemperer, work in progress.
17. K.I. Peterson and W. Klemperer, J. Chem. Phys., manuscript in preparation.
18. P.D. Aldrich, A.C. Legon and W.H. Flygare, J. Chem. Phys. 75, 2126 (1981).
19. A.J. Fillery-Travis, A.C. Legon and L.C. Willoughby, Chem. Phys. Lett. 98, 369 (1983).
20. C.R. Quade, J. Chem. Phys. 47, 1073 (1967).
21. M.J. Frisch, J.A. Pople and J.E. Del Bene, J. Chem. Phys. 78, 4063 (1983).
22. G.T. Fraser, K.R. Leopold and W. Klemperer, J. Chem. Phys. 80, 1423 (1984).
23. G.T. Fraser, K.R. Leopold, D.D. Nelson, Jr., A. Tung and W. Klemperer, J. Chem. Phys. 80, 3073 (1984).
24. B.J. Howard, private communication.
25. G.T. Fraser, K.R. Leopold and W. Klemperer, J. Chem. Phys. 81, 2577 (1984).
26. G.T. Fraser, D.D. Nelson, Jr. and W. Klemperer, work in progress.
27. Z. Latajka and S. Scheiner, J. Chem. Phys. 81, 407 (1984) and references therein.
28. R.L. DeLeon and J.S. Meunter, J. Chem. Phys. 72, 6020 (1980).
29. The 10 K rotational temperature of the van der Waals dimers is suggested by previous microwave studies of weakly bound complexes. See, for example, K.R. Leopold, G.T. Fraser and W. Klemperer, J. Chem. Phys. 80, 1039 (1984).

VIBRATIONAL AND ROTATIONAL TRANSITIONS OF HYDROGEN BONDED COMPLEXES FROM THEORY AND EXPERIMENT

Clifford E. Dykstra and James M. Lisy
Department of Chemistry
University of Illinois
Urbana, Illinois 61801

ABSTRACT. Exploiting the interface between spectroscopy and *ab initio* investigation can be of real benefit in understanding the subtle properties of hydrogen bonded complexes. However, what theory and experiment determine most directly are not precisely corresponding values, and so the differences are important in a detailed comparison. This point is considered here for certain of the primary questions that arise in studying weak complexes, such as bond strengths, equilibrium structures, vibrational frequencies and interconversions. Particular reference is made to the hydrogen fluoride dimer where *ab initio* calculations and experiment have been able to provide spectroscopic parameters to high accuracy.

1. INTRODUCTION

Comparison of *ab initio* predictions and spectroscopic parameters of molecules at the finest level of detail is the ultimate test of the electronic structure theory that yielded the potential energy surface. However, such a comparison implicitly tests dynamical approximations made on both sides of the comparison. In the case of what we term a hard molecule -- one whose force constants are large enough to preclude large-amplitude oscillation in the ground vibrational state -- agreement between theory and spectroscopy can be excellent by most standards. But for floppy molecules, the situation is different.

In the case of the hard molecule $HCNH^+$, the first spectrosco-

R. J. Bartlett (ed.), Comparison of Ab Initio Quantum Chemistry with Experiment for Small Molecules, 245–266.

pic information was recently obtained from the experiments of Oka and coworkers [1]. The CH stretch vibrational band was resolved in this work and the band center frequency of 3188 cm^{-1} agreed nicely with the _ab initio_ value of 3201 cm^{-1} obtained by Lee and Schaefer [2] prior to the experimental work. In addition, the measured spectrum provided even more dramatic agreement with a much earlier spectroscopic prediction of the $J = 1 \rightarrow 0$ transition frequency. The 1980 _ab initio_ study of Dardi and Dykstra [3] predicted that this transition would occur at 74.07 GHz with a roughly estimated accuracy of $\pm$ 0.15 GHz. The experimental value that was obtained was 74.11 GHz which is within 0.05% of the prediction. Several other _ab initio_ calculations [4-6] with differences in approaches have been within 0.2 to 1.0% as well. Most recently, spectroscopic study of a vibrational band of $DCNH^+$ by Amano [7] has led to a value of 30.79 GHz for the B_0 rotational constant; Dardi and Dykstra's prediction [3] was 30.77 GHz.

The success of _ab initio_ calculations in predicting these vibrational and rotational constants is of course related to the achieved sophistication in electronic structure theory. Conversely, the fact that the predictions could be tested for such elusive species testifies to the sophistication achieved in contemporary molecular spectroscopy. But the comparisons themselves are meaningful because the species are hard molecules; quite simply, the nuclear motion complications are minimized for a system at the bottom of a steep potential well. Thus, _harmonic_ vibrational frequencies from theory could be compared with measured _transition_ frequencies, while for the $J = 1 \rightarrow 0$ rotational transition, the theoretical effort consists of just finding the equilibrium structure and thus the rotational constant since vibrational averaging effects are small. High accuracy in the equilibrium structural parameters, though, is essential and Dardi and Dykstra [3] expected their equilibrium bond lengths to be accurate to better than 0.002 Angstroms.

At the fine level of detail, one must keep in mind that even the concept of an equilibrium structure of a molecule is something that is approached directly only with explicit knowledge of a potential energy surface. Structures based on observed transition frequencies are vibrationally averaged so there is an inherent mismatch if theory and experiment are to be compared with no further analysis. For floppy molecules, the mismatch can be exacerbated because of the large amplitude motions. Hydrogen bonded species are quite often in the category of floppy molecules since most hydrogen bond strengths are only a few kcal [8]. These complexes can undergo large amplitude motion about the hydrogen bond even in the ground vibrational state. Consequently, an effective comparison of spectroscopic and theoretical results for these systems requires a means of extending equilibrium theoretical values to vibrationally-averaged experimental values or vice-versa.

2. VIBRATIONAL SPECTROSCOPY

The vibrational spectroscopy of hydrogen-bonded systems can be separated into intermolecular and intramolecular vibrations. The intramolecular vibrations, i.e. within the molecular units, have been observed by long path length absorption [9-13], matrix isolation [14] and vibrational predissociation [15,16] spectroscopy. Intermolecular vibrations have been observed in long-path absorption [12,13], and matrix isolation [14] spectra. Also, the gas-cell microwave spectra of hydrogen-bonded clusters contain vibrational satellites [17] corresponding to excited intermolecular vibrational states. The vibrational transition frequencies can be compared directly with theoretically determined, isolated molecule transition frequencies, while for matrix isolation spectra, the matrix perturbation of the intramolecular and intermolecular force fields must be considered. The comparison between theory and experiment for the vibrational transition frequencies is a key test of accuracy from the experimental perspective. Table 1 contains the vibrational frequencies obtained spectroscopically and theoretically for a number of hydrogen-bonded systems.

The complete theoretical determination of intermolecular and

Table 1. Vibrational Transition Frequencies of Certain Hydrogen Bonded Complexes

		Experiment		
Complex	Vibration[a]	Method[b]	Frequency	Theory
HCN-HF	ν_{HF}	LP	3710.5(2)[c]	3990[h]
	ν_{HF}	LP	3710.20(2)[d]	
	ν_{B}	LP	555(3)[c]	
	ν_{β}	LP	70(24)[c]	
	ν_{β}	LP	84(19)[d]	86[h]
	ν_{β}	MRI	91(20)[e]	
	ν_{σ}	LP	197(15)[e]	193[h]
	ν_{σ}	MRI	155(10)[c]	160[i]
OC-HF	ν_{HF}	LP	3844.029(5)[f]	
	ν_{β}	LP	75(12)[f]	
HF(1)-HF(2)	$\nu_{HF(1)}$	LP	3930.68[g]	3963[j]
	$\nu_{HF(2)}$	LP	3867.86[g]	3898[j]
	ν_{σ}			157[j]
DF(1)-DF(2)	$\nu_{DF(1)}$	LP	2882(1)[g]	2908[j]
	$\nu_{DF(2)}$	LP	2834.58[g]	2868[j]
HF-DF	ν_{σ}			158[j]

[a] ν_B - high-frequency hydrogen-bond bend; ν_β - low frequency doubly degenerate hydrogen bond bend; ν_σ - hydrogen bond stretch.
[b] LP - long path absorption; MRI - microwave relative intensities.
[c] Ref. 12.
[d] Ref. 13.
[e] Ref. 17.
[f] Ref. 11.
[g] Ref. 10.
[h] Ref. 20.
[i] Ref. 18. (160 cm^{-1} is twice the zero point energy and not a calculated transition frequency.)
[j] Ref. 19.

intramolecular vibrational transition frequencies requires that one treat a nuclear motion Schrödinger equation, and have in hand a potential energy surface rather than just harmonic force constants at the equilibrium. That potential energy surface must be properly shaped; the anharmonicity must be well-described by the level of treatment employed to generate the surface. As discussed more fully in the next section, the incorporation of electron correlation effects is very important in achieving an accurately shaped surface.

The theoretically determined vibrational transition frequencies for $(HF)_2$ and $(DF)_2$ in Table 1 were found from a pseudo-diatomic model with a Numerov-Cooley numerical vibrational analysis of the manifold of vibrational state wavefunctions [21]. Hydrogen bond stretching is well described in this way because physically it is the weak motion between two rigid molecules that are the "atoms" of the pseudo-diatomic. From a perturbative standpoint, however, the procedure can be applied as well to the intramolecular stretches because the intramolecular force constants are much greater than the hydrogen bond force constant, suggesting little coupling. Nonetheless, in carrying out this diatomic analysis of intramolecular stretching [19] one-dimensional potential curves were obtained that could fully account for that part of the coupling which may be associated with the potential. These curves were generated as a function of one or the other intramolecular coordinate by relaxing all other coordinates to minimize the energy at each point along the curve. For even large changes (0.3 Angstroms) in the intramolecular coordinates, the changes in the other structural parameters amounted to thousandths of Angstroms and tenths of degrees. The coupling through the potential surface is small, therefore, and in this sense, a separated or local mode treatment of the intramolecular stretching is justified. We have also approximately incorporated dynamical coupling, but again the effects are small, amounting to a few cm^{-1} [19]. The agreement between the theoretical and measured vibrational frequencies of

$(HF)_2$ and $(DF)_2$, as shown in Table 1, is about 1%. When the shifts in the local mode frequencies from the isolated monomer are compared, the agreement is extremely good, 31 (theory) vs. 32 cm^{-1} (expt.) for the free HF and 88 vs. 93 cm^{-1} for the bonded HF [19].

3. HYDROGEN BOND STRENGTHS

The *ab initio* determination of a hydrogen bond strength differs in certain respects from the determination of the strength of a conventional chemical bond. At the outset one anticipates that a molecular orbital picture or SCF wavefunction dominates throughout the entire hydrogen bond potential curve; there is no orbital occupancy rearrangement of the type often seen in breaking a chemical bond. Undoing this apparent simplification of the problem, however, is the unavoidable importance of intermolecular electron correlation effects which contribute significantly to the bond strength. This can be shown by comparing SCF bond strengths with experimental values, or by comparing SCF with well-correlated theoretical values provided there has been careful consideration given to basis set effects. Table 2 gives illustrative results for the N_2-HF complex. An *ab initio* prediction of the dissociation energy is achieved by showing convergence with respect to basis set effects in the SCF and correlated values of D_e. The double-zeta (DZ) basis is seen to be seriously in error because of how it disagrees with the larger basis set results: it exaggerates the bond strength. This, in turn, is associated with a basis set superposition effect as tested by a counterpoise or ghost orbital calculation [22]. Part of the deficiency is the lack of polarization functions (P), but part is also the need for greater valence flexibility as is shown with the use of a triple-zeta (TZ) valence set. In the end one sees that correlation effects contribute significantly to the intermolecular attraction. For NN-HF, correlation is responsible for over 450 cm^{-1} of the bond strength at the DZP level. This is reduced to a still significant 250 cm^{-1} with the more appropriate TZP basis. Table 3 lists correlation contributions to the bond strengths of other systems and shows that a range of effects is possible.

Given the potential role of electron correlation effects in hydrogen bonding, its incorporation into the ab initio treatment must be fairly complete. In particular, a size-extensive wavefunction is quite important in order to properly compare the complex's energy to the energies of the separated products. Furthermore, higher-order electron correlation effects, those that go beyond pair correlations, may afford useful refinements. Size-extensivity and higher-order electron correlation can both be built into a wavefunction with coupled cluster (CC) approaches [23-29]. In CC, higher-order correlating configurations are included in the wavefunction but with expansion coefficients given by products of the expansion coefficients of the lower-order substitutions. Thus, CC with double substitutions, CCD, includes not only the primary pair substitutions (the "doubles"); it also includes all quadruple substitutions, hextuple substitutions and so on. Benzel and Dykstra [18] used a CC approach, designated ACCD, for the N_2-HF results in Table 2 and for certain of the results in Table 3. ACCD takes advantage of an approximate near cancellation of matrix element terms in CCD [30-33]. A matrix-oriented method [34] was employed for the CC calculations and the similarly-oriented self-consistent electron pairs (SCEP) method [35-37] was used to obtain CI wavefunctions with pair substitutions (no higher-order effects). The higher-order/size-extensivity effects that distinguish the ACCD from SCEP results in Table 3 show the value of using the more sophisticated of the correlated wavefunctions. Otherwise, an often good-sized fraction of the correlation contribution to the bond strength can be missed. The ultimate check, though, is not the extensiveness of the treatment but the comparison with measured values.

The spectroscopic determination of the hydrogen bond strength is somewhat difficult, since in general the dissociation path is not related to a single vibrational coordinate of the complex. A standard approximation is to assume that the hydrogen bond stretching coordinate matches the dissociation coordinate. In effect, one

Table 2. Calculational Level Dependence in the Dissociation Energy of N_2-HF.[a]

	Dissociation Energy (cm^{-1})	
Basis Set	SCF	With Electron Correlation[b]
DZ	1460	
DZ(G)[c]	896	
DZP	737	1188
DZP(G)[c]	493	
TZP	540	795

[a] Ref. 18.
[b] ACCD calculations (see text).
[c] Counterpoise or ghost orbital corrected results (Ref. 22).

Table 3. Electron Correlation Energy Contributions to Hydrogen Bond Dissociation Energies, D_e (cm^{-1}).

	Total D_e	Pair Correlation Contributions (from SCEP)	Higher Order Correlation Contributions (from ACCD)
NN-HF[a]	795	200	55
OC-HF[a]	1066	307	11
CO-HF[b]	603	-57	-49
HCCH-HF[c]	1319	226	-8
HF-HF[d]	1593	120	-138
HCN-HF[a]	2397	190	-15

[a] Ref. 18.
[b] Ref. 42.
[c] Ref. 43.
[d] Ref. 19. The size-extensivity error of a CIDS wavefunction, calculated as the sum of the monomer's energies less the energy of a composite, supersystem of the two non-interacting monomers, is actually much greater than even the total D_e value for the HF-dimer. The pair correlations given in this table are all based on supersystem calculations

treats the hydrogen bond as a radial interaction and fixes the relative orientation of molecular subunits to the vibrationally-averaged ground state geometry. A standard form for the potential (e.g. Lennard-Jones, Morse, etc.) is used to fit the spectoscopic information to the potential parameters. The dissociation energy for the radial potential is the approximate hydrogen bond strength.

In molecular beam microwave spectroscopy, precise measurements of the rotational transitions have led to extremely accurate determinations of the rotational constant $\overline{B}_o = 1/2(B_o+C_o)$ and the centrifugal distortion constant, D_J, for a large number of binary molecular complexes. By assuming that the stretching of the intermolecular hydrogen bond is responsible for the magnitude of D_J, and invoking the pseudo-diatomic approximation [38,39], a value for the harmonic stretching frequency, ν_s, of the hydrogen bond can be determined by the relation:

$$D_J = 4\overline{B}_o^3/\nu_\sigma^2 \qquad (1)$$

where

$$\nu_\sigma^2 = (2\pi)^{-1}\ (k_\sigma/\mu_{PD}) \quad \text{and} \quad \mu_{PD} = M_B M_{HX}/M_B + M_{HX} \qquad (2)$$

for the hydrogen-bonded complex B•••HX. Using a Lennard-Jones 6-12 potential for the radial interaction:

$$U(r) = \varepsilon[(r_e/r)^{12} - 2(r_e/r)^6] \qquad (3)$$

the force constant, k_σ, can be related to the potential parameters:

$$k_\sigma = 36\varepsilon/r_e^{\ 2} \qquad (4)$$

An excellent review of this procedure has been recently published by Legon [40]. A note of caution must be added in that this approximation has the effect of averaging in the zero point energy of the hydrogen bond bending vibrations. Thus the dissociation energy from the pseudo-diatomic analysis will be a lower limit to the true hydrogen bond equilibrium dissociation energy.

Hydrogen bond dissociation energies can also be obtained from conventional gas-phase microwave spectra of gas mixtures, by absolute intensity measurements of the components in an equilibrium mixture of B, HA and B•••HA [16,41]. Analysis of the temperature dependence yields the zero-point dissociation energy, D_o. When combined with the zero-point vibrational energies due to the intermolecular modes, a value for the equilibrium dissociation energy can be obtained. A comparison of available experimental and theoretical results for a number of binary complexes involving HF is summarized in Table 4.

Table 4. Comparison of Hydrogen Bond Strengths

Complex	Dissociation energy, D_o (cm^{-1})		
	Experiment/Method[a]		Theory
HCN-HF	2070[b]	MB	
HCN-HF	2180[c]	ABS	2307[f]
OC-HF	987[d]	MB	1007[f]
NN-HF	618[e]	MB	742[f]
HF-HF			1297[g]

[a] MB - molecular beam; ABS - absolute intensity.
[b] L.W. Buxton, E.J. Campbell, M.R. Keenan, A.C. Legon, W.H. Flygare, unpublished results.
[c] Ref. 17.
[d] Ref. 44.
[e] Ref. 45.
[f] Ref. 18. Values corrected only for zero-point energy of ν .
[g] Ref. 19. Values corrected only for zero-point energies of all but out-of-plane torsional vibrations. Zero-point energies for in-plane bends used normal mode frequencies.

4. EQUILIBRIUM STRUCTURES

An equilibrium structure of a molecule is uniquely determined by a global minimum on a potential energy surface. *Ab initio* electronic structure theory provides a means of generating such a surface and its prediction of an equilibrium structure is therefore direct. The accuracy of the predicted structure, though, depends on the level of treatment used to generate the surface. The factors discussed in the last section for calculating the bond strength are just as important in calculating the hydrogen bond length because both demand an accurate potential energy curve along the hydrogen bond coordinate. The parameters associated with the monomer's structures typically show small differences from their values in the isolated species, but this difference is an important feature of hydrogen bonding. Thus, the calculational level employed must be sufficient to accurately describe the bonds of hard molecules, too. The essential requirement here is a balanced incorporation of electron correlation effects, which elongate bond length predictions, and extension of the basis set, which tends to shorten bonds [46]. Table 4 lists bond length predictions obtained at roughly comparable levels for a number of systems.

The development of an equilibrium structure for a hydrogen-bonded complex from spectroscopic data involves a number of approximations. The initial assumption is that the geometries of B and HA are essentially unchanged by complex formation, and so far this has been borne out by calculations. The pseudo-diatomic approximation of treating the molecular units as point masses is assumed and the distance, r, between the centers of mass of the molecular units is used as the intermolecular coordinate. Changes in the rotational spectra due to isotopic substitution, centrifugal distortion or vibrational excitation of the hydrogen-bond stretch can be analyzed to give the equilibrium intermolecular distance, r_e. However, it is a requisite of such analysis that a model intermolecular potential for the radial interaction be selected, and this can influence the determination of r_e.

Two different methods using a Lennard-Jones potential have been developed. Using the vibrationally-averaged intermolecular distance, r_o, and the harmonic stretching frequency, ν_σ, within the pseudo-diatomic approximation, the potential parameters, ε and r_e, are adjusted [38] to reproduce ν_σ and r_o. The best fit yields the value for r_e. Alternatively the equilibrium rotational constant, B_e, can be obtained by using the vibration-rotation interaction constant, $\alpha_e = 36(B_e^2/\nu_\sigma)$, in the pseudo-diatomic approximation by solving the quadratic [39]:

$$B_o = B_e - 1/2\alpha_e = B_e - 18(B_e^2/\nu_\sigma) \tag{5}$$

and then

$$r_e = (8\pi^2 \mu_{PD} B_e/h)^{1/2} \tag{6}$$

If sufficient isotopic information is available, one can fit mass-reduced structural constants:

$$B_e' = B_e \mu_{PD} \qquad D_e' = D_e \mu_{PD}^2 \qquad \alpha_e' = \alpha_e \mu_{PD}^{3/2} \tag{7}$$

to the following transition frequency expression [47]:

$$\frac{E_{J+1\leftarrow J}}{h} = 2\overline{B}_e' \left(\frac{J+1}{\mu_{PD}}\right) - 4D_e' \frac{(J+1)^3}{\mu_{PD}^2} - \alpha_e' \frac{(J+1)}{\mu_{PD}^{3/2}} \tag{8}$$

These structural constants can then be directly related to Morse potential parameters. While the above technique has yet to be applied to hydrogen-bonded systems, it does lead [47] to shorter values for r_e, and lower binding energies in ArHF than obtained using earlier treatments with the Lennard-Jones potential and centrifugal distortion [38,48]. Table 5 contains a set of experimental equilibrium distances for hydrogen-bonded complexes involving HF obtained with these approaches. The agreement between theory and experiment is good, for the most part. However, there is enough disagreement to leave uncertainty about lingering errors in the _ab initio_ surfaces as well as in the analysis of the spectra

Table 5. Equilibrium Bond Lengths (Å) of Hydrogen Bonded Complexes.

		Intermolecular Separation	
		Expt.	Theory
HF-HF	(F-F)	2.72[a]	2.7675[b]
HCCH-HF	(F-|||)	3.121[c]	3.258[d]
HCN-HF	(N-F)	2.802[e]	2.829[f]
OC-HF	(C-F)	3.176[g]	3.080[f]
NN-HF	(N-F)	3.031[f]	3.072[f]
HCℓ-HCℓ		3.797[j]	3.81[i]

[a] Ref. 49.
[b] Ref. 19. The calculated changes in the monomer bond lengths were $\delta r_1 = +0.0024$ and $\delta r_2 = +0.0037$.
[c] Ref. 58. Distance is vibrationally averaged, not equilibrium.
[d] Ref. 55.
[e] L.W. Buxton, E.J. Campbell, M.R. Keenan, A.C. Legon, W.H. Flygare, unpublished results.
[f] Ref. 18. H-||| was reported to be 2.341 Å.
[g] Ref. 44.
[h] Ref. 45.
[i] Ref. 59.
[j] Ref. 60. Vibrationally averaged center-of-mass separation.

Table 6. Rotational Constants of $(HF)_2$.[a]

	Expt.[b]	Theory[c]
HF-HF	6487	6487
HF-DF	6475	6500
DF-DF	6177	6252

[a] Values are (B+C)/2 and are in MHz.
[b] Ref. 49.
[c] Ref. 19. Vibrational averaging of only the intermolecular stretching was included.

and its effective averaging of the zero point motion for the hydrogen-bond bending vibrations.

In an attempt to provide a more meaningful check of theoretical structure parameters, a more direct comparison has been made in the case of the HF-dimer [19]. Here theory has been carried to the point of evaluating $\overline{B}_0$, and results are given in Table 6. The theoretical rotational constant was obtained by vibrationally averaging the intermolecular separation using the one-dimensional Numerov-Cooley method [21]. For $(HF)_2$, the averaged F-F distance obtained is 2.789 Å which is 0.021 Å longer than the equilibrium distance. The corresponding value obtained from microwave spectra [49] is 2.791 Å. As shown in Table 6, the calculated rotational constant agrees nicely with the measured value. Deuterium substituted complexes show good, but not as good agreement between the theoretical and measured values. This is probably due in part to the neglect of bending averaging in the theoretical determination since the difference of the averaged bending angle from the equilibrium angle with the heavier deuteriums will have a more noticeable effect on $\overline{B}_0$. At the same time, the theoretical study serves to show unequivocably that the difference between isotopes for the vibrationally averaged F-F distance are just too small to be important.

5. TRANSITION INTENSITIES

Vibrational transition enhancements for intramolecular stretches are often found in hydrogen bonded systems. They can arise in two ways, a change in the potential and a change in the dipole moment function upon hydrogen bond formation. The HF-dimer is one of the few cases where this has been studied [19] and the results are interesting. As shown by the values given in Table 7, the transition moment for the bonded HF's stretch increases sizably over the monomer. The transition moment for the other HF stretch increases, but less. The overtone transition moments decrease relative to the monomer, and again the greatest effect is for the bonded HF.

Table 7. Absolute Vibrational Transition Moments[a] of $(HF)_2$.

	v=0→1	v=1→2	v=0→2
HF monomer	0.1006	0.1408	-0.0131
HF monomer (expt.)[b]	0.0985	0.138	-0.0127
Free HF stretch	0.1108	0.1585	-0.0108
Bonded HF stretch	0.1925	0.2949	-0.0033
DF monomer	0.0859	0.1207	-0.0094
Free DF stretch	0.0940	0.1341	-0.0078
Bonded DF stretch	0.1604	0.2402	-0.0025
$(HF)_2$ intermolecular	-0.0560	0.0736	0.0150

[a] These values are in Debyes and are calculated as $\langle v'|\mu|v''\rangle$.
[b] Ref. 51.

Sandorfy has discussed the effect of a changing dipole moment function on transition intensities and how this can be important in hydrogen bonded systems [50]. Since the dipole moment function of a molecule like HF in the region of the equilibrium bond length is close to a linear function [51-53], the important part of the transition moment is that arising from a linear component followed next by the part arising from the quadratic component of the dipole moment function. For an anharmonic oscillator, both terms are non-zero for overtone transitions, and according to Sandorfy [50], the typical situation is that these terms are of opposite sign. A steepening of the linear component of the dipole moment function may cancel a greater share of the quadratic (or electrical anharmonicity) part of the transition moment with a resulting loss of overtone transition probability. If this happens, the steepening of the dipole moment function will manifest itself as an increase in the transition moment of the fundamental transition, since that is dominated by the linear part of the dipole moment function. This seems to be what is happening in the HF-dimer according to the results in Table 7 and the steepening of the dipole moment function that is actually calculated (Fig. 1).

Quantitative comparisons of the intramolecular stretching vibration intensities for the fundamental and overtone between the

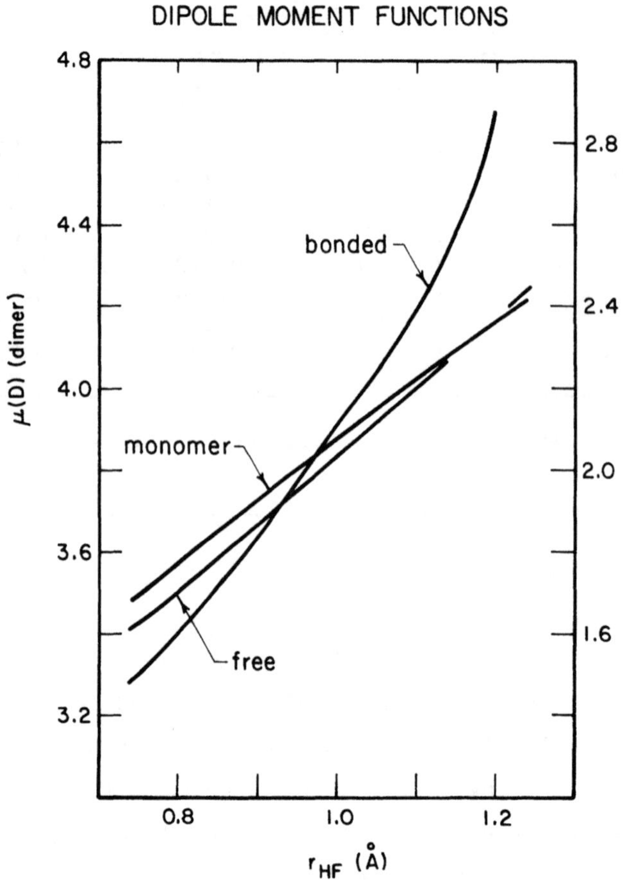

Figure 1. The dipole moment function of HF and one-dimensional cuts along HF corrdinates through the dipole moment surface of $(HF)_2$.

monomer and binary complexes have yet to be determined experimentally in the gas phase. In many cases such as $(HF)_2$, the identification of the dimer vibrational bands has only recently been made. Additional problems such as overlapping bands are also present. However, qualitative observations of enhancement of the fundamental and diminution of the overtone intensities have been made in liquids and solutions [54].

5. INTERCONVERSIONS

The low energy motions of hydrogen bonded complexes and the long-range electrostatic attractions that are a component of the interaction in the more weakly held systems afford the opportunity for facile interconversion. In the HF dimer, motion along a coordinate that for the most part rotates each monomer can smoothly change the bonded hydrogen into the free hydrogen and the vice versa. The result, of course, is an equivalent structure. A more complicated interconversion path is anticipated for the complex of CO with HF [42], and probably for NN with HF, too. As shown in Fig. 2, the orientation of HF and CO at the T-shaped midpoint on the interconversion path requires an extra turn of HF. What dictates this orientation to some extent is the dipole-quadrupole interaction which would be unfavorable if the HF did not undergo the extra 360° rotation. The HCCH-HF complex has features that are similar [55], except the sign of the quadrupole moment of HCCH is opposite to that of CO.

Tunneling splitting associated with interconversion in $(HF)_2$ and $(DF)_2$ are extremely useful experimental data, providing valuable information on the potential barrier to interconversion. Inverting the experimental data and adjusting the barrier height so as to reproduce the observed splittings, Barton and Howard [56] have determined a barrier height of 302 cm^{-1}, for a C_{2h} transition state. Alternatively, by choosing a form for the potential energy barrier and an appropriate interconversion coordinate, e.g. $\rho = \theta_2 = \theta_1$, the barrier height is again adjusted in a simple one-dimensional treatment to yield the observed splittings for each

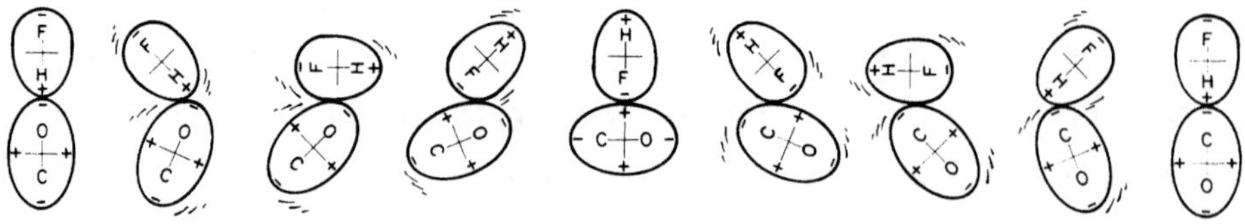

Figure 2. Qualitative representation of the minimum-energy interconversion pathway from CO–HF to OC–HF. This path requires that HF rotate by 540° as seen from the CO molecule in order to have an energetically favorable dipole-quadrupole interaction at the T-shaped structure.

particular vibrational state [57,10]. All of these treatments are highly dependent on the form of the potential energy barrier. A calculation [19] of the barrier, neglecting all zero-point corrections, yields a value of 379 cm^{-1}.

7. IDEAL COMPARISON AND CONCLUSIONS

From looking at the floppy molecules that arise from hydrogen bond formation, it appears that the ideal comparison of theory and experiment -- the best way to check the validity of increasingly sophisticated levels of electronic structure calculations -- is to continue the application of theory to the point where it calculates what experiment measures directly. The reason for this is simple: The interface between theory and experiment is the potential energy surface. Good quality calculations can be applied to generate whole potential surfaces whereas mapping out a surface from spectroscopic data is a very indirect task.

What is seen here as the ideal means for comparison of calculational results with real measurement is being approached in steps as the treatment of the nuclear motion problem becomes more complete. Because of the nature of the HF-dimer as well as certain other complexes, certain practical conclusions can already be made regarding the levels of electronic structure theory that achieve high accuracy in their description of these molecules. These conclusions are that very well correlated, size-extensive-wavefunction treatments using polarized triple-zeta (or better) basis sets can provide meaningful structural, vibrational and energetic predictions for hydrogen-bonded complexes. At the same time, seemingly small steps to lower-level calculations can lead to seriously different hydrogen bond lengths, exaggerated bond strengths, and inaccurate vibrational frequencies.

ACKNOWLEDGEMENTS

Several graduate students at the University of Illinois are responsible for many of the calculations discussed here and for analysis work at the theory/spectroscopy interface. We therefore wish to thank Mr. Kirk Kolenbrander, Mr. Daniel Michael, Ms. Shi-yi Liu, Dr. Mark A. Benzel and Dr. William G. Read. We also thank the National Science Foundation for grants to J.M.L. and C.E.D. that have supported our work in this area.

References:

1. R.S. Altman, W.M. Crofton and T. Oka, J. Chem. Phys. 80, 3911 (1984).
2. T.J. Lee and H.F. Schaefer, J. Chem. Phys. 80, 2977 (1984).
3. P.S. Dardi and C.E. Dykstra, Ap. J. Lett. 240, 171 (1980).
4. P.K. Pearson and H.F. Schaefer, Ap. J. 192, 33 (1974).
5. D.J. DeFrees, G.H. Loew and A.D. McLean, Ap. J. 257, 376 (1982).
6. D.J. DeFrees, J.S. Binkley and A.D. McLean, J. Chem. Phys. 80, 3720 (1984).
7. T. Amano, J. Chem. Phys. 81, 3350 (1984).
8. M.D. Joesten and L. J. Schaad, "Hydrogen Bonding" (Marcel Dekker, New York, 1974).
9. A.S. Pine and W.J. Lafferty, J. Chem. Phys. 78, 2154 (1983).
10. A.S. Pine, W.J. Lafferty and B.J. Howard, J. Chem. Phys. 81, 2939 (1984).
11. E.K. Kyrö, P. Shoja-Chaghervand, K. McMillan, M. Eliades, D. Danzeiser, and J.W. Bevan, J. Chem. Phys. 79, 78 (1983).
12. R.K. Thomas, Proc. R. Soc. London A325, 133 (1971).
13. E. Kyrö, R. Warren, K. McMillan, M. Eliades, D. Danzeiser, P. Shoja-Chaghervand, S.G. Lieb and J.W. Bevan, J. Chem. Phys. 78, 5881 (1983).
14. L. Andrews, J. Phys. Chem. 88, 2940 (1984).
15. J.M. Lisy, A. Tramer, M.F. Vernon and Y.T. Lee, J. Chem. Phys. 75, 4733 (1981).
16. M.P. Cassasa, C.M. Western, F.G. Celii, D.E. Brinza and K.C. Janda, J. Chem. Phys. 79, 3227 (1983).
17. A.C. Legon, D.J. Millen and S.C. Rogers, Proc. R. Soc. Lond. A370, 213 (1980).
18. M.A. Benzel and C.E. Dykstra, J. Chem. Phys. 76, 1602 (1982); 78, 4052 (1983); 80, 3510 E (1984).
19. D.W. Michael, C.E. Dykstra and J.M. Lisy, J. Chem. Phys. 81, 5998 (1984).
20. L.A. Curtiss and J.A. Pople, J. Molec. Spect. 48, 413 (1973).
21. J.W. Cooley, Math. Comp. 15, 363 (1961).
22. S.F. Boys and F. Bernardi, Mol. Phys. 19, 553 (1970).
23. J. Cizek, J. Chem. Phys. 45, 4256 (1966); Adv. Chem. Phys. 14, 35 (1960).
24. J. Cizek, J. Paldus and L. Sroubkova, Int. J. Quantum Chem. 3, 149 (1969).
25. A.C. Hurley, "Electron Correlation in Small Molecules" (Academic Press, New York, 1976).

26. J.A. Pople, R. Krishnan, H.B. Schlegel and J.S. Binkley, Int. J. Quantum Chem. 14, 545 (1978).
27. J. Paldus, J. Chem. Phys. 67, 303 (1977).
28. R.J. Bartlett and G.D. Purvis, Int. J. Quantum Chem. 14, 561 (1978); Physica Scripta 21, 255 (1980).
29. R.J. Bartlett, C.E. Dykstra and J. Paldus, in "Advanced Theories and Computational Approaches to the Electronic Structure of Molecules," ed. C.E. Dykstra (Reidel, Dordrecht, Holland, 1984).
30. R.A. Chiles and C.E. Dykstra, Chem. Phys. Lett. 80, 69 (1981); S.M. Bachrach, R.A. Chiles and C.E. Dykstra, J. Chem. Phys. 75, 2270 (1981).
31. K. Jankowski and J. Paldus, Int. J. Quantum Chem. 18, 1243 (1980).
32. J. Paldus, J. Cizek and M. Takahashi, submitted.
33. J. Paldus, M. Takahashi and R.W.H. Cho, submitted.
34. R.A. Chiles and C.E. Dykstra, J. Chem. Phys. 74, 4544 (1981).
35. W. Meyer, J. Chem. Phys. 64, 2901 (1976).
36. C.E. Dykstra, H.F. Schaefer and W. Meyer, J. Chem. Phys. 65, 2740 (1976).
37. W. Meyer, R. Ahlrichs and C.E. Dykstra in "Advanced Theories and Computational Approaches to the Electronic Structure of Molecules," ed. C.E. Dykstra (Reidel, Dordrecht, Holland, 1984).
38. S.J. Harris, S.E. Novick and W. Klemperer, J. Chem. Phys. 60, 3208 (1974).
39. T.J. Balle, E.J. Campbell, M.R. Keenan and W.H. Flygare, J. Chem. Phys. 72, 922 (1980).
40. A.C. Legon, Ann. Rev. Phys. Chem. 34, 275 (1983).
41. R.F. Curl, T. Ikeda, R.S. Williams, S. Leavell and L.H. Scharpen, J. Am. Chem. Soc. 95, 6182 (1973).
42. M.A. Benzel and C.E. Dykstra, Chem. Phys. 80, 273 (1981).
43. S.-Y. Liu and C.E. Dykstra, unpublished results.
44. A.C. Legon, P.D. Soper and W.H. Flygare, J. Chem. Phys. 74, 4944 (1981).
45. P.D. Soper, A.C. Legon, W.G. Read, and W.H. Flygare, J. Chem. Phys. 76, 292 (1982).
46. C.E. Dykstra, Ann. Rev. Phys. Chem. 32, 25 (1981).
47. B.L. Cousins, S.C. O'Brien and J.M. Lisy, J. Phys. Chem. 88, 5142 (1984).
48. M.R. Keenan, L.W. Buxton, E.J. Campbell, A.C. Legon and W.H. Flygare, J. Chem. Phys. 74, 2133 (1981); M.R. Keenan, Ph.D. Thesis, University of Illinois, 1981.
49. B.J. Howard, T.R. Dyke and W. Klemperer, J. Chem. Phys., in press.
50. T. DiPaola, C. Boaurderon and C. Sandorfy, Can. J. Chem. 50, 3161 (1972); C. Sandorfy, Topics Curr. Chem. 120, 41 (1984).
51. R.N. Sileo and T.A. Cool, J. Chem. Phys. 65, 117 (1976).
52. H.-J. Werner and P. Rosmus, J. Chem. Phys. 73, 2319 (1980).
53. P.G. Jasien and C.E. Dykstra, Int. J. Quant. Chem. S17, 289 (1983).

54. W.A.P. Luck in "Water, A Comprehensive Treatise," Vol. 2, ed. F. Franks (Plenum Press, New York, 1973).
55. M.J. Frisch, J.A. Pople and J.E. Del Bene, J. Chem. Phys. 78, 4064 (1983).
56. A.E. Barton and B.J. Howard, Faraday Discuss. Chem. Soc. 73, 45 (1982).
57. I.M. Mills, J. Phys. Chem. 88, 532 (1984).
58. W.G. Read and W.H. Flygare, J. Chem. Phys. 76, 2238 (1982).
59. Chr. Votava, R. Ahlrichs and A. Geiger, J. Chem. Phys. 78, 6841 (1983).
60. N. Ohashi and A.S. Pine, J. Chem. Phys. 81, 73 (1984).

AB INITIO CALCULATIONS OF RADIATIVE TRANSITION PROBABILITIES IN DIATOMIC MOLECULES

Hans-Joachim Werner* and Pavel Rosmus**,
Fachbereich Chemie der Universität Frankfurt/M,
D-6000 Frankfurt, West Germany
*1984-1985: Visiting Fellow, Chemical Laboratory,
University of Cambridge, Cambridge CB2 1EW, England
**1984-1985: Visiting Fellow, JILA and
University of Colorado, Boulder, Colorado 80309

ABSTRACT. Multiconfiguration self consistent field (MCSCF) and multiconfiguration reference configuration interaction (MCSCF-CI) techniques have been employed to calculate electric dipole moment and electronic transition moment functions of diatomic molecules and molecular ions. From these data radiative transition probabilities and lifetimes have been evaluated. It is demonstrated that the results are strongly dependent on electron correlation effects. Using highly correlated MCSCF-CI wavefunctions the radiative lifetimes are obtained with an accuracy of 10-15 percent, which is comparable to the uncertainties of the best experimental values. Often the calculated predictions appear to be more reliable than measured values. In some cases it has been found that empirical transition moment functions, which have been derived from measured relative intensities, exhibit an incorrect variation with the internuclear distance.

1. INTRODUCTION

The knowledge of accurate radiative transition probabilities is of considerable interest for the quantitative interpretation of spectra from atmospheric and space environments, and for the investigation of elementary chemical processes by spectroscopic

R. J. Bartlett (ed.), Comparison of Ab Initio Quantum Chemistry with Experiment for Small Molecules, 267–323.

techniques. For instance, in the reactions summarized in Fig. 1, the product molecules are formed in vibrationally or electronically excited states, and fluorescence from these initially populated states can be observed. Assuming the transition probabilities are known, the populations in the various states may be derived from the measured intensities. Thus, such experiments provide valuable information about reaction mechanisms and kinetics.

The radiative emission probabilities are given by the Einstein coefficients of spontaneous emission. Provided the electronic, vibrational, and rotational motions can be separated, the Einstein coefficient $A_{v's'v''J''}$ for a transition between two rotational levels J', J" in two different vibronic states v', v" takes the form

$$A_{v'J'v''J''} = \text{const.}\ \bar{\nu}^3_{v'J'v''J''} |\langle v'_{J'}|R_e(r)|v''_{J''}\rangle|^2 \times S_{J'J''}/(2J'+1), \tag{1}$$

where $\bar{\nu}_{v'J'v''J''}$ is the energy difference between the initial and final states and $S_{J'J''}$ is the rotational line intensity factor (Hönl-London factor). For a rovibrational transition within one electronic state, $R_e(r)$ is the electric dipole moment function. When the transition occurs between different electronic states, $R_e(r)$ is the electronic transition moment function. $\langle v'_{J'}|$ and $|v''_{J''}\rangle$ are the upper and lower state vibrational wavefunctions derived from effective potential energy functions for the rotating system. Average vibrational transition probabilities can be obtained by summing over all initial and final rotational states, which yields

$$A_{v'v''} = \text{const.}\ \bar{\nu}^3_{v'v''} g |\langle v'|R_e(r)|v''\rangle|^2 . \tag{2}$$

Here, g is a degeneracy factor, which for pure vibrational transition takes the value 1 and for electronic transitions

$HF + HF_2 \rightarrow HF^* + F$	chemical HF laser
$F + H_2 \rightarrow HF^* + H$	
$CS + O \rightarrow CO^* + S$	chemical CO laser
$H + O_3 \rightarrow OH^* + O_2$	atmosphere
$N^+ + O_2 \rightarrow NO^{+*} + O$	
$C^+ + H_2 \rightarrow CH^{+*} + H$	interstellar gases
$C^+ + O_2 \rightarrow CO^{+*} + O$	

Figure 1. Some simple chemiluminescence reactions.

$$g=(2-\delta_{0,\Lambda'+\Lambda''})/(2-\delta_{0,\Lambda'}) \quad . \tag{3}$$

The radiative lifetime of a vibrational level is given by

$$\tau_{v'} = 1/\Sigma_{v''} A_{v'v''} \quad . \tag{4}$$

It is noted that in the literature other definitions of the electronic transition moments have been used, in which part of the degeneracy factor was incorporated. This has led to considerable confusion, and it is therefore recommended to use the definitions given by Whiting et al [1]. Explicit conversion factors between transition moments and various other dynamical variables in diatomic molecules which are consistent with these definitions have been summarized by Larsson [2].

From the above formulae the transition probabilities can be evaluated once the potential energy function(s) and the dipole or transition moment functions are known. Usually only the intensities of a small subset of all possible transitions can be measured. Therefore, it has often been attempted to derive the dipole moment or transition moment functions from a limited database to evaluate the radiative rates for other arbitrary transitions. Generally, the problem with such procedures arises due to the fact that the shape of the functions depend sensitively on relative intensities of very strong and very weak transitions. For instance, in the analysis of rovibrational line intensities the dipole moment function is usually approximated as

$$R_e(r) = M_o + M_1X + M_2X^2 + \ldots \tag{5}$$

with $X=(r-r_e)/r_e$. It is well known, that (to first order in $\gamma=2B_e/\omega_e$) the vibrational matrix elements $\langle v'|R_e(r)|v'+n\rangle$ depend on the expansion coefficients M_i, $i\leq n$ [3]. Hence, in order to find M_i, intensities of the fundamental and at least the first i-1 overtone transitions have to be determined. Since measurements of the weak higher overtone intensities are difficult, only relatively few accurate dipole moment functions have been determined experimentally. These are usually accurate only over a relatively small region of internuclear distances. A further complication is that the signs of the coefficients M_i can be obtained only via an analysis of rotational line intensities.

In principle, the same problems arise for electronic transitions. Since in this case the potential energy functions of the initial and final states are different, the situation is even more complicated. Frequently, the r-centroid approximation [4] is employed in the analysis of intensities of electronic transitions. In this approach the transition matrix element is approximated as

$$|\langle v'|R_e(r)|v''\rangle|^2 = q_{v'v''}R_e^2(r_{v'v''}) \ , \qquad (6)$$

where $R_e(r_{v'v''})$ is the electronic transition moment at the distance

$$r_{v'v''}=\langle v'|r|v''\rangle/\langle v'|v''\rangle \qquad \text{(r-centroid)}, \qquad (7)$$

and $q_{v'v''}=|\langle v'|v''\rangle|^2$ is the Franck-Condon factor. Provided the potential energy functions are known, $q_{v'v''}$ and $r_{v'v''}$ can easily be calculated numerically. The absolute values of the transition moments at the distances $r_{v'v''}$ can then be derived from the measured intensities of the corresponding transitions. In order to obtain transition moments for a sufficiently large range of internuclear distances, again very weak intensities often have to be measured. Obviously, small absolute errors in the weak intensities can lead to large relative errors in the intensity ratios and thus to an incorrect variation of the transition moments with the internuclear distance. Examples of such cases will be given in Section 5. It is noted that some criticism on the validity of the r-centroid approximation has been made [5,6]. In all cases reported in the present paper, however, we have checked its accuracy numerically and found that it is a very good approximation except for some very weak transitions.

In contrast to the experimental determination of dipole or transition moment functions, the calculation of these quantities from ab initio wavefunctions is in principle straightforward. However, as will be shown for several examples below, electron correlation effects often have a large influence on the calculated transition probabilities. Therefore, highly correlated electronic wavefunctions and large basis sets are generally needed in order to obtain reliable results. Unfortunately, there is no intrinsic criteron for the accuracy of the calculated data (as, e.g., provided by the variational principle for the total energy), but there are three indirect possibilities which may be used to investigate the reliability of the theoretical results: (i) Similar calculations for closely related molecules or other electronic states can be performed, for which experimental data are available. It is reasonable to assume that analogous calculations for similar molecules yield results of comparable quality. (ii) Sometimes information about the reliability of the results can be obtained indirectly by checking whether they are consistent with related experimental data, e.g. by comparing relative transition probabilities rather than the absolute values. (iii) The stability of the results with respect to improvements of the wavefunction can be studied. This implies that an open ended method is employed as is the case for the multiconfiguration reference CI procedure used in this work. Although some uncertainty always remains, since the number and the extent of the variations is limited, it is our experience that such investigations are very useful. They give an estimate for the sensitivity of the calculated results and may help to reveal, understand, and overcome intrinsic difficulties. In this paper examples of the application of all three of the above approaches will be given.

In Section 2 the computational methods will be briefly outlined. In Section 3 the accuracy of the computed potential energy functions and dipole moments is discussed. Section 4 deals with calculations of vibrational transition probabilities, while in Section 5 calculations of radiative rates and lifetimes for electronic transitions are presented.

2. COMPUTATIONAL METHODS

For most of the calculations reviewed in this paper, the multiconfiguration self-consistent field (MCSCF) [7-10] and the self-consistent electron pairs method (SCEP) [11-13] have been employed. In the MCSCF method the orbitals and coefficients of a configuration expansion are fully optimized according to the variational principle. This allows one to calculate the potential energy and the property functions which are qualitatively correct over the whole range of internuclear distances. Even though current MCSCF wavefunctions may utilize a very large number of configurations ($\sim 10^5$) [10], the accuracy obtained with this approximation is often not sufficient. The reason for this is that the number of orbitals optimized in the MCSCF wavefunctions is usually quite small, and therefore dynamical correlation effects are at most partly accounted for. Furthermore, the energy optimization process often yields wavefunctions which are biased towards certain regions of the molecule in which electron correlation is most important.

A dramatic example of this is shown in Table 1 for the FeO molecule [10], for which the dipole moment plays an important role in the interpretation of its negative ion dipole bound states [10a]. In this case the SCF approximation yields a very polar structure ($Fe^{+}O^{-}$) as indicated by the large dipole moment. The dipole moment is drastically reduced in a complete active space (CASSCF) calculation in which one more σ orbital is included. This orbital is essentially of iron d_σ character, and the reduction of the dipole moment can therefore be explained as a $p \rightarrow d$ electron back transfer. The dipole moment changes only very little if a very extensive multireference CI calculation is performed, and it can, therefore, be assumed to be close to the correct value. If one tries to account for electron correlation effects by optimizing further orbitals in the MCSCF wavefunction, the dipole moment strongly increases. Inspection of the orbitals shows that the additional optimized σ and π orbitals are of localized oxygen

Table 1. Comparison of some results for the FeO $^5\Delta$ ground state at R=3.0538 bohr.

Method	CSFs[a]	Energy	Dipole moment[b]	Percent correlation[c]
SCF	1	-1337.137524	3.524	0
CASSCF[d]	260	-1337.220934	1.286	20
CASSCF[e]	49140	-1337.300164	1.826	40
CASSCF[f]	178910	-1337.323532	2.005	45
MC-CI[g]	207212	-1337.548794	1.322	100

[a] In C_{2v} symmetry.
[b] In a.u., polarity Fe^+O^-.
[c] Relative to the MC-CI calculation.
[d] Active space: 8σ-10σ, 3π-4π, 1δ (12 electrons in 9 orbitals).
[e] Active space: 8σ-11σ, 3π-5π, 1δ (12 electrons in 12 orbitals).
[f] Active space: 8σ-11σ, 3π-5π, 1δ, $2\delta_{xy}$ (12 electrons in 13 orbitals. The $1\delta_{xy}$ orbital is doubly occupied in the SCF determinant.)
[g] For details see Ref. [10].

3p character. Hence, the oxygen 2p orbitals are lowered relative to the iron 3d orbitals, and thus too much charge is bound on the oxygen atom, which results in a dipole moment which is too large. It appears to be quite difficult to reduce such errors by extending the MCSCF wavefunctions. It is more reasonable (and cheaper) to include only the most important configurations (e.g. those with coefficients >0.06) into the MCSCF wavefunction and to account for the major part of the correlation effects in a subsequent multireference CI calculation. Usually, all configurations which are singly or doubly excited relative to any of the reference configurations are included in such configuration expansions. The coefficients of these configurations can be efficiently optimized with the SCEP procedure [12-13]. A noteworthy feature of this method is that the configurations may be generated by applying pair excitation operators to the total MCSCF function. In this way configurations which are generated by application of the same excitation operators to different reference configurations are contracted with the coefficients of the MCSCF wavefunction. In cases with many reference configurations this "internal contraction" drastically reduces the number of variational parameters and the computational effort. Except for some cases near avoided crossings, the influence of the contraction on the results is negligible [13,14].

In cases in which one configuration strongly dominates the total wavefunction over the range of internuclear distances of interest it appears to be cost effective to consider in the CI expansion only all singly and doubly excited configurations out of the SCF reference function, and to account for the important "unlinked cluster" effects of higher excitations by means of the CEPA approximation [15,11]. In many applications it has been shown that around the equilibrium geometries of small molecules the SCEP-CEPA results are of comparable accuracy as the MCSCF-SCEP ones. In all calculations reported here the CEPA-1 version [11] has been used. It has been found to give very similar results to more extensive coupled cluster (CCSD) calculations. It is noted that no configuration selection, except for the selection of the reference configurations and the restriction to single and double excitations, is performed. Thus smooth potential energy and property functions are always obtained, and uncertainties in the calculated properties, which may be more sensitive to the omission of certain configurations than the energy, are avoided.

Electric dipole moments can be calculated in two different ways. They are either obtained as the expectation value of the dipole moment operator, or they are evaluated as the first energy derivative with respect to an applied electric field. With SCF or MCSCF wavefunctions, for which the Hellmann-Feynman theorem is valid, both methods yield the same results. For CI wavefunctions, however, this is not the case. It has been found in various applications that the results obtained with the latter method are less sensitive with respect to variations of the wavefunction and are often in better agreement with experiment. This finding can also be justified theoretically. It is due to the fact that the energy derivative method allows for a relaxation of the orbitals with an applied electric field. The deviations of the dipole moments obtained with both methods become smaller as the quality of the wavefunction is improved and they vanish completely for full-CI wavefunctions. Since the computational effort to obtain the energy derivative is two to three times larger than to calculate the expectation value, it is often useful first to check for small deviations at several geometries. Provided this is the case one calculates only the expectation values at further geometries.

Electronic transition moments can be evaluated from MCSCF-SCEP wavefunctions as described in Ref. 14. This method requires that both states are described by the same orbital basis. Suitable orbitals for this purpose can be obtained by minimizing the energy average of two or more reference wavefunctions in the MCSCF procedure [8]. The errors of the MCSCF and MCSCF-SCEP energies and dipole moments due to the use of such orbitals are usually small (see, e.g., [16,17]), provided the selection of the reference configurations has been performed using the same orbitals. This is to be expected, since the singly and doubly excited configurations can account to second order for a relaxation of the internal orbitals. Transition moments calculated with averaged or fully optimized orbital sets have been compared within the MCSCF approach (see, e.g., [8,17-19]. The deviations were usually small. Since the effect should be considerably reduced in MCSCF-SCEP wavefunctions, it is unlikely that large errors in the SCEP transition moments arise by using a common orbital set. Nevertheless, the development of efficient methods which allow one to use different orbital sets for both states appears to be very desirable. This not only eliminates one source of uncertainty, but often also reduces the number of reference configurations needed. Further work in this direction has recently been reported [20,21,22].

Finally, it is noted that we generally use the dipole length operator to calculate the electronic transition moments. The values obtained with this operator have been found to be much less sensitive to basis set effects or changes of the CI wavefunctions than those calculated with the momentum operator.

3. POTENTIAL ENERGY FUNCTIONS, SPECTROSCOPIC CONSTANTS, AND DIPOLE MOMENTS

Illustrative examples of equilibrium distances and $\Delta G_{1/2}$ values obtained from MCSCF-SCEP potential energy functions are shown in Table 2. In all cases the reference functions properly described the dissociation of the bonds and contained some further configurations which were found to be important at short or intermediate distances (details are given in the original publications [23-27,17]). It is found that the equilibrium distances are often too long by around 0.002 to 0.005Å. The harmonic constants ω_e or the $\Delta G_{1/2}$ values are usually 10 to 20 cm^{-1} too small. These errors are probably due partly to basis set deficiencies and partly to the neglect of higher order excitations. In particular, unlinked cluster contributions are not fully accounted for. However, the agreement is still excellent. It is more difficult to obtain excitation and dissociation energies correctly, since these quantities may be strongly influenced by electron correlation effects. Some such examples are shown in Tables 3 and 4. For molecules with single bonds an accuracy of about 0.1 eV can be achieved, but for

Table 2. Comparison of calculated and experimental spectroscopic constants[a]

Molecule	State	Ref. Conf.	r_e(Å)	error[a]	$\Delta G_{1/2}$(cm^{-1})	error[a]
BH	$X^1\Sigma^+$	3	1.235	0.003	2268	0
OH	$X^2\Pi$	3	0.971	0.001	3561	4
OH^+	$X^3\Sigma^-$	7	1.031	0.002	2944	-10
HF^+	$X^2\Pi$	6	1.002	0.001	2928	-4
HCl^+	$X^2\Pi$	7	1.319	0.004	2579	10
SH	$X^2\Pi$	12	1.340	-0.001	2588	-3
	$A^2\Sigma^+$	10	1.428	0.005	1760	-24
SiO	$X^1\Sigma^+$	8	1.515	0.005	1230	-1
C_2^-	$X^2\Sigma_g^+$	10	1.276	0.008	1756	-2
	$B^2\Sigma_u^+$	12	1.231	0.008	1951	12
CN	$X^2\Sigma^+$	12	1.174	0.002	2053	10
	$A^2\Pi$	19	1.233	0.005	1788	-16
N_2	$A^3\Sigma_u^+$	11	1.294	0.007	1415	-18
	$B^3\Pi_g$	21	1.216	0.003	1701	-4

[a] As compared to experimental values. Internally contracted MCSCF-SCEP calculations.

molecules with multiple bonds errors of 0.3-0.5 eV may remain even if quite extensive basis sets are used. A typical case is that of the triplet states of the N_2 molecule [17]. At the Hartree-Fock level, these states are not bound, or only very weakly bound. In our calculations, more than 90 percent of the correlation contributions to the binding energies have been accounted for. These estimates result from the experimental and SCF D_e values. Nevertheless, due to large correlation contributions to the dissociation energies, considerable absolute errors in the dissociation energies remain. It is likely that the major part of the errors are due to basis set deficiencies rather than to shortcomings of the CI expansions. This is indicated by the fact that calculations with two d functions on each atom yield nearly twice the error obtained from calculations (Tables 3 and 4) with three d and one f-functions on each atom. It is noted that the large error of the dissociation energy calculated for the ground state of CN (cf. Table 3) can be ascribed to the fact that no f-functions were included in the basis set.

Dipole moments in the vibrational ground states can usually be calculated with an accuracy of one percent. Some examples which have been obtained with either SCEP-CEPA or MCSCF-SCEP wavefunctions are shown in Table 5. It is interesting to note that for a long time there was a large discrepancy between theoretical and experimental values for the difference of the dipole moments of OH and OD in their vibrational ground states. Experimentally, an isotope effect of 0.0145D had been found [29], while all theoretical calculations predicted only a difference of 0.001-0.002D. This disparity has recently been resolved by new measurements of Peterson et al [30]. They found a smaller dipole moment for OH, and an isotope effect of only 0.001D. This finding is in better agreement with the theoretical prediction.

4. VIBRATIONAL TRANSITION PROBABILITIES

In this section, we present some selected calculations of electric dipole moment functions and transition probabilities for rotational-vibrational transitions. As a first example, we consider the

Table 3. Comparison of some calculated and experimental dissociation energies D_e (in eV).

Molecule/State		MCSCF-SCEP	Experiment	Error
BH	$X^1\Sigma^+$	3.65	3.57	0.08
OH	$X^2\Pi$	4.52	4.62	-0.10
OH^-	$X^1\Sigma^+$	5.03	4.98	0.05
CN	$X^2\Sigma^+$	7.17	7.78	-0.61
N_2	$A^3\Sigma_u^+$	3.30	3.70	-0.40
	$B^3\Pi_g$	4.38	4.90	-0.52
	$W^3\Delta_u$	4.55	4.87	-0.32
	$B'^3\Sigma_u^-$	5.06	5.26	-0.20

Table 4. Comparison of some calculated and experimental excitation energies T_e (in eV).

Molecule/Trans.		MCSCF-SCEP	Experiment	Error
CH	$A^2\Delta-X^2\Pi$	2.91	2.87	0.04
OH	$A^2\Sigma^+-X^2\Pi$	4.07	4.05	0.02
SH	$A^2\Sigma^+-X^2\Pi$	3.85	3.88	0.03
HF^+	$A^2\Sigma^+-X^2\Pi$	3.19	3.16	0.03
HCl^+	$A^2\Sigma^+-X^2\Pi$	3.57	3.55	0.02
C_2^-	$A^2\Pi_u-X^2\Sigma_g^+$	0.44	0.52	-0.08
	$B^2\Sigma_u^--X^2\Sigma_g^+$	2.35	2.28	0.07
CN	$A^2\Pi-X^2\Sigma^+$	1.13	1.15	0.02
N_2	$B^3\Pi_g-A^3\Sigma_u^+$	1.45	1.17	0.28
	$W^3\Delta_u-B^3\Pi_g$	-0.18	0.02	-0.20
	$B'^3\Sigma_u^--B^3\Pi_g$	0.66	0.82	-0.16
	$C^3\Pi_u-B^3\Pi_g$	3.70	3.66	0.04

Table 5. Comparison of calculated and experimental dipole moments in vibrational ground states (in Debye).

Molecule	Calculated	Experimental	Difference
OH	1.642[a]	1.655	0.013
OD	1.641[a]	1.653	0.012
HF	1.807[b]	1.8275	-0.019
DF	1.801	1.8188	-0.018
HCl	1.120[b]	1.1085	0.012
DCl	1.116	1.1033	0.013
HBr	0.829[b]	0.828	0.001
HI	0.472[c]	0.447	0.025
SH	0.754[d]	0.757	-0.003
CO	0.117[e]	0.1098	0.007
SiO	3.056[e]	3.098	-0.042
	3.067[f]	3.098	-0.031

[a] MCSCF-SCEP calculations, Ref. [24].
[b] SCEP-CEPA calculations, Ref. [42].
[c] SCEP-CEPA calculations, Ref. [28].
[d] MCSCF-SCEP calculations, Ref. [25].
[e] MCSCF-SCEP calculations, Ref. [23].
[f] SCEP-CEPA calculations, Ref. [56b].

transpositions within the ground state of the OH radical [24]. This radical is of atmospheric importance, since it is formed by the reaction of ozone with hydrogen. The emission from the excited vibrational states of OH is observed in the atmospheric twilight and night glow and is known as Meinel bands [31]. OH is also of interest in combustion processes [32] and many other elementary reactions. The determination of the radiative transition probabilities has, therefore, been the subject of several investigations [33-36]. However, absolute experimental transition probabilities [33-35] vary by several hundred percent, and previous calculations [37-40] yielded very different results (see Table 6).

Table 6. Comparison of rotationless Einstein transition probability coefficients for the $X^2\Pi$ state of OH (in s^{-1}).

	Experiments			Calculations[a]		
	b	c	d	e	f	MCSCF-SCEP[g]
A_0^1	33	8.5	42.5	6.20	18.3	12.2
A_1^2		13	62.5	4.87	22.5	15.6
A_2^3			66.7	1.00	17.7	13.0
A_0^2	5.4		21.4	7.75	13.5	8.43
A_1^3			52.4	15.6	38.9	23.4

[a] All values have been calculated with the RKR potential energy function from Ref. [41].
[b] Benedict, Plyler [33].
[c] D'Incan et al [34].
[d] Roux et al [35].
[e] MC-CI dipole moment function of Chu et al [39].
[f] MCSCF(17) dipole moment function of Stevens et al [37].
[g] Ref. [24].

The difficulty in calculating reliable transition rates for OH originates from the fact that the dipole moment function has a maximum close to the equilibrium distance. Small changes of the dipole moment function in this region can lead to considerable changes in the vibrational transition matrix elements. Fig. 2 shows a comparison of our calculated dipole moment function with two previously calculated ones. The varying shapes of these functions are responsible for the differences of the theoretical Einstein coefficients in Table 6. Since no reliable experimental data are available for comparison, only indirect methods can be utilized to determine which dipole moment function is the most accurate one. All three approaches mentioned in the introduction to investigate the reliability of the results were applied to this case.

Previously, similar calculations to those performed on OH were carried out for hydrogen fluoride [42,23], for which accurate experimental Einstein coefficients are known. It was found that SCEP-CEPA or MCSCF-SCEP wavefunctions with similar reference configurations as used for OH yielded the transition probabilities with an accuracy of about 5 percent. On the other hand, less extensive MCSCF-SCEP calculations, which included just those two reference configurations which are needed to describe the dissociation properly, yielded too flat a dipole moment function and the maximum at too short a distance. This is also the case for the similar MC-CI calculations for OH of Chu et al. [39] shown in Fig. 2. MCSCF calculations for HF yielded quite strongly varying results depending on the choice of configurations. In particular, if the fluorine 2s orbital was correlated in the MCSCF wavefunction, the slope of the dipole moment function was too steep. This is also the case for the MCSCF dipole moment function calculated for OH by Stevens et al [37].

Next, the stability of our results with respect to variations of the number and structure of the reference configurations was tested. The results are shown in Fig. 3. It is found that the dipole moments obtained from the MCSCF reference wavefunctions show large variations depending on the choice of configurations, particularly at longer internuclear distances. On the other hand, the MCSCF-SCEP values are very stable to such variations, provided they are calculated as derivatives of the energy with respect to an external field. Somewhat larger variations have been obtained if they are calculated as expectation values, but for the final MCSCF(7)-SCEP wavefunctions the

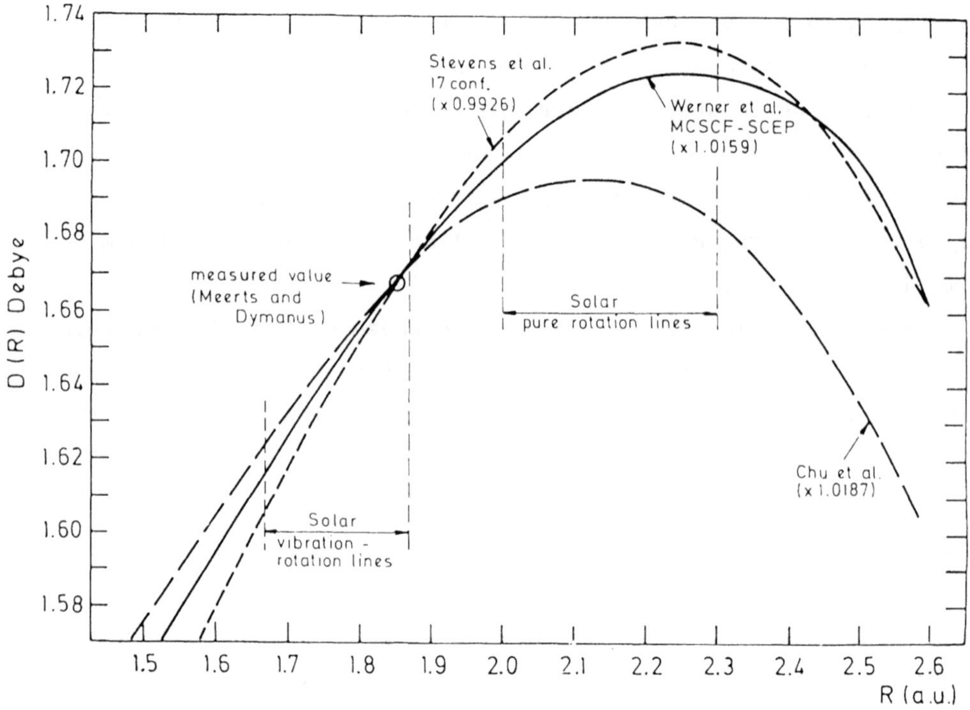

Figure 2. Comparison of calculated dipole moment functions for the ground state of OH. The regions of the dipole moment functions which are tested by the solar observations are indicated (from Ref. [45]).

differences are very small (cf. Ref. 24).

Thirdly, various indirect checks of the calculated Einstein coefficients can be made. One possibility is to compare the calculated and measured ratios of Einstein coefficients rather than the absolute values. The intensity ratios for transitions with the same initial level can be measured much more accurately than the absolute values. For this comparison to be meaningful it is important to consider the same rotational lines which have been measured, because for OH rotation-vibration coupling and spin decoupling effects are not negligible (see the work of Mies [43]). This is demonstrated by a comparison of probability ratios calculated for the P and Q branches of the bands in Table 7. A comparison of measured and calculated ratios is shown in Table 8. The values obtained from our dipole moment function are in excellent agreement with the empirical values except for A_2^3/A_1^3. The experimental value for this ratio was obtained from unresolved Q_1 and Q_2 branches, and is considered to be much less reliable than the other values. It is noted that the ratios depend sensitively on the shape of the dipole moment function. A different indirect comparison of various dipole moment functions was recently made by Sauval et al [44,45]. They derived the oxygen abundance in the solar atmosphere from measured intensities in the solar spectrum and calculated transition probability coefficients. It was found that transition probabilities derived from the MCSCF-SCEP dipole moment function yielded the same abundance independently of the transition considered. Other dipole moment functions gave a considerable variation of the abundance with the wavelength (see Fig. 4). Also, the analysis of purely rotational spectra and vibration-rotation spectra yielded very consistent results, even though different regions of the dipole moment function are important for the corresponding transition matrix elements (see Fig. 3). Finally, in an investigation by Argawalla et al. [48] the rate constant for the reaction $H+NO_2$ was determined from measured relative emission intensities, and compared to values obtained in other independent measurements. For their analysis they used a PNO-CEPA dipole moment function calculated by Meyer [38] and

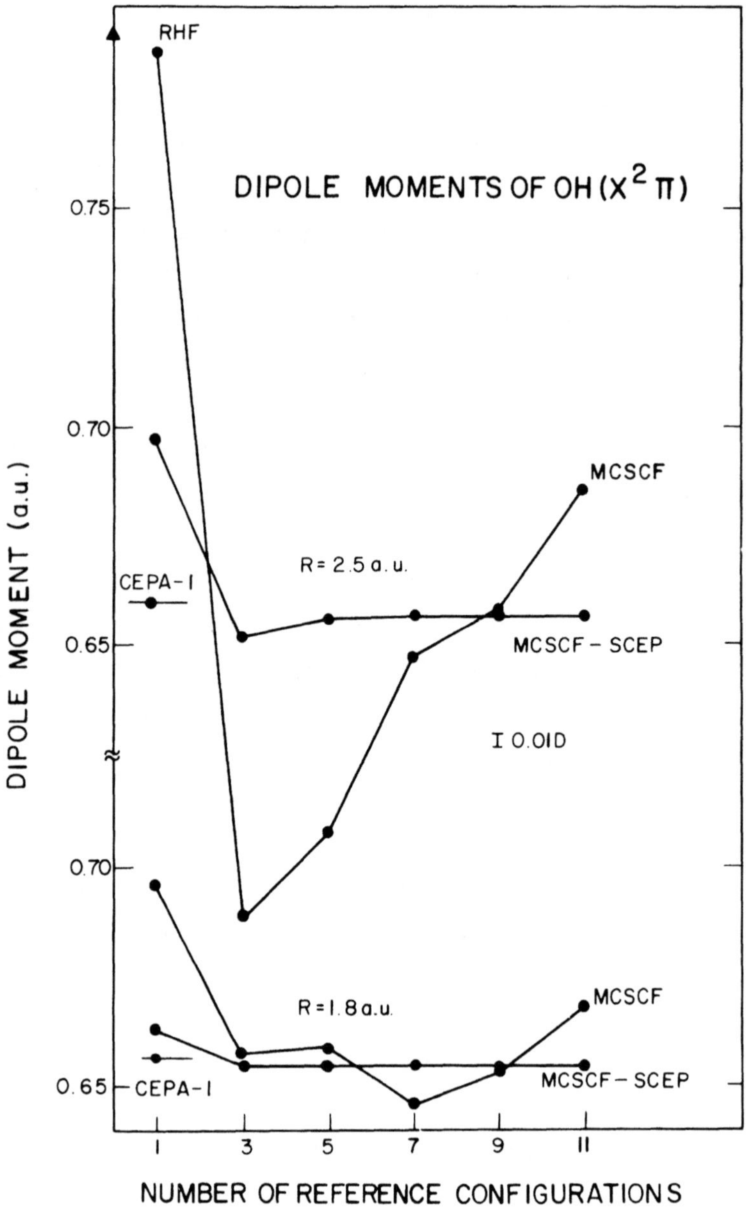

Figure 3. Calculated dipole moments of OH at r=1.8 a.u. and r=2.5 a.u. as a function of the number of reference configurations in MCSCF-SCEP calculations. The dipole moments of the reference wavefunctions are also shown. All values were obtained as first energy derivatives with respect to an electric field.

Table 7. Comparison of relative transition probabilities in P and Q branches for OH ($X^2\Pi$)[a]

Branch	A_0^2/A_1^2 (K=2-7)	A_1^3/A_2^3 (K=2-7)	A_2^4/A_3^4 (K=1-3)
P_1	0.38±0.06	1.10±0.37	3.15±0.58
Q_1	0.56±0.02	1.86±0.09	5.60±0.17

[a] Calculated from MCSCF-SCEP dipole moment function and RKR potential energy function. The error bounds are due to the averaging over several rotational lines.

Table 8. Comparison of measured and calculated relative radiative transition probabilities for OH ($X^2\Pi$)[a]

	experimental		theoretical	
	Murphy[36]	Roux et al. [35]	Mies[43][b]	this work[c]
A_0^2/A_1^2	0.44±0.03	0.41	0.46±0.06	0.38±0.06
A_1^3/A_2^3	1.15±0.05	1.25	1.62±0.28	1.10±0.37
A_2^4/A_3^4	3.3±0.7		7.8±0.21	5.6±0.58

[a] The error bounds in the theoretical values are due to averaging over several rotational lines (for the first two ratios K=2-7 in P branches, for the last ratio K=1-3 in Q branches).

[b] Using MCSCF(17) dipole moment function of Stevens et al [37].

[c] Using MCSCF-SCEP dipole moment function and RKR potential.

the MCSCF dipole moment function of Stevens et al [37]. Only with the PNO-CEPA function, which yields similar transition rates to the MCSCF-SCEP function, were correct rate constants obtained. Since all the above mentioned indirect theoretical and experimental tests of the dipole moment functions yield consistently satisfactory results with the MCSCF-SCEP data, it is reasonable to assume that this dipole moment function is the most accurate one presently available. S.R. Langhoff has used it to calculate a large number of transition probability coefficients for rovibrational transitions, which are available on request [49].

The SH radical is expected to play an important role in processes such as coal combustion or the atmospheric sulphur cycle. So far, this radical has not been investigated as a product of chemical reactions by IR-chemiluminescence techniques, because of a very weak emission in its electronic ground state. Recently, Bernath, Amano and Wong [89] succeeded in observing the fundamental vibration-rotation band of SH in absorption at Doppler-limited resolution. Winkel and Davis [90] also detected three infrared bands in emission. These results suggest that it should be possible to monitor SH by its IR-chemiluminescence, even though the rates of spontaneous emission are calculated [25] to be very small (SH(v=1): 1 sec^{-1} and SD(v=1): 0.3 sec^{-1}) (cf. Table 9b).

We have also calculated the dipole moment functions of the ions OH^-, SH^- and OH^+, SH^+ [24], which are compared with the functions of the neutral radicals in Fig. 5a and 5b. It is observed that the dipole moment functions of the ions are much steeper which results in considerably larger probabilities for the fundamental transitions (cf. Table 9). The dipole moments of molecular ions are origin dependent. In order to calculate vibrational transition probabilities they have to be referred to the center of mass. Hence, formally an infinite dipole moment is obtained at infinite internuclear distance. The strong increase or decrease of the dipole moment function with internuclear distance is due to the removal of the positive or negative charge from the center of mass. Therefore, large transition probabilities have been predicted, for instance, for HF^+, and HCl^+ [20], and

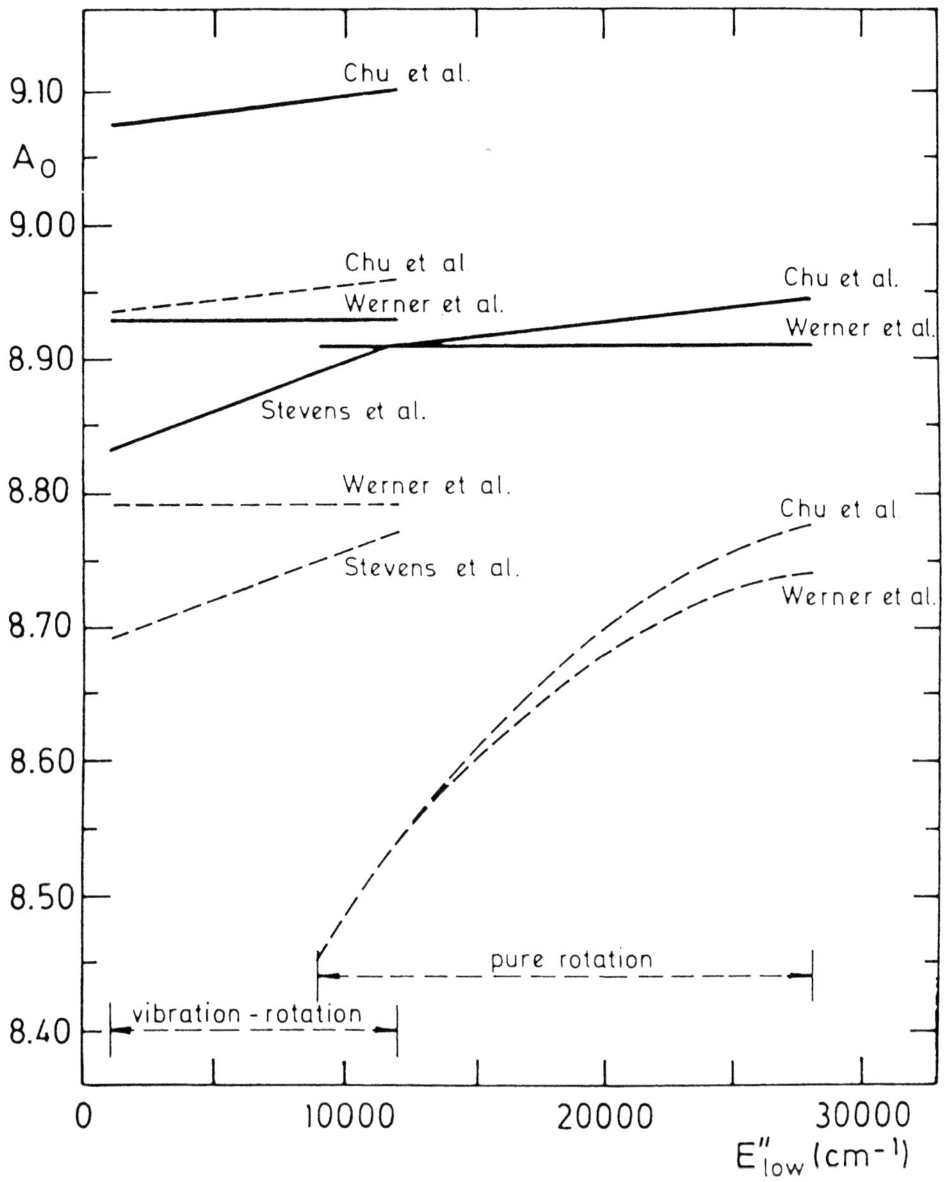

Figure 4. Least squares fits to the solar oxygen abundance inferred from the OH vibration-rotation lines as a function of the excitation energy of the lower level of the pure rotation lines of OH [44] and the vibration-rotation lines [45], using different dipole moment functions (as indicated) and two different photospheric models: Holweger Müller [46] (solid lines) and Vernazza (dashed lines) et al [47]. The results favor the first model (from Ref. [45]).

Table 9a. Comparison of infrared transition rates for OH, OH^-, and OH^+ (all values from MCSCF-SCEP calculations; in sec^{-1})

	OH	OH^-	OH^+
A_0^1	12.2	137	263
A_1^2	15.6	283	471
A_0^2	8.4	0.2	16

Table 9b. Comparison of infrared transition rates for SH, SH^-, and SH^+ (all values from MCSCF-SCEP or CEPA calculations, in sec^{-1})

	SH	SH^-	SH^+
A_0^1	1.1	57	54
A_1^2	2.6	115	100
A_0^2	0.07	0.6	0.3

particularly for the protonated rare gas atoms HeH^+, NeH^+, ArH^+, KrH^+ and XeH^+ [50] (cf. Table 10). In fact, this prediction assisted the first observation of infrared emission in a molecular ion (ArH^+) by Brault and Davis [51]. Meanwhile, only the XeH^+ and RnH^+ have not been investigated by IR-techniques, other protonated rare gas atom spectra are known [50].

In contrast to the hydrides, the MCSCF dipole moment functions for less polar molecules containing two first row atoms seem to be less sensitive to the choice of configuration. MCSCF calculations for the CO molecule [52] yielded surprisingly accurate vibrational transition probabilities (cf. Table 11). Similar calculations were also performed for the (valence) isoelectronic species NO^+ [54], BF, AℓF [55] and SiO [56] as well as for the 13 electron systems CO^+ [57], SiO^+ [56], BF^+ and $A\ell F^+$ [55]. Fairly strong infrared intensities have been predicted for BF and BF^+, which exhibit extremely steep dipole moment functions. The dipole moment of BF increases by about 12 Debye as the atoms are separated by 1.5Å. Since BF is very strongly bound (D_o=7.8eV) the boron atom may abstract fluorine from most fluorine containing compounds, thereby generating BF in highly excited vibrational states [58]. These pumping reactions can be used to produce efficient laser action in the infrared region.

The calculated transition rates for NO^+ have been used by Smith, Bierbaum and Leone [59] to determine initial populations of vibrational states of NO^+ formed in the reaction $N^+ + O_2 \rightarrow NO^+ + O$ at low oxygen pressures. The number of vibrational states which can be populated depends on the electronic state in which the oxygen atom is formed. Interestingly, the results of Smith, et. al. indicate that the predominant channel is the formation of the highest possible oxygen state (^1S) and low vibrational levels of NO^+ (cf. Fig. 6). Due to the low resolution of the measured chemiluminescence spectra further investigations appear to be necessary to clarify these findings.

5. ELECTRONIC TRANSITIONS

In this section some MCSCF-SCEP calculations of electronic

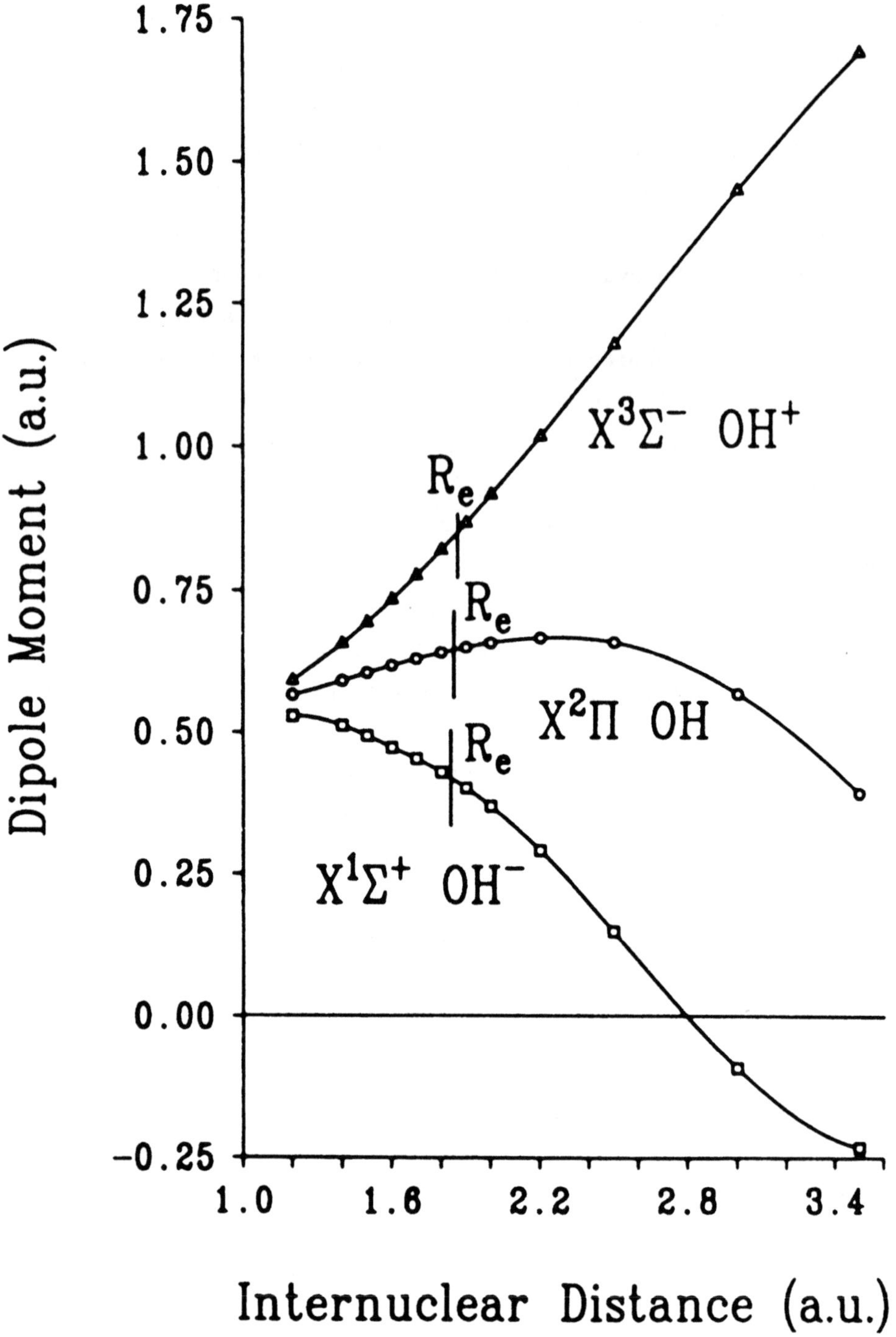

Figure 5a. Comparison of the dipole moment functions of OH, OH^- and OH^+.

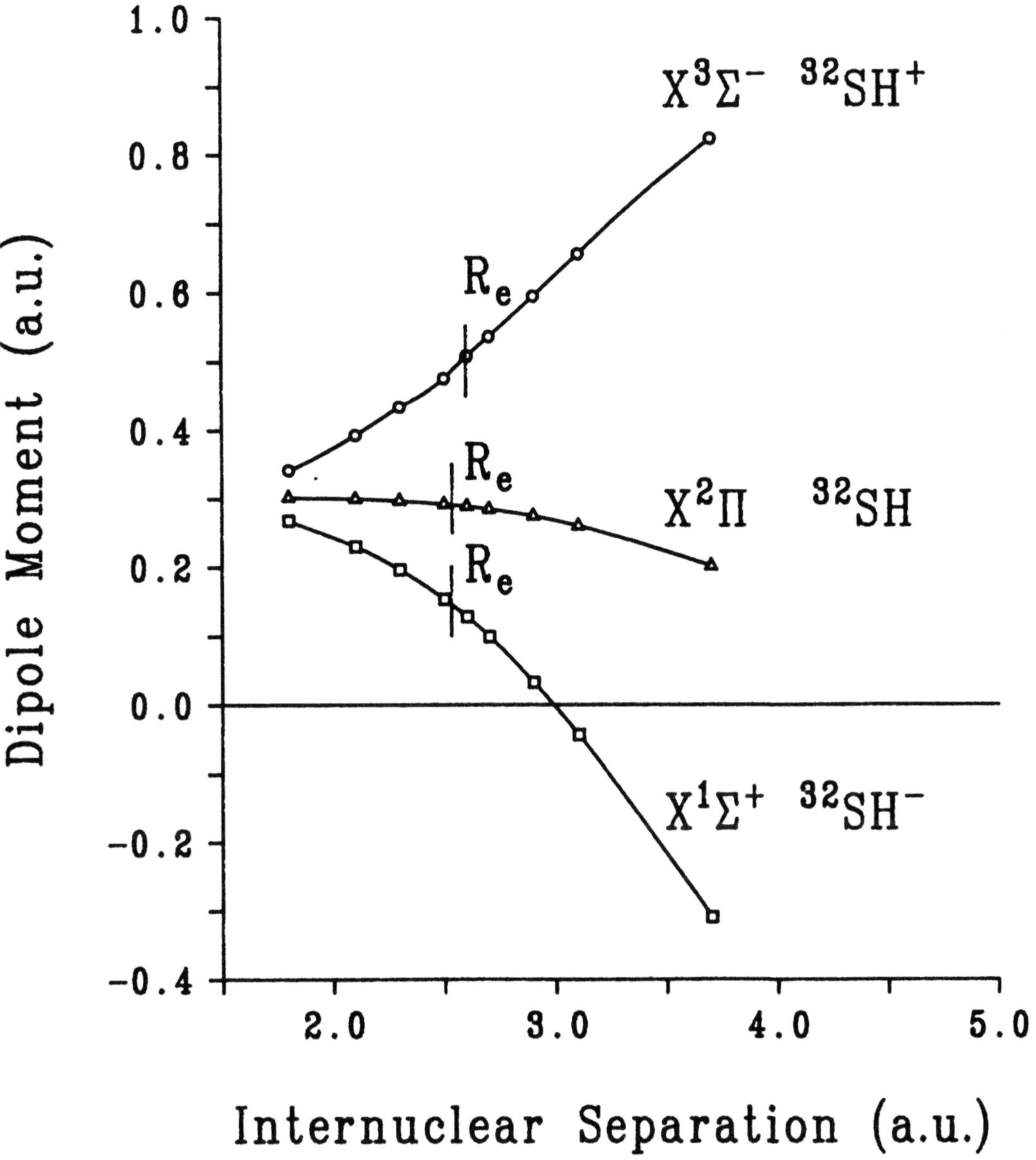

Figure 5b. Comparison of the dipole moment functions of SH, SH^- and SH^+.

Table 10. Comparison of calculated Einstein coefficients A_0^1 for the fundamental vibrational transition in some diatomic molecules and molecular ions (in sec^{-1})

				HeH^+	837
HF	189	HF^+	618	NeH^+	756
HCl	35	HCl^+	217	ArH^+	438
HBr	5.6			KrH^+	285
HI	0.1			XeH^+	125
CO	35	CO^+	31		
BF	43	BF^+	77		
AℓF	9	$AℓF^+$	13		
CN	26	NO^+	11		
SiO	8.6(10.5)[a]	SiO^+	10		

[a] Ref. [56b].

Table 11. Comparison of calculated and measured dipole matrix elements for the vibrational fundamental sequence of CO (in D)

v'	J'	v"	J"	MCSCF[a]	Experimental[b]
1	0	0	0	0.107	0.1055
5	11	4	12	0.238	0.238 0.003
6	10	5	11	0.260	0.259 0.002
7	8	6	9	0.280	0.277 0.002
8	7	7	8	0.299	0.295 0.002
9	10	8	11	0.316	0.313 0.003
10	11	9	12	0.332	0.326 0.004
11	9	10	10	0.347	0.344 0.005

[a] Evaluated with RKR potential energy function.

[b] Ref. [53].

transition moments and radiative lifetimes of electronically excited states are reviewed. It will be demonstrated that these quantities may be very strongly influenced by electron correlation effects. A first example of this is shown in Fig. 7 for the $A^2\Delta - X^2\Pi$ transition of CH [13]. The lifetime of the $A^2\Delta$ state has been measured accurately by two groups [60,61], and may, therefore, be used as a reliable check of the theoretical transition moments. Furthermore, Larsson et al. [19] recently systematically investigated the convergence of the calculated transition moments as a function of the basis set and the number of orbitals included in complete active space SCF (CASSCF) wavefunctions. They found that not only large basis sets with diffuse d functions are required to obtain accurate results, but also the convergence of the calculated lifetimes towards the experimental value with increasing length of the MCSCF expansion is very slow. Relatively compact MCSCF wavefunctions, which describe the potential energy function qualitatively correctly, yielded lifetimes which were too short by about a factor of two. For comparison, we have performed MCSCF-SCEP calculations which comprised all singly and doubly excited configurations relative to the most important configurations of the CASSCF wavefunctions. As demonstrated in Fig. 7, the stability of the MCSCF-SCEP values is much better than the CASSCF values. Even with the simplest reference wavefunction the calculated lifetime is accurate to within about five percent. It turned out that in order to obtain more accurate values, configurations containing a δ orbital had to be included into the reference function. Such configurations also have the largest effect on the CASSCF results [19].

Quite similar results were obtained for the $A^2\Sigma^+ - X^2\Pi$ transition moments of OH, HF^+ and HCl^+ [20]. The transition moment functions calculated in the SCF, MCSCF and MCSCF-SCEP wavefunctions are shown in Figs. 8 and 9. It is observed that in all cases the electron correlation effects are very large. Only a very small part of these effects is accounted for with compact MCSCF wavefunctions. Many other choices of configurations and orbitals optimized in the MCSCF wavefunctions have been tested for OH [23], but thus far it is not possible to get

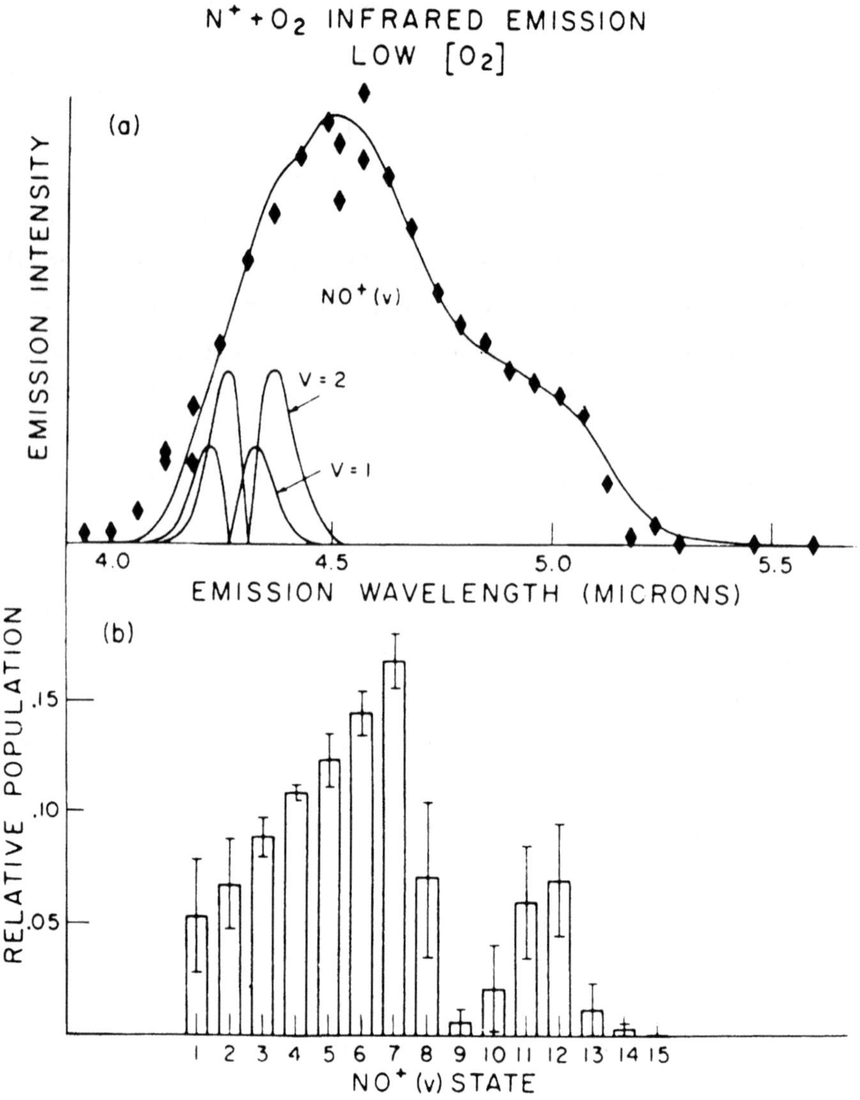

Figure 6. a) The observed fluorescence spectrum due to $NO^+(X^1\Sigma^+,v)$ from the reaction $N^+ + O_2$ at oxygen pressures less than 0.8 Pa. The solid line is a least-squares fit to the data. b) The resulting NO^+ (v) distribution from three independent spectra (from Ref. [59]). If the oxygen atom is formed in the ^1S state, only the NO^+ $X^1\Sigma^+$, v=0-8 levels are energetically accessible, while higher levels (up to v=18) can be reached if O is formed in the ^1D state.

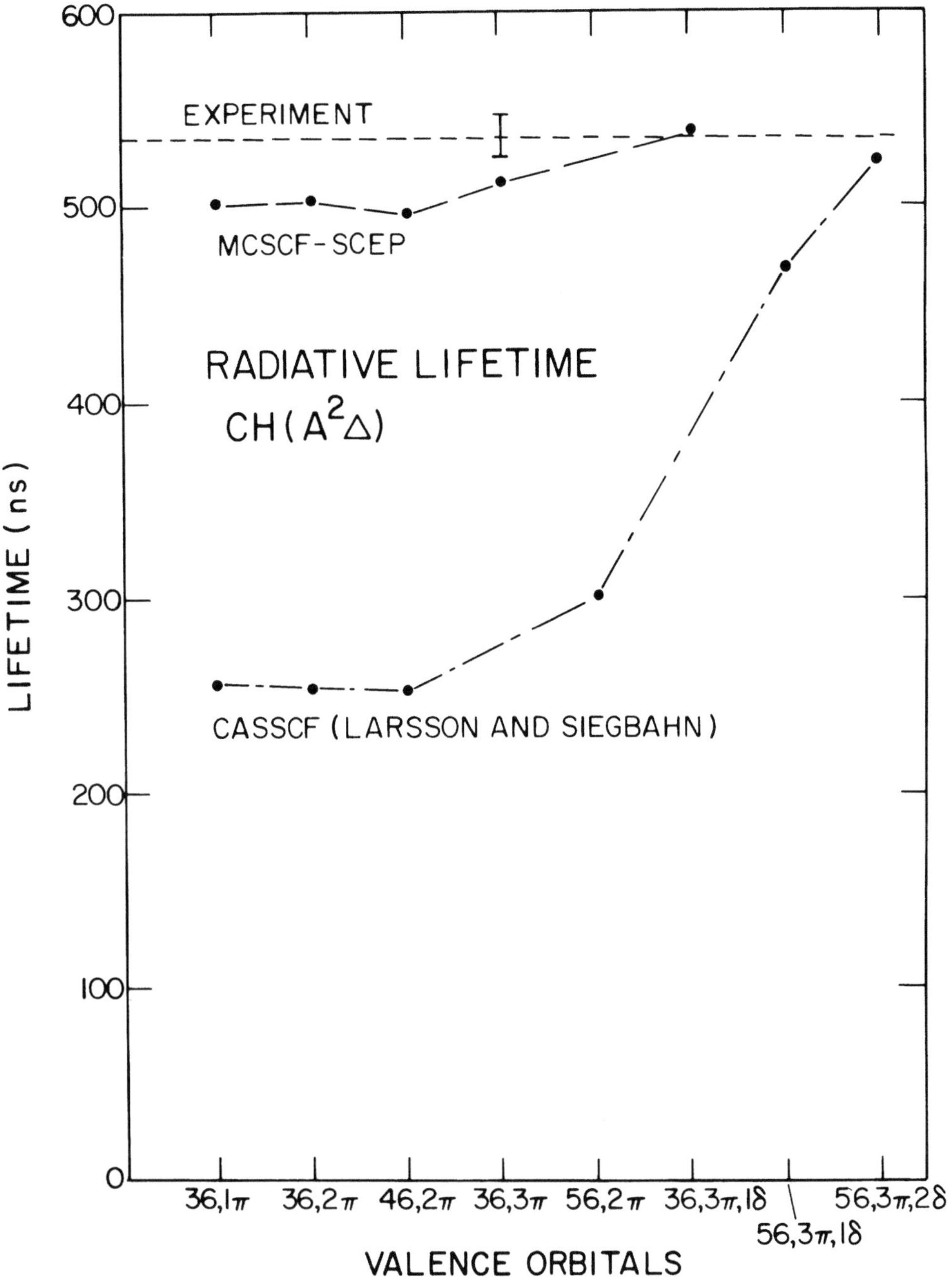

Figure 7. Comparison of calculated and measured radiative lifetimes of CH ($A^2\Delta$, v'=0) using various CASSCF (from Ref. 19) and MCSCF-SCEP wavefunctions.

an accurate lifetime in this approximation. The calculated radiative lifetimes for OH ($A^2\Sigma^+$,v'=0) are shown in Table 12, in comparison with other theoretical and experimental data. Our MCSCF-SCEP value is in close agreement with previous less extensive MCSCF-CI calculations [39, 40]. Older experimental values show considerable variations [62], but the most reliable recent values [63-67] lie close to 700 ns. It can be noted that the lifetime for OH can also be obtained from independently measured oscillator strengths [68,69] for the 0-0 transition, which by far dominates the decay from v'=0. This yielded a lifetime of ~650 ns. Thus, the available experimental data are not fully consistent. It appears that the calculated MCSCF-SCEP lifetime might be 10-15 percent too low. In view of the extremely large electron correlation effects, an error of this size is not surprising.

Similar results were obtained for HCl^+ [20,70,71] (cf. Fig. 10). As was found for OH, the MCSCF-SCEP lifetime appears to be somewhat too low. The MCSCF method, however, yields values which are too short by a factor of two. In contrast to OH, the calculated lifetimes for HF^+ are much longer than the measured ones [72] (cf. Fig. 11). It is, therefore, very likely that in this case the experimental values are in error. It should be noted that the HF^+ ($A^2\Sigma^+$) lifetimes are rather long (ca. 20 μsec), and that it is difficult to measure directly such long lifetimes for molecular ions.

Recently, Friedl, Brune and Anderson [91] observed that predissociation occurs throughout the v'=0 level of the $A^2\Sigma^+$ state of SH and SD. This result was based on LIF measurements obtained in conjunction with the determination of the absolute concentrations of both radicals. The authors calculated the predissociative and radiative lifetimes to be 3 and 820 ns for SH and 260 and 730 ns for SD, respectively. Since the A-X transition is important as an analytical diagnostic for SH we we have performed calculations [25] of the radiative transition probabilities for this transition. The theoretical lifetimes and absorption oscillator strengths have been compared with the experimentally determined values. Using the MCSCF-SCEP potentials and transition moment functions displayed in Figure 12, we calculate a

Table 12. Comparison of calculated and experimental radiative lifetimes for OH ($A^2\Sigma^+$, v'=0)

Method/Authors	Lifetime (ns)
Calculations:	
MCSCF [20]	327
MCSCF-SCEP [20]	590
PNO-CEPA [20]	637
MR-CI, Langhoff et al. [40]	580
Lifetime measurements:	
German [67]	693±10
Dimpfl and Kinsey [65]	684±14
Dermid and Laudenslager [63]	720±9
Brzozowski et al. [66]	760±20
Bergeman et al. [64]	755±40
From oscillator strengths:	
Wang and Huang [68]	650
Smith and Crosley [69]	650

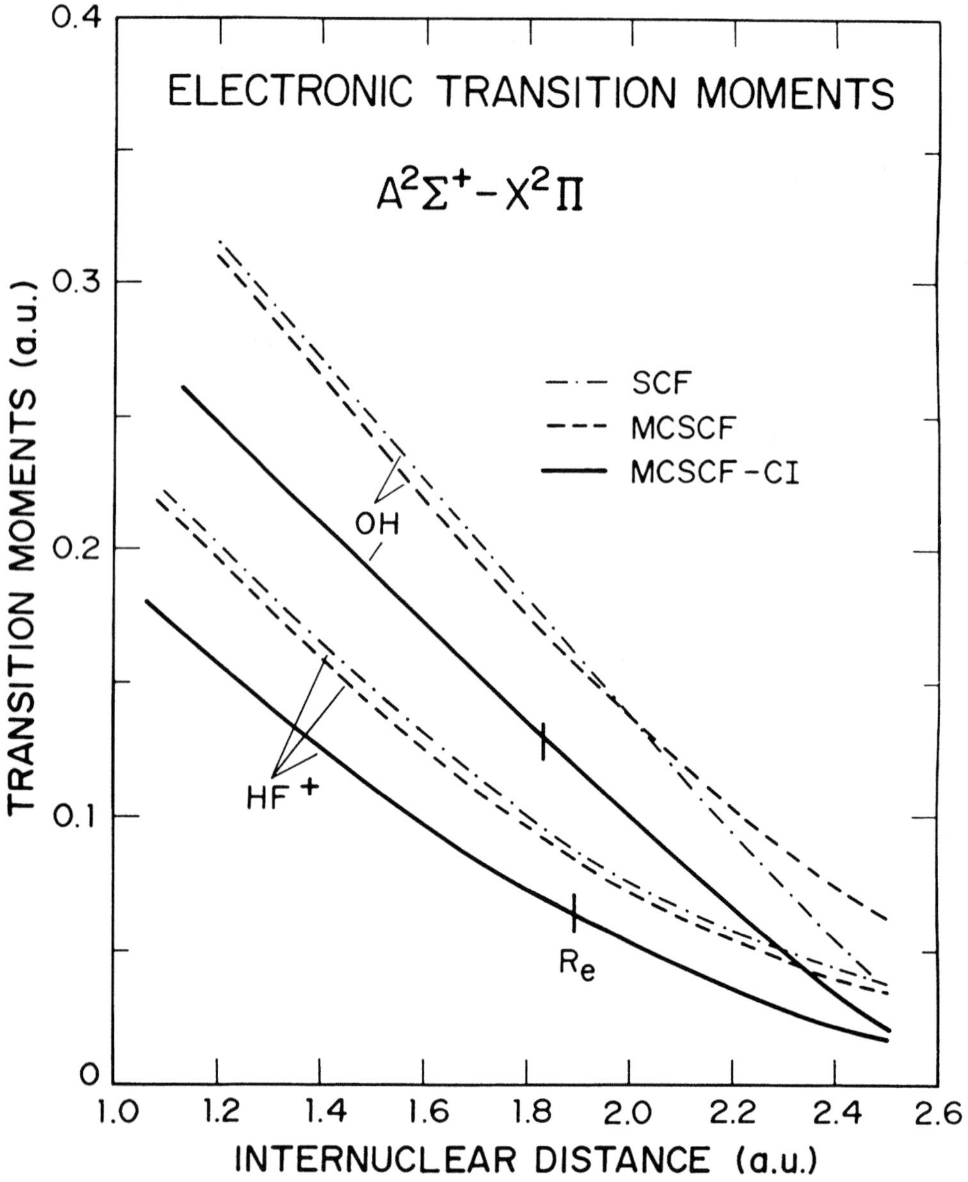

Figure 8. Electronic transition moment functions for the $A^2\Sigma^+-X^2\Pi$ transitions in OH and HF^+.

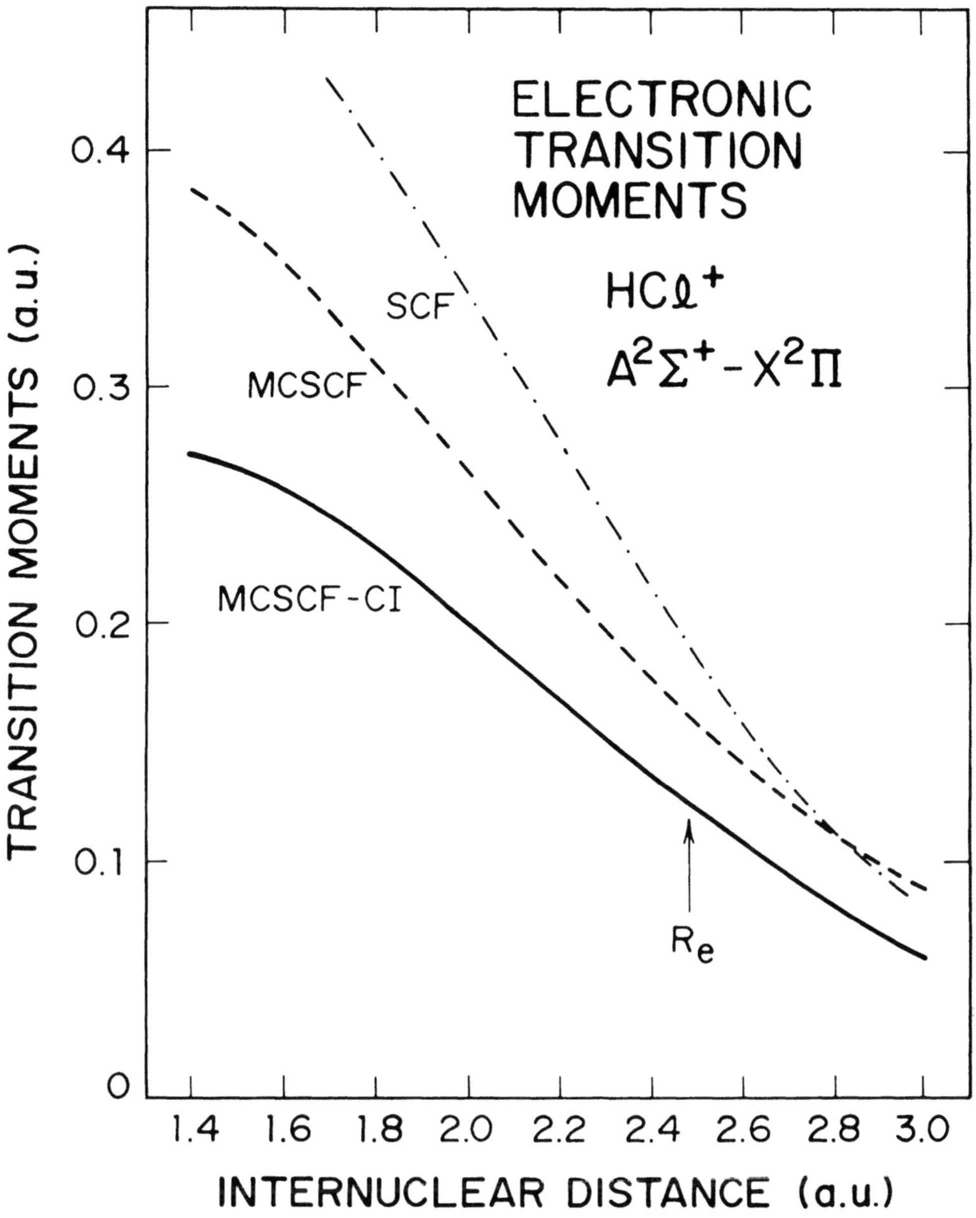

Figure 9. Electronic transition moment functions for the $A^2\Sigma^+-X^2\Pi$ transition in HCl^+.

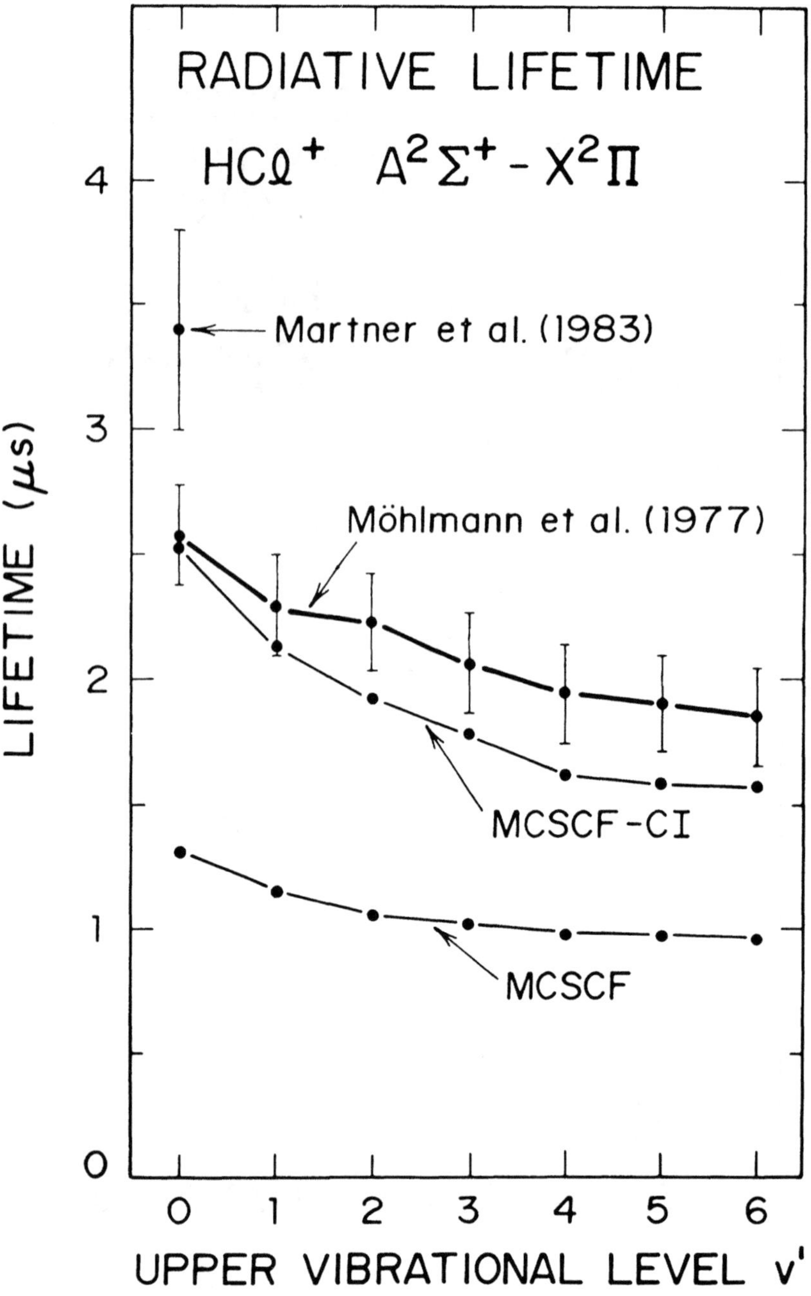

Figure 10. Comparison of calculated and measured radiative lifetimes of the $A^2\Sigma^+$ state of HCl^+ as a function of the initial vibrational level.

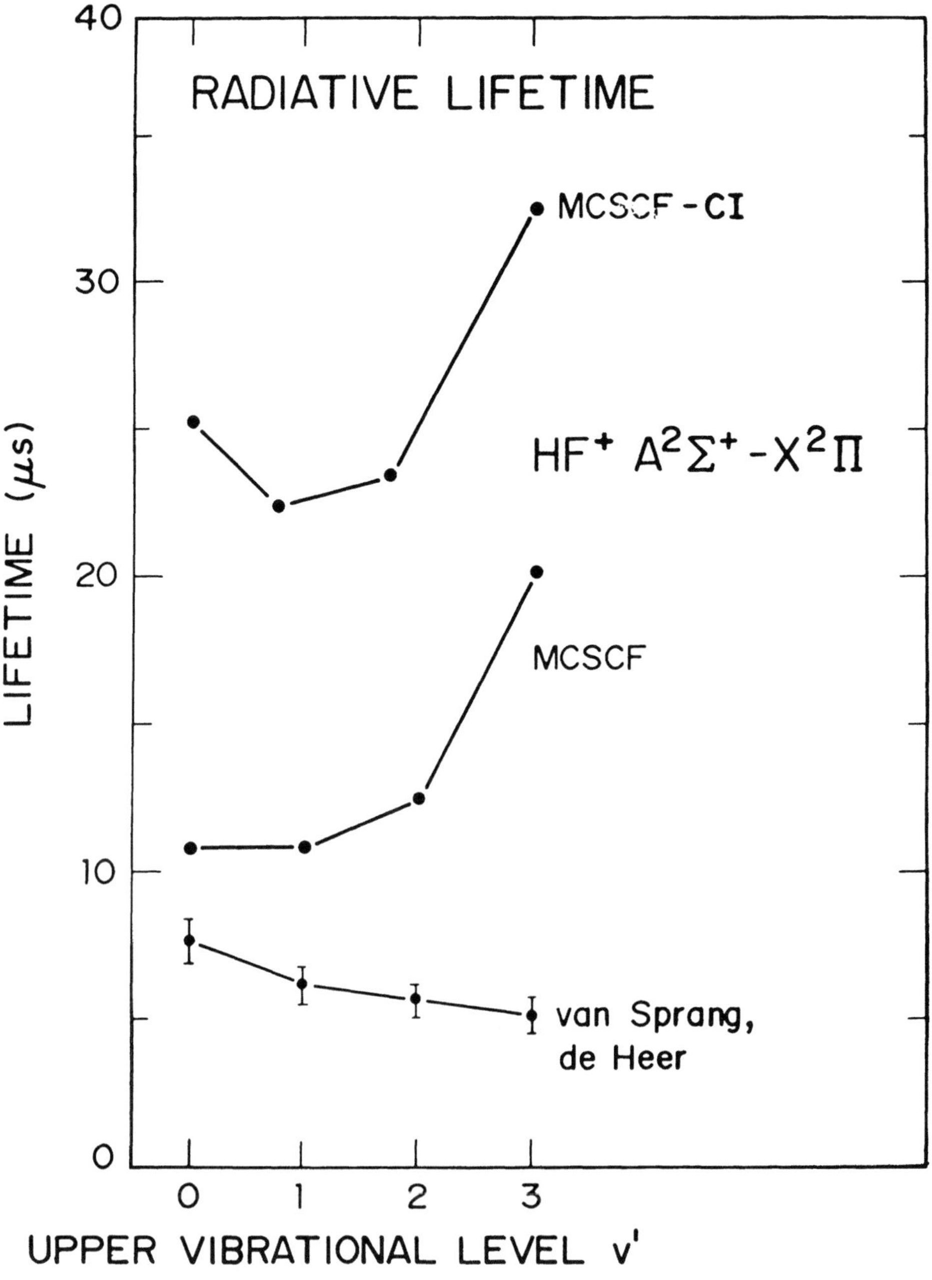

Figure 11. Comparison of calculated and measured radiative lifetimes of the $A^2\Sigma^+$ state of HF^+ as a function of the initial vibrational level.

lifetime for v'=0 in the A state of SH to be 704 ns (exp. 820± 240 ns) and for SD 688 ns (exp. 730±180 ns). The absorption oscillator strengths f_{00} have been calculated to be 10.2 x 10^{-4} (SH) and 9.6 x 10^{-9} (SD), respectively.

An example of the use of calculated transition rates for the analysis of electronic chemiluminescence spectra is the work done for BH^+ [16] and $A\ell H^+$. These ions are formed in the $A^2\Pi$ state or the $B^2\Sigma^+$ state from the reaction of B^+ or $A\ell^+$ ions with H_2 [73]. The calculated transition probabilities for BH^+ are shown in Fig. 13. It is noteworthy that there are two frequency regions in which strong B-X transitions are predicted. This can readily be explained by the overlap of the vibrational wavefunctions. In fact, the short wavelength part of this spectrum was first overlooked due to a strongly decreasing sensitivity of the spectrometer below 3000 Å. It was later discovered as a consequence of the theoretical prediction [74]. The calculated transition rates have been used by Ottinger and Reichmuth [74] to determine initial vibrational populations as a function of the B^+ impact energy. Some of these results are shown in Fig. 14. Presently, calculations of the relevant potential energy surfaces are being performed in our and other groups, in order to rationalize these results. Very recently, stimulated by our theoretical predictions Ottinger and coworkers [73] discovered a chemiluminescence of $A\ell H^+$ resulting from the reaction $A\ell^+ + H_2$; the spectra are in very good agreement with the theoretically predicted ones.

Another interesting ion that was studied is C_2^- [26]. It has two bound electronically excited states (see Fig. 15). To our knowledge, it is the only negative ion in the gas phase in which emission from an excited state has been observed. The lifetime for the $B^2\Sigma_u$ state has recently been measured [75]. In high vibrational levels of this state (v'>5) autodetachment takes place. The rates for this process have been determined from ultrahigh resolved resonances [76]. Our calculated lifetimes and oscillator strengths are shown in Table 13. The values obtained for the $B^2\Sigma_u v'=0,1$ states are in excellent agreement with the experimental values of Leutwyler et al [75]. The predictions

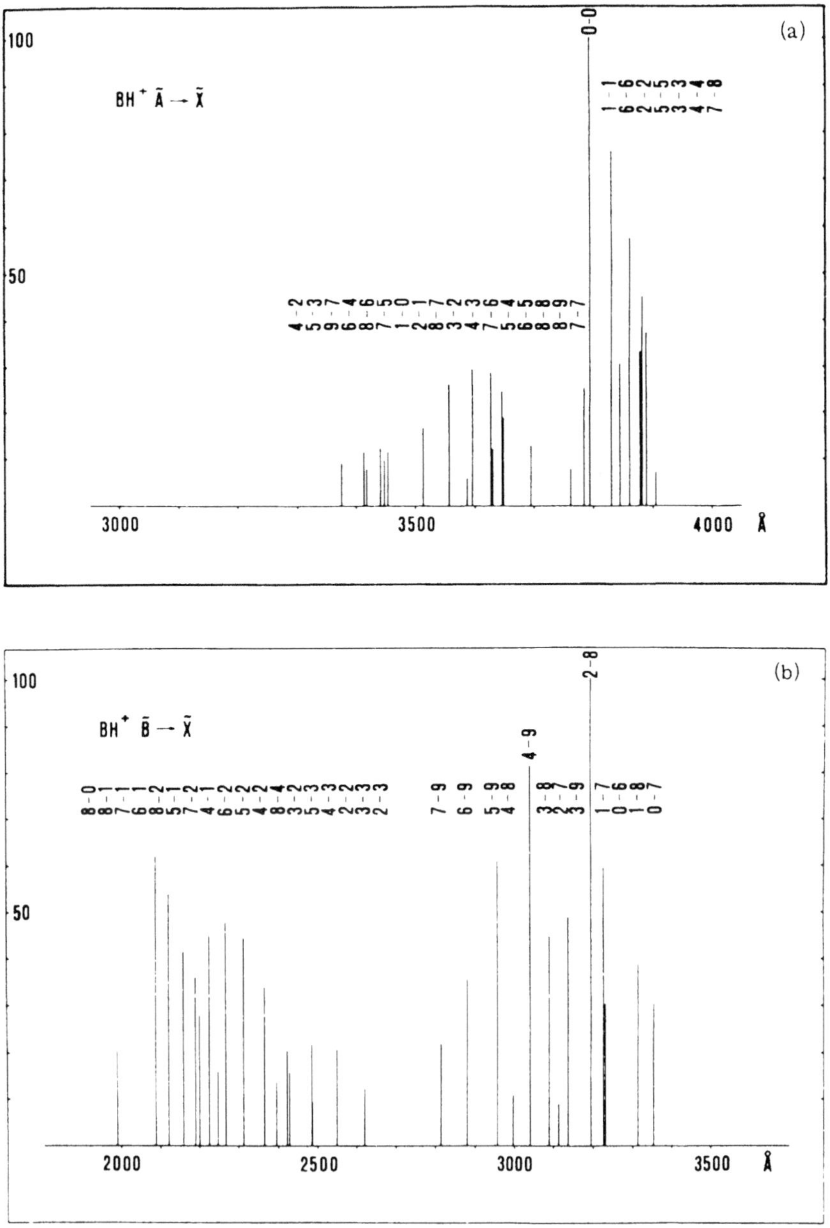

Figure 12. Potential energy functions including the vibrational levels of the $A^2\Sigma^+$, $a^4\Sigma^-$ and $X^2\Pi$ states of SH and the $A^2\Sigma^+$ - $X^2\Pi$ transition moment function as calculated from the MCSCF-SCEP wavefunctions.

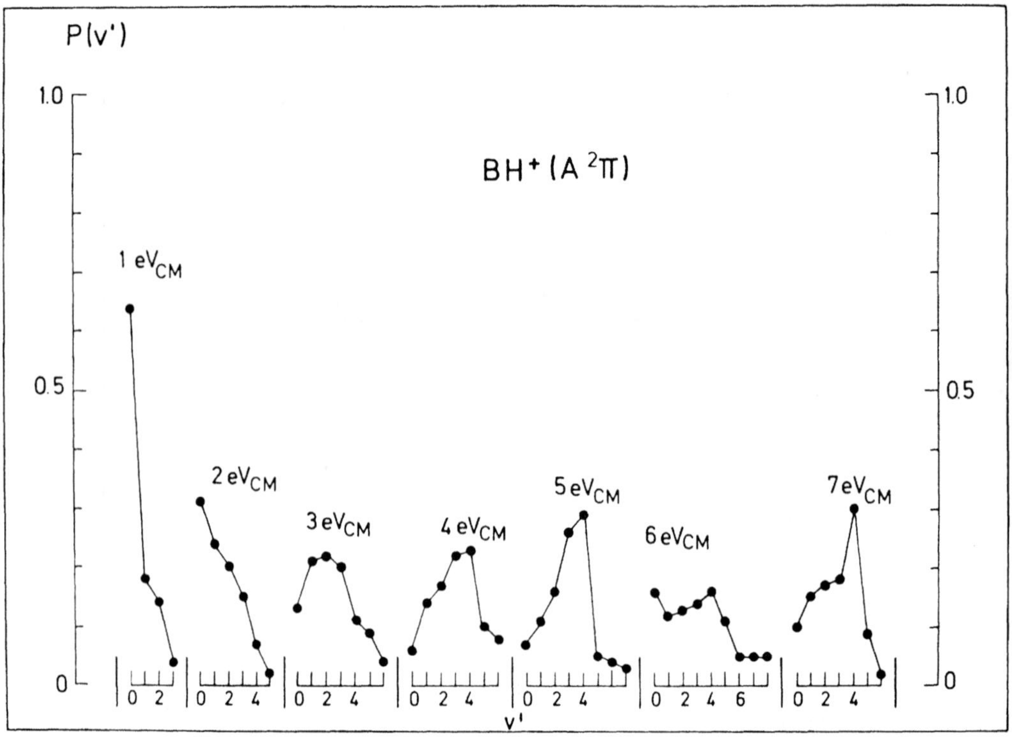

Figure 13. Calculated relative band emission intensities for the $A^2\Pi$-$X^2\Sigma^+$ and $B^2\Sigma^+$-$X^2\Pi$ transitions in BH^+.

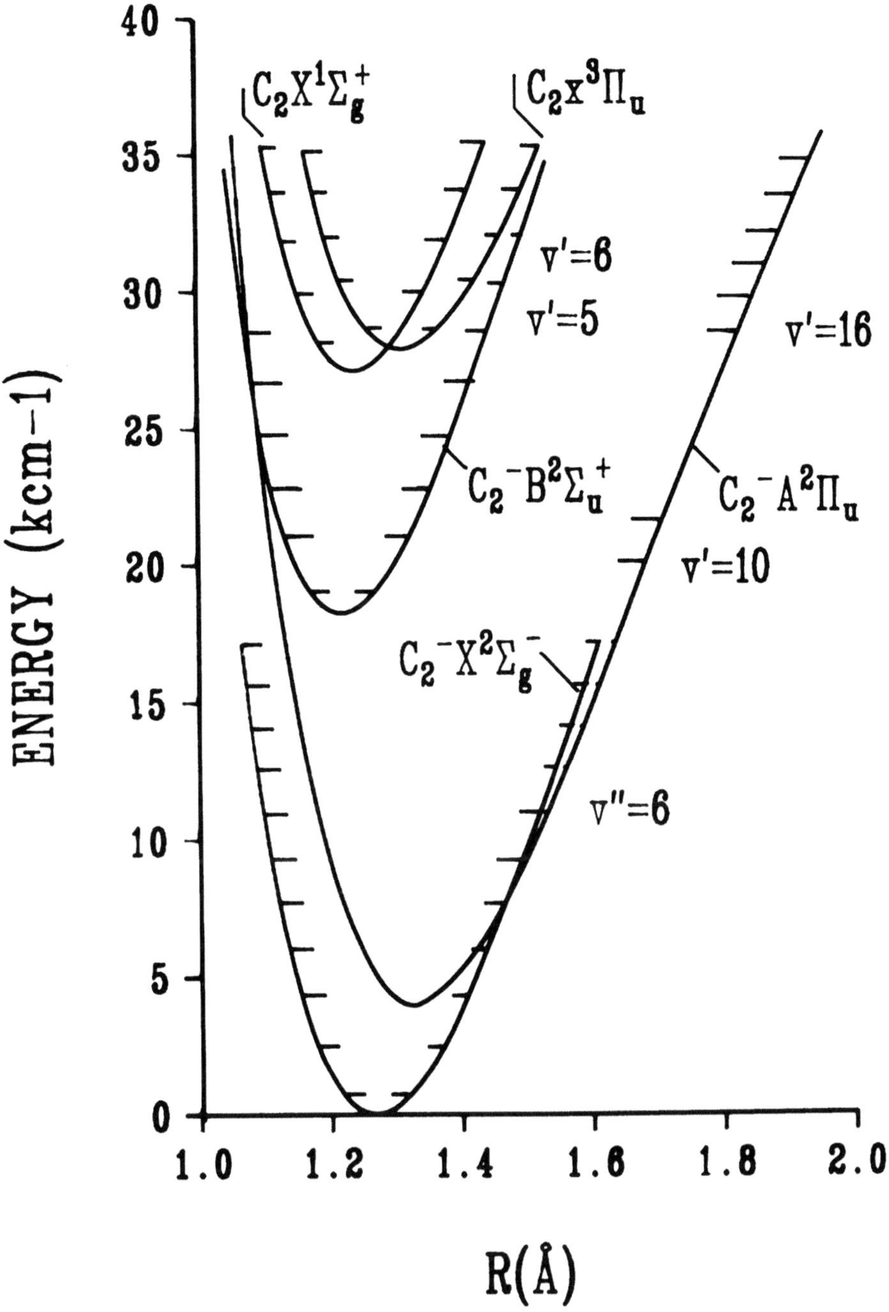

Figure 14. Vibrational product distribution in the reaction B^+ + H_2 as a function of the B^+ impact energy (from Ref. [74]).

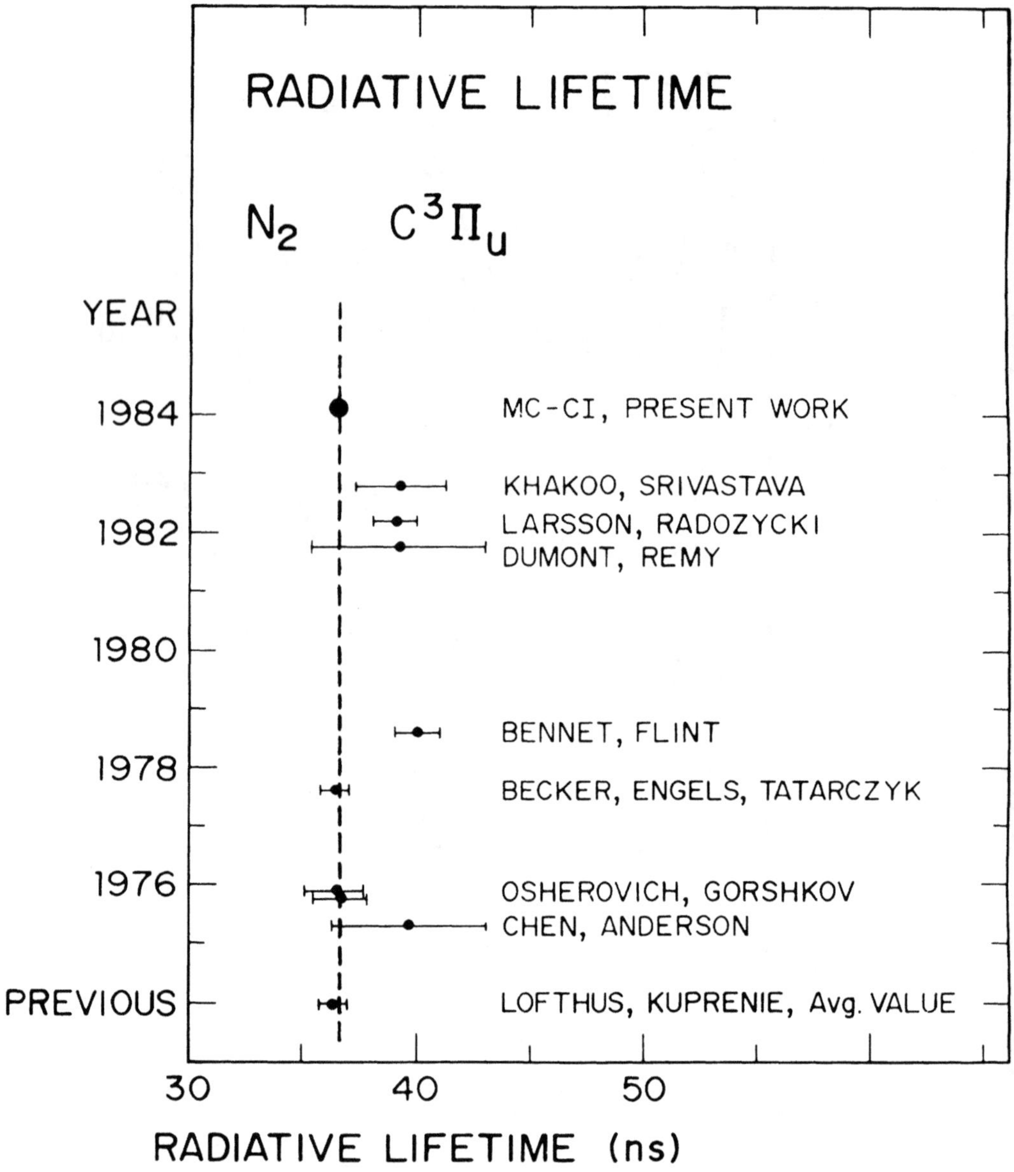

Figure 15. Potential energy functions of C_2^- and low lying states of C_2.

Table 13. Radiative lifetimes and oscillator strengths of the $A^2\Pi_u$ and $B^2\Sigma_u^+$ states of C_2^-.

	$A^2\Pi_u$	$B^2\Sigma_u^+$	
	MCSCF-SCEP	MCSCF-SCEP	Experiment[a]
Lifetimes:			
V'=0	49.9 μs	76.5 ns	77±8 ns
1	40.6	75.8	73±7
2	34.6	75.3	
3	30.5	75.1	
4	27.4	75.3	
Oscillator Strengths:			
	$f_{00}=0.340\text{x}10^{-2}$	$f_{00}=0.436\text{x}10^{-1}$	$f_{00}=(0.44\pm0.04)\text{x}10^{-1}$
	$f_{01}=0.129\text{x}10^{-2}$	$f_{00}=0.144\text{x}10^{-1}$	

[a] Ref. [75].

made for the $A^2\Pi_u$ state are of considerable interest for IR spectroscopy. In the (0,0) band origin at 4160 cm^{-1} the transition matrix element R_0^o was calculated to be 0.94 D, which leads to a transition rate of $1.9\times10^4 s^{-1}$ and to radiative lifetimes in the microsecond range. This transition rate is about a hundred times larger than the 1-0 vibrational transition rate in hydrogen fluoride, which is known to be a good infrared emitter [77]. Thus, the C_2^- ion should be an excellent candidate for studying emission processes in the infrared spectral region.

The C_2^+ ion is a probable constituent of many space environments. Its spectroscopic constants, however, are not yet established experimentally. Very recently, O'Keefe, Derai and Bowers [92] observed the translational energy spectra of C_2^+ with low resolution. Several transitions have been tentatively assigned as excitations of the lowest $^2\Pi_u$ state of C_2^+. In addition, two transitions were attributed to the excitations of the quartet states. The only assignment of a rotationally resolved absorption band to the C_2^+ ion proposed by Meinel [93] has been questioned in a thorough theoretical study by Petrongolo, Bruna, Peyerimhoff and Buenker [94].

We have performed MCSCF-SCEP and CASSCF calculations [95] for the nine lowest electronic states of C_2^+ in order to obtain its radiative transition probabilities needed for astrophysical purposes. The results for the three lowest quartet states are displayed graphically in Figures 16 and 17. The calculated spectroscopic constants are: $X^4\Sigma_g$: v_e = 1.411r, ω_e = 1335 cm^{-1}; $A^4\Pi_g$: r_e = 1.259Å, ω_e = 1874 cm^{-1} and $B^4\Sigma_u$: r_e=1.352Å, ω_e = 1507 cm^{-1}. In the electric dipole approximation the A-X transition is not allowed. The radiative decay of the B state can populate both the X and the A state. The B-X transition energy T_e is calculated to be 2.47 eV, i.e., this band system will lie very close to the Swan system of C_2. The MCSCF-SCEP v'=0 lifetime in the B state is calculated to be 148 ns for radiation into the X state and 8.7 μs into the A state. The former emission system will, therefore, be much stronger. The CASSCF method, with all valence orbitals active, yields longer lifetimes (171 ns for v'=0) for the B state than

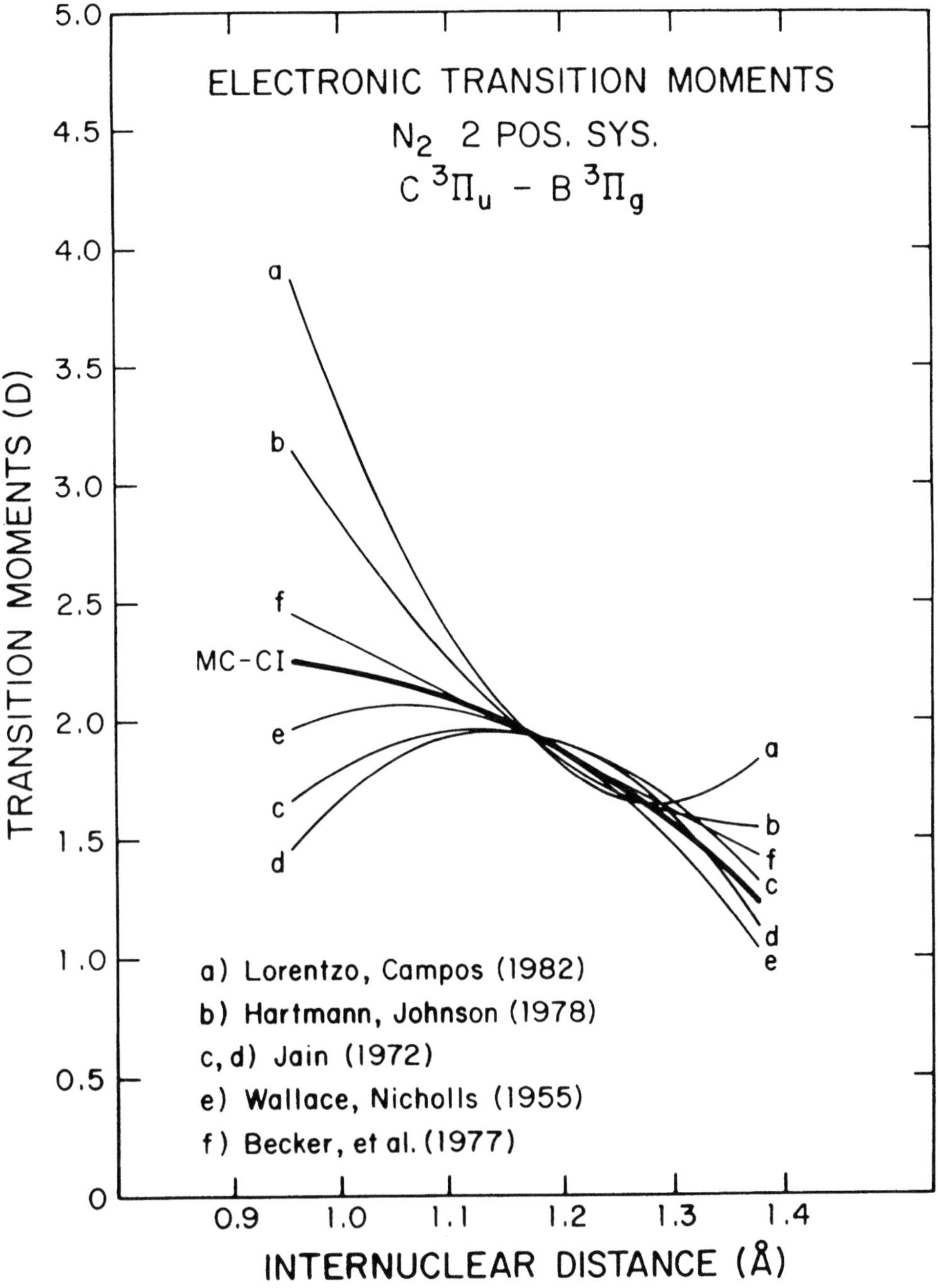

Figure 16. MCSCF-SCEP potential energy functions of the three lowest quartet states of C_2^+ (values in a.u.).

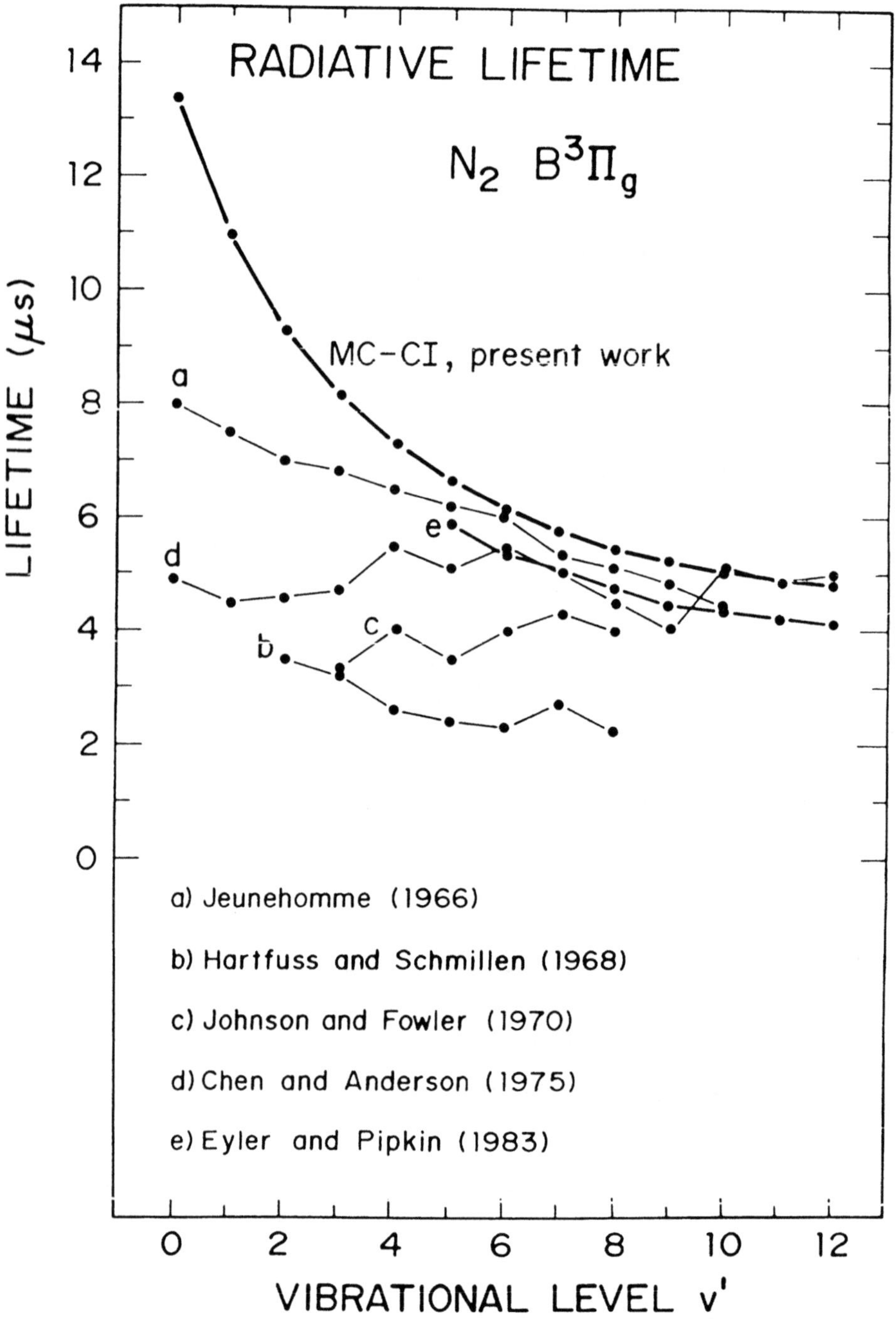

Figure 17. MCSCF, MCSCF-SCEP and CASSCF transition moment functions for the quartet transitions in C_2^+ (values in a.u.).

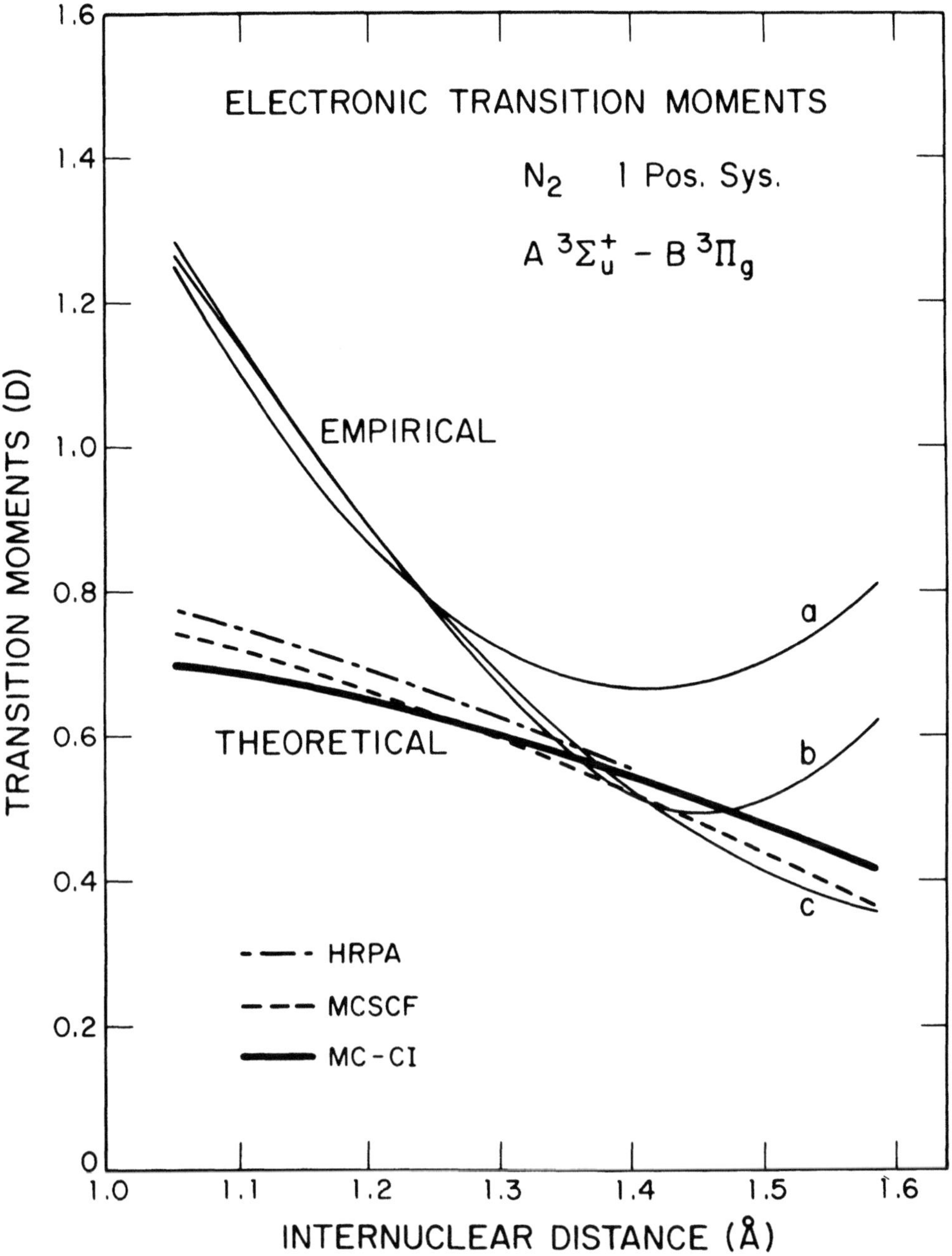

Figure 18. Comparison of calculated and measured radiative lifetimes of the $C^3\Pi_u$, v'=0 state of N_2.

the more reliable MCSCF-SCEP method. The absorption oscillator strengths f_{00} have been calculated to be 1.56 x 10^{-2} (B-X) and 4.4 x 10^{-4} (B-A), respectively. All the considered transitions within the doublet series exhibit lifetimes on the microsecond time scale and these transitions will be, therefore, much weaker.

Finally, we present calculations for the lowest five triplet states of N_2 [17]. The transition probabilities between vibronic levels of these states are of importance, e.g., for technological exploitation of lasing bands [78] and for the understanding of auroral phenomena [79]. Reliable experimental radiative lifetimes are available only for the $C^3\Pi_u$ state. Several recent measurements yielded values in close agreement (cf. Fig. 15) [80-85]. In the first positive system ($B^3\Pi_g$-$A^3\Sigma_u^+$) there are numerous measurements of relative intensities and radiative lifetimes, but the results of these studies differ considerably [86-88]. For the Wu-Benesch system $W^3\Delta_u$-$B^3\Pi_g$ and for the infrared afterglow system, $B'^3\Sigma_u^-$-$B^3\Pi_g$ only rough estimates of the transition probabilities have been available [86].

The well known lifetime of the $C^3\Pi_u$ state can serve to check the accuracy of the calculated data. As seen in Fig. 18, our calculated value is in excellent agreement with the various measured lifetimes. Apparently, there are two groups of experiments which yielded differing results. Although our calculated value is in better agreement with the lower values, we do not consider our calculations accurate enough to discriminate between the experimental differences. However, the good agreement indicates that the calculated lifetime is accurate to about 10 percent. A similar agreement can also be expected for the other transitions.

Several attempts have been made to derive the transition moment function for the second positive system ($C^3\Pi_u$-$B^3\Pi_g$) from measured relative intensities. Most of the published functions were represented by quadratic polynomials. None of these functions is in reasonable agreement with the calculated transition moment function (cf. Fig. 19). Some of them showed convex or concave variation with the internuclear distance. These discrepancies are probably due to

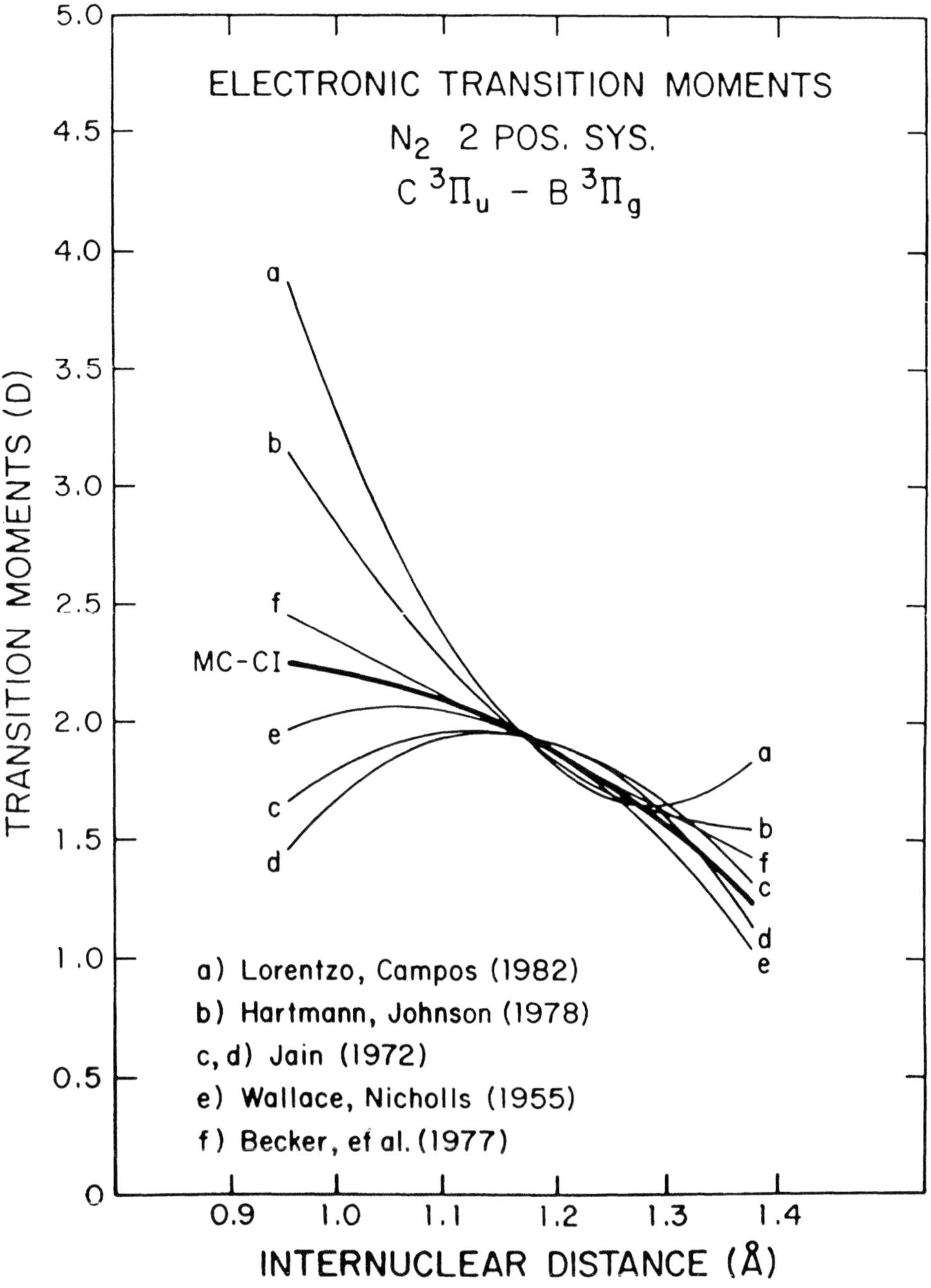

Figure 19. Comparison of computed and empirical transition moment functions for the second positive system ($C^3\Pi_u - B^3\Pi_g$) of N_2.

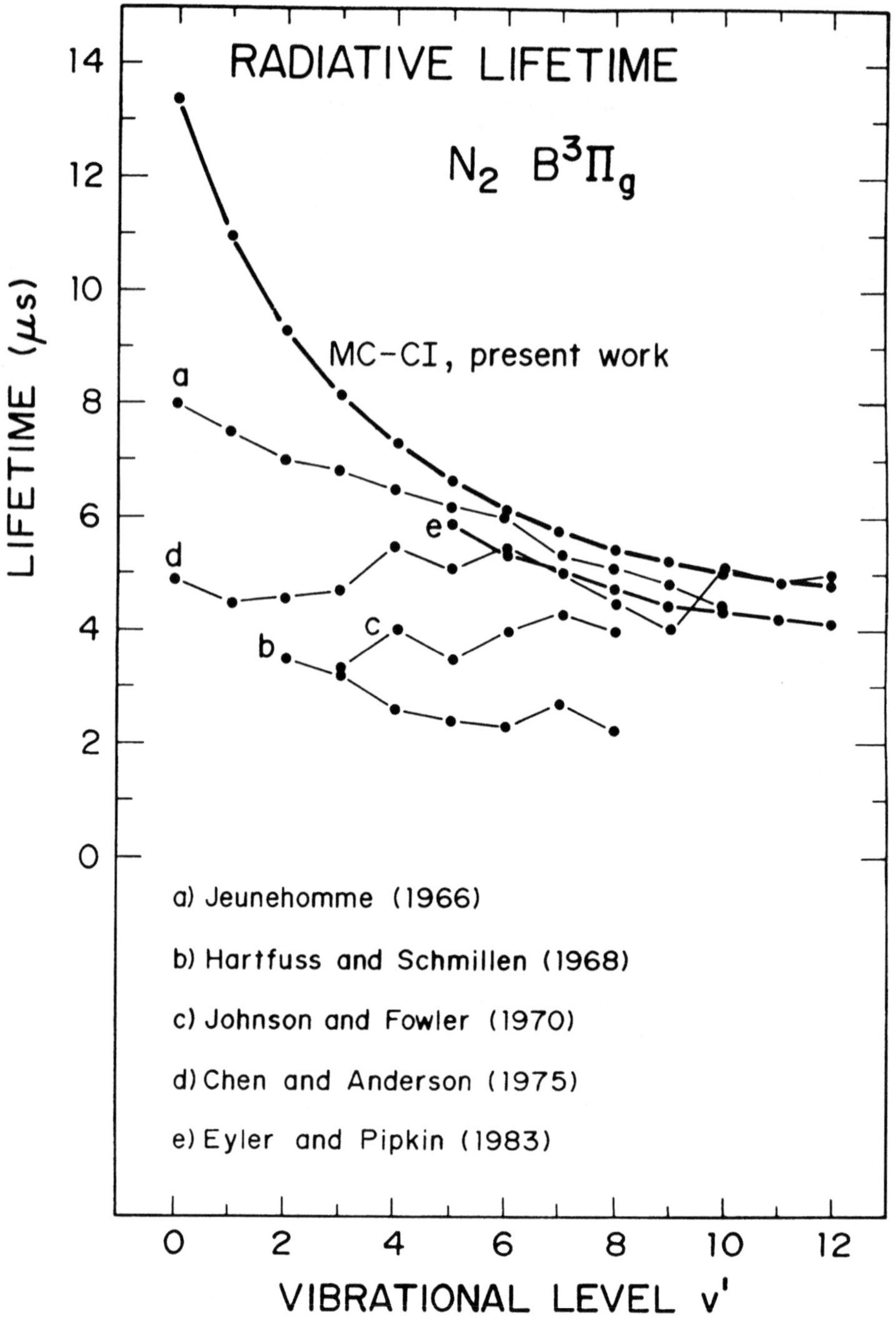

Figure 20. Comparison of calculated and measured radiative lifetimes of the $B^3\Pi_g$ state of N_2 as a function of the initial vibrational level.

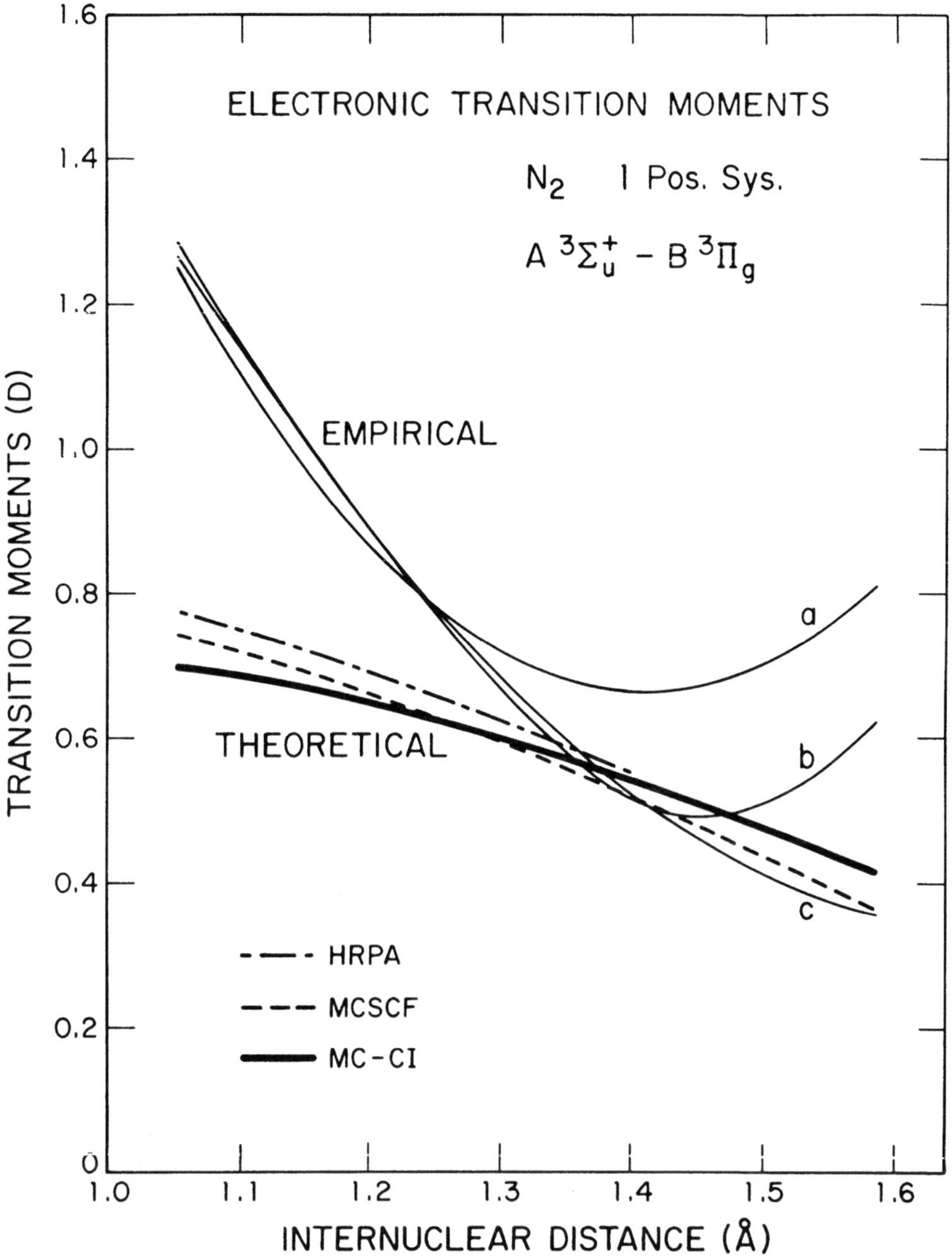

Figure 21. Comparison of calculated and empirical transition moments for the first positive system ($B^3\Pi_g - A^3\Sigma_u^+$) of N_2.

Table 14. Calculated and empirical radiative lifetimes of the $W^3\Delta_u$ and $B'^3\Sigma_u^-$ states of N_2 (in μs).

	$^3\Delta_u$		$^3\Sigma_u^-$	
v'	MCSCF-SCEP	a	MCSCF-SCEP	b
0	31.6×10^6	16.7×10^6	46	↑
1	45.2×10^2	20.0×10^2	36.1	
2	12.2×10^2	489	30.1	
3	607	234	25.9	
4	380	138	23.0	25–52
5	268	93	20.8	
6	203	69	19.0	
7	162	54	17.7	
8	134		16.5	
9	115		15.6	↓
10	100		14.8	

[a] Covey et al. [86], error estimate 50%.
[b] Wentink et al. [86].

uncertainties in the weak transitions, which were used to determine the transition moments at short distances. Some authors have used a linear approximation to the transition moment function. The slopes of these functions are in rather good agreement with each other and with the slope of the calculated function in the region of the r-centroids for the strongest transitions (around 1.2Å).

Fig. 20 shows the calculated and experimental lifetimes for the $B^3\Pi_g$ state. In this case the experimental values show large variations. The most reliable measurements appear to be those of Carlson et al. [87] and of Eyler and Pipkin [88]. In the latter case the laser induced fluorescence technique has been used to excite individual rovibronic levels of the B state, and single exponential fluorescence decay was observed. The variation of the measured lifetimes with the initial vibrational state closely parallels the calculated ones. However, the theoretical lifetimes are consistently larger by about 15 percent. Since errors of this magnitude can neither be excluded from the experimental nor from the theoretical data, it is difficult to decide which values are more accurate. Accurate measurements of the lifetimes from low vibrational levels (v'=0-5) are desirable, because previous experimental values appear to be too low (cf. Fig. 17). Fig. 21 displays calculated and empirical transition moment functions for the first positive system ($B^3\Pi_g$-$A^3\Sigma_u$). As in the case of the second positive system, the variation of the empirical transition moments with the internuclear distance is much too strong. It is very unlikely that the calculated transition moments at short distances are in error by 100 percent, since the correct variation of the lifetimes with the initial vibrational state indicates that the shape of the calculated function at larger distances is accurate. The predicted lifetimes for the $W^3\Delta_u$ and $B'^3\Sigma_u^-$ states are shown in Table 14 and compared to previous estimates. The lifetimes of low lying vibrational levels of the $W^3\Delta_u$ state are extremely long, because this state is nearly degenerate with the $B^3\Pi_u$ state. In this case rotational and spin orbit effects may extensively influence the lifetimes of the low lying vibrational levels. The corresponding data in Table 14 are to be considered as estimates. Tables with predicted transition probability coefficients for the B-A, W-B, and B'-B systems of N_2 can be found in Ref. [17].

6. CONCLUSIONS

For small diatomic molecules and molecular ions radiative transition probabilities for vibrational and electronic transitions can be calculated with an accuracy of 10-15 percent, provided highly correlated electronic wavefunctions are employed. In many cases the theoretical data appear to be more reliable than experimental values. Hence, such calculations are a valuable complement to experimental studies as illustrated by chemical dynamics studies and astrophysical investigations. It is hoped that this work stimulates further cooperative experimental and theoretical work in these areas.

Acknowledgements

The authors are indebted to S.R. Langhoff for the calculations of the rovibrational transition probabilities of OH from the MCSCF-SCEP dipole moment function, and to E.F. van Dishoeck, Ch. Ottinger, and K.I. Peterson for providing their results prior to publication. The critical reading of the manuscript and many valuable suggestions by M. McCarthy are gratefully acknowledged.

References:

1. E.E. Whiting, A. Schadee, J.B. Tatum, J.T. Hougen and R.W. Nicholls, J. Molecular Spectr. 80, 249 (1980).
2. M. Larsson, Astronomy annd Astrophysics 128, 291 (1983).
3. see, e.g. R.H. Tipping, J. Molec. Spectr. 43, 31 (1976).
4. P.A. Frazer, Can. J. Phys. 32, 515 (1954).
5. see, e.g. J.C. McCallum, J. Quant. Spectrosc. Radiat. Transfer 21, 563 (1979).
6. S.M. Yazykova and E.V. Butyrskaya, J. Phys. B13, 3361 (1980).
7. H.-J. Werner and W. Meyer, J. Chem. Phys. 73, 2342 (1980) and references therein.
8. H.-J. Werner and W. Meyer, J. Chem. Phys. 74, 5794 (1981).
9. H.-J. Werner and P.J. Knowles, J. Chem. Phys., in press. For other recent methods see references therein.
10. P.J. Knowles and H.-J. Werner, Chem. Phys. Letters 115, 259 (1985).

0a. W.C. Lineberger, private communication.
1. W. Meyer, J. Chem. Phys. 64, 2901 (1976).
2. H.-J. Werner and E.A. Reinsch, J. Chem. Phys. 76, 3144 (1982).
3. H.-J. Werner and E.A. Reinsch, in "Advanced Theories and Computational Approaches to the Electronic Structure of Molecules," p. 79 (edited by C.E. Dykstra, D. Reidel Publishing Company, 1984).
4. H.-J. Werner, J. Chem. Phys. 80, 5080 (1984).
5. W. Meyer, Int. J. Quantum Chem. 55, 341 (1971).
6. R. Klein, P. Rosmus and H.-J. Werner, J. Chem. Phys. 77, 3559 (1982).
7. H.-J. Werner, J. Kalcher and E.A. Reinsch, J. Chem. Phys. 81, 2420 (1984).
8. M. Larsson, P.E.M. Siegbahn and H. Ågren, Astrosphys. J. 272, 369 (1983).
9. M. Larsson and P.E.M. Siegbahn, J. Chem. Phys. 79, 2270 (1983).
20. H.-J. Werner, P. Rosmus, W. Schätzl, and W. Meyer, J. Chem. Phys. 80, 831 (1984).
21. W. Schätzl and W. Meyer, to be published.
22. P.Å. Malmquist, in press.
23. H.-J. Werner, unpublished results.
24. H.-J. Werner, P. Rosmus and E.A. Reinsch, J. Chem. Phys. 79, 905 (1983).
25. J. Senekowitsch, P. Rosmus and H.-J. Werner, E.-A. Reinsch and S. O'Neil, to be published.
26. P. Rosmus and H.-J. Werner, J. Chem. Phys. 80, 5085 (1984).
27. H.-J. Werner, P.J. Hay and D. Cartwright, to be published.
28. H.-J. Werner, E.A. Reinsch and P. Rosmus, Chem. Phys. Lett. 78, 311 (1981).
29. W.L. Meerts and A. Dymanus, Chem. Phys. Lett. 23, 45 (1973).
30. K.I. Peterson, G.T. Frazer and W. Klemperer, Canad. J. Phys. 62, 1502 (1984).
31. A.B. Meinel, J. Astrophys. 111, 555 (1950).
32. see, e.g., J.A. Coxon and S.C. Foster, Canad. J. Phys. 60, 41 (1982) and references therein.
33. W.S. Benedict and E.K. Plyler, Energy Transfer in Hot Gases, Natl. Bur. Stand. 523 (U.S. GPO, Washington, D.C., 1954, p. 54).
34. J. D'Incan, C. Effantin and F. Roux, J. Quant. Spectrosc. Radiat. Transfer 11, 1215 (1971).
35. F. Roux, J. D'Incan and D. Cerny, J. Astrophys. 186, 1141 (1973).
36. R.E. Murphy, J. Chem. Phys. 54, 4852 (1971).
37. W.J. Stevens, G. Das, A. Wahl, M. Krauss and D. Neumann, J. Chem. Phys. 61, 3686 (1974).
38. W. Meyer, Theor. Chim. Acta 35, 277 (1974).
39. S.I. Chu, M. Yoshimine and B. Liu, J. Chem. Phys. 61, 5389 (1974).
40. S.R. Langhoff, E.F. van Dishoeck, R. Wetmore and A. Dalgarno, J. Chem. Phys. 77, 1379 (1982).

41. R.J. Fallon, I. Tobias and J.T. Vanderslice, J. Chem. Phys. 34, 167 (1961).
42. H.-J. Werner and P. Rosmus, J. Chem. Phys. 73, 2319 (1980).
43. F.H. Mies, J. Mol. Spectrosc. 53, 150 (1974).
44. A.J. Sauval, N. Grevesses, J.W. Brault, G.M. Stokes and R. Zander, Astrophys. J. 282, 330 (1984).
45. N. Grevesse, A.J. Sauval and E.F. van Dishoeck, Astronomy and Astrophysics, 141, 10 (1984).
46. H. Holweger and E.A. Müller, Solar Physics 39, 19 (1974).
47. J.E. Vernazza, F.H. Avrett and R. Loeser, Astrophys. J. Suppl 30, 1 (1976).
48. B.S. Argawalla, A.S. Manocha and D.W. Setser, J. Phys. Chem. 85, 2873 (1981).
49. S.R. Langhoff, private communication.
50. P. Rosmus, E.A. Reinsch and H.-J. Werner, in "Molecular ions," edited by J. Berkowitz and K.O. Groeneveld, Plenum Publishing Corp., 1983; R. Klein and P. Rosmus, Z Naturforsch 39a, 348 (1984).
51. J.W. Brault and S.P. Davis, Phys. Scr. 25, 268 (1982).
52. H.-J. Werner, Mol. Phys. 44, 111 (1981).
53. M.F. Weisbach and C. Chackerian, Jr., J. Chem. Phys. 59, 4272 (1973).
54. H.-J. Werner and P. Rosmus, J. Mol. Spectrosc. 96, 362 (1982).
55. P. Rosmus, H.-J. Werner and M. Grimm, Chem. Phys. Letters 92, 250 (1982); R. Klein and P. Rosmus, Theor. Chim. Acta 66, 21 (1984).
56a. H.-J. Werner, P. Rosmus and M. Grimm, Chem. Phys. 73, 169 (1982);
b. P. Botschwina and P. Rosmus, J. Chem. Phys., in press.
57. P. Rosmus and H.-J. Werner, Mol. Phys. 47, 661 (1982).
58. see, e.g., W.W. Rice, W.H. Beattie, R.C. Oldenborg, S.E. Johnson and P.B. Scott, Appl. Phys. Letters 28, 444 (1976).
59. M.A. Smith, V.M. Bierbaum and S.R. Leone, Chem. Phys. Letters 94, 398 (1983).
60. J. Brzozowski, P. Bunker, N. Elander and P. Erman, Astrophys. J. 207, 414 (1976).
61. K.H. Becker, H.H. Brenig and T.T. Tatarcyk, Chem. Phys. Letters 71, 242 (1980).
62. see review of K. Schofield, J. Chem. Phys. Ref. Data 8, 723 (1979).
63. I.S. Dermid and J.B. Laudenslager, J. Chem. Phys. 76, 1824 (1982).
64. T. Bergeman, P. Erman, Z. Haratym and M. Larsson, Phys. Scr. 23, 45 (1981).
65. W.L. Dimpfl and J.L. Kinsey, J. Quant. Spectrosc. Radiat. Transfer 21, 233 (1979).
66. J. Brzozowski, P. Erman and M. Lyyra, Phys. Scr. 17, 507 (1978).
67. K.R. German, J. Chem. Phys. 63, 5252 (1976).
68. C.C. Wang and C.M. Huang, Phys. Rev. A 21, 1235 (1980).
69. G.P. Smith and D.R. Crosley, Eighteenth International Symposium on Combustion, 1511 (1981).

70. G.R. Möhlmann, K.K. Bhutani and F.I. de Heer, Chem. Phys. 21, 127 (1977).
71. C. Martner, J. Pfaff, H. Rosenbaum, A. O'Keefe and J. Saykally, J. Chem. Phys. 78, 7074 (1983).
72. H.A. van Sprang and F.E. de Heer, Chem. Phys. 33, 73 (1978).
73. Ch. Ottinger and J. Reichmuth, J. Chem. Phys. 74, 928 (1981).
74. Ch. Ottinger, private communication. See J. Reichmuth, Diplomarbeit, Max-Planck Institut für Strömungsforschung, Göttingen (West Germany), 1981.
75. S. Leutwyler, J.P. Meyer and L. Misev, Chem. Phys. Lett. 91, 206 (1982).
76. U. Hefter, R.D. Mead, P.A. Schulz and W.C. Lineberger, Phys. Rev. A 28, 1429 (1983).
77. R.N. Sileo and T.A. Cool, J. Chem. Phys. 65, 117 (1976).
78. see, e.g., D. Cerny, R. Bacis, R.W. Field and R.A. McFarlane, J. Phys. Chem. 85, 2626 (1981) and references therein.
79. see, e.g., W. Benesch, J. Chem. Phys. 78, 2978 (1983).
80. A.L. Osherovich and V.N. Gorshkov, Opt. Spectrosc. 41, 92 (1976).
81. K.H. Becker, H. Engels and T. Tatarczyk, Chem. Phys. Lett. 51, 111 (1977).
82. W.R. Bennet, Jr. and J. Flint, Phys. Rev. A 18, 2527 (1978).
83. M.N. Dumont and F. Remy, J. Chem. Phys. 76, 1175 (1982).
84. M. Larsson and T. Radozycki, Phys. Scr. 25, 627 (1982).
85. M.A. Khakoo and S.K. Srivastava, J. Quant. Spectrosc. Radiat. Transfer 30, 31 (1983).
86. see review of A. Lofthus and P.H. Kuprenie, J. Phys. Chem. Ref. Data 6, 113 (1977)and references therein.
87. T.A. Carlson, N. Duric, P. Erman and M. Larsson, Phys. Scr. 19, 25 (1979).
88. E.E. Eyler and F.M. Pipkin, J. Chem. Phys. 79, 3654 (1983).
89. P.F. Bernath, T. Amano and M. Wong, J. Mol. Spectrosc. 99, 20 (1983).
90. R.J. Winkler and S.P. Davis, Canad. J. Phys. 62, 1420 (1984).
91. R.R. Friedl, W.H. Brune and J.G. Anderson, J. Chem. Phys 79, 4227 (1983).
92. A. O'Keefe, R. Derai and M.T. Bowers, Chem. Phys. 91, 161 (1984).
93. H. Meinel, Canad. J. Phys. 50, 158 (1972).
94. C. Petrongolo, P.J. Bruna, S.D. Peyerimhoff and R.J. Buenker, J. Chem. Phys. 74, 4594 (1981).
95. C. Januschewski, P. Rosmus, H.-J. Werner and M. Larsson, to be published.

EXCITED STATES OF Li_2 AND THE GROUND ELECTRONIC STATE OF Li_2^+

R.A. Bernheim, L.P. Gold and C.A. Tomczyk
Department of Chemistry
The Pennsylvania State University
152 Davey Laboratory
University Park, PA 16802

ABSTRACT. Because the Li_2 dimer is the least complex stable homonuclear diatomic molecule besides hydrogen, it is an interesting and important testing ground for comparison between theory and experiment. In particular, the elucidation of the excited state structure of Li_2 has strongly benefitted from the close interaction between the experimental and ab initio theoretical investigations. In this work the experimental and theoretical results are reviewed and compared. Recent results for the very interesting $3^1\Sigma_g^+$ double minimum potential state are presented.

1. INTRODUCTION AND BACKGROUND

Theoretical and spectral interest in the electronic structure of the lithium dimer dates back to the 1930's [1,2]. Part of this interest is due to the relatively simple electronic structure of Li_2. However, it was also recognized at an early stage that the bonding in the lithium dimer differed in an unusual way from that of the hydrogen molecule. For example, it was found that the bond energy of the lithium dimer ion is actually larger than the bond energy of the neutral molecule [3,4]. This is contrary to what is seen for the hydrogen molecule where the ion has the smaller bond energy [5], and is contrary to the simple concept of an increase in bond energy with the number of "bonding" electrons. Indeed, all of the homonuclear alkali dimers exhibit an increased bond energy for the ion compared with that of the neutral molecule [5]. These and

R. J. Bartlett (ed.), Comparison of Ab Initio Quantum Chemistry with Experiment for Small Molecules, 325–337.

other aspects of the electronic structure of Li_2 and Li_2^+ have received recent theoretical attention [6].

With the increased availability of tunable dye lasers which operate in the visible spectral range, interest in the excited electronic state structure of the alkali diatomic molecules has grown dramatically over the past decade. The coincidences between the wavelengths of molecular transitions and the different ion laser lines has resulted in numerous laser induced fluorescence studies which have revealed in new information for the electronic ground and excited states. Another spur to experimental investigation comes from the fact that the vapor pressures of the alkali dimers are high enough that spectroscopic absorption and fluorescence experiments may be performed on the gaseous substances at reasonably accessible temperatures in vapor cells or molecular beams. As a consequence, the experiments on alkali vapors have generated interest in their possible application as laser media.

The conventional spectroscopic investigations that have been performed on the excited states of Li_2 include detailed studies of the $A^1\Sigma_u^+$ and $B^1\Pi_u$ states [7,8] and other measurements of the $C^1\Pi_u$ and $D^1\Pi_u$ states [4,9,10]. Work in our own laboratory has used pulsed optical-optical double resonance (OODR) techniques to study the *gerade* excited states [11-19]. These recent studies have revealed an additional three dozen excited electronic states for which the pulsed OODR spectra have been observed and assigned, molecular constants determined, RKR potentials derived, and Franck-Condon factors calculated. The present paper summarizes these experimental findings and compares them with *ab initio* treatments where available.

2. THE IONIZATION POTENTIAL OF Li_2 AND THE GROUND ELECTRONIC STATE CONSTANTS OF Li_2^+

The spectroscopic constants of the ground state of the lithium dimer ion, including its energy above the ground electronic state of the neutral lithium dimer, can be obtained from an extrapolation of the properties of the Rydberg states of Li_2. This has been done in our laboratory for the $ns\sigma\,^1\Sigma_g^+$, $nd\sigma\,^1\Sigma_g^+$, and $nd\pi\,^1\Pi_g$ Rydberg series using the pulsed OODR method [18,19]. This extrapolation was performed with the lowest vibrational levels (v = 0-7) of the Rydberg states and rotational levels in the J = 0-40 range. The most accurate value for the ionization potential comes from an extrapolation of the term energies of the $nd\pi\,^1\Pi_g^-$ components and yields $T_o(\infty) = 41496 \pm 4$ cm^{-1} [18]. No rotational correction of this value is required since the virtual v = 0, J = 0 levels of the $^1\Pi_g^-$ components correlate with the v = 0, N = 0 level of the $X^2\Sigma_g^+$ state of Li_2^+ [22]. The ionization potential has been determined by other workers in a variety of experiments including photoionization [33], electron impact ionization [24,25], and OODR with two argon ion lasers [26]. There are also two other studies [20,27] that use the autoionization of Rydberg states with vibrational levels v ⩾ 1. These studies have been discussed and compared with our own result [19]. In Table 1 the two best experimental values [18-20] are compared with several results from theory. A comprehensive discussion and comparison of the different theoretical treatments can be found in reference [28].

The experimental spectroscopic constants for the molecular ion are also displayed in Table 1 and compared with theory. The appropriate deviations between the highest level <u>ab initio</u> theory and best experiment are as follows: 0.5% or 0.02Å for R_e, 0.2% or 20 cm^{-1} for D_e, 0.5% or 1 cm^{-1} for ω_e, and 15 cm^{-1} or 0.03% for the ionization potential of the neutral dimer.

The number of molecular constants (usually Dunham coefficients) that are used to characterize an excited electronic state of a diatomic molecule vary according to the extent and accuracy of the known field of data and depend upon the complexity of the shape of the molecular potential as a function of internuclear separation. On the other hand, the molecular constants derived from theory rarely go beyond the values of B_e, α_e, ω_e, $\omega_e x_e$, D_e, and $T_e(\infty)$ for a description of the state. For many electronic states

Table 1. Comparison between *ab initio* theory and experiment for the ionization energies of Li_2 and the molecular constants for the ground $X^2\Sigma_g^+$ state of Li_2^+.[a]

Reference	$T_o(\infty)$	R_e (Å)	B_e (cm^{-1})	α_e (cm^{-1})
Experiment				
19	41 496(4)	3.11	0.496(2)	0.0052(17)
20	41 475(8)			
Theory				
5,21	41 480	3.127	0.491	
20		3.099	0.500	
28	41 450	3.096	0.501	0.00533
29		3.09	0.503	

Reference	ω_e (cm^{-1})	$\omega_e x_e$ (cm^{-1})	D_e (cm^{-1})
Experiment			
19	262.2(1.5)	1.7(0.5)	10 464(6)
20	263		10 485(24)
Theory			
5,21	264	1.67	10 324
20	263.5	1.7	10 445
28	265.5	1.89	10 445
29	268		10 470

[a] The numbers in parentheses give the experimental error in the last digit(s).

it is sufficient and meaningful to compare these first few experimental Dunham coefficients with the theoretical values as was done above for Li_2^+. However, when a large fraction (>50%) of the potential well is known experimentally, or when there exist extensive homogeneous perturbations or double minimum or "shelf" states, a comparison between theory and experiment of just five or six molecular constants can be misleading and is inappropriate. A more meaningful procedure is obtained by comparing an experimentally derived potential, such as the RKR, with the corresponding theoretical potential, point-by-point. Alternatively, spectroscopic information, such as the vibrational terms, G_v, or vibrational term splitting ΔG_v, can be compared, level-by-level with the corresponding information from theory. Those parameters that depend only weakly upon the shape of the potential, such as $T_o(\infty)$ and D_o can be compared directly. The zero-point energy correction to obtain the equilibrium values $T_e(\infty)$ and D_e are usually not strongly affected by the higher order Dunham coefficients and comparison of these quantities with theory is usually appropriate.

2.1 The Electronic Structure of the Excited States of Li_2

Two general experimental approaches have been used to investigate the excited electronic states of Li_2. Conventional absorption spectroscopy from the $X^1\Sigma_g^+$ ground state has been used to obtain data for the ungerade excited states, and optical-optical double resonance has been used to study the gerade excited states. Some spectroscopic information for one of the triplet states is also available from perturbations detected by measuring radiative lifetimes. Of course, the ground electronic state of Li_2 has been extensively characterized, including the portions of the potential near the dissociation limit [29,30]. The states of Li_2 that have received theoretical and experimental treatment are summarized in Table 2. Not surprisingly, the theoretical investigations are mainly confined to the lower electronic states. The experimental studies have been constrained by the diatomic molecular spectral

Table 2. A survey of the electronic states of Li_2 that have received theoretical study, experimental study, or both. Except where indicated, all states except the Rydberg and the $C^1\Pi_u$ and $D^1\Pi_u$ states have received theoretical treatment. Those states which have been experimentally studied are indicated by the letter (e), and those which are predicted to be unbound are indicated by (*).

Separated Li Atom States	Singlets	Triplets
$2^2S + 2^2S$	$X^1\Sigma_g^+$(e)	$1^3\Sigma_u^+$
$2^2S + 2^2P$	$2^1\Sigma_g^+$, $1^1\Pi_g$, $1^1\Sigma_u^+$, $1^1\Pi_u$	$1^3\Sigma_g^+$, $1^3\Pi_g$, $1^3\Pi_u$, $2^3\Sigma_u^+$
$2^2S + 3^2S$	$3^1\Sigma_g^+$, $2^1\Sigma_u^+$	$2^3\Sigma_g^+$, $3^3\Sigma_u^+$
$2^2P + 2^2P$	$2^1\Pi_g$, $4^1\Sigma_g^+$, $1^1\Delta_g$, $5^1\Sigma_g^+$	$2^3\Pi_g$, $1^3\Sigma_g^-$, $2^3\Pi_u$, $4^3\Sigma_u^+$,
	$2^1\Pi_u$, $1^1\Sigma_u^-$(*)	$1^3\Delta_u$, $5^3\Sigma_u^+$(*)
?	$C^1\Pi_u$, $D^1\Pi_u$, experiment only	
$2^2S + n^2S$	$ns\sigma$ $^1\Sigma_g^+$ Rydberg series ($n \leq 10$) experiment only	
$2^2S + n^2D$	$ns\sigma$ $^1\Sigma_g^+$ Rydberg series ($n \leq 10$) experiment only	
	$nd\pi$ $^1\Pi_g$ Rydberg series ($n \leq 15$) experiment only	

transition probabilities and experimentally convenient spectroscopic equipment.

A rough comparison between ab initio theory and experiment can be made by an examination of the first several Dunham coefficients for each state. In making this comparison it must be realized that, experimentally, there are often 10-20 Dunham coefficients required to adequately describe the rovibrational energy level positions. This is usually the case when a very large fraction of the potential well depth has been characterized and especially when homogeneous perturbations are present. States with double minimum potentials present special problems. Theoretically, the molecular constants D_e and R_e come directly from the minimization procedure, while ω_e and $\omega_e x_e$ can come from a fit of the Morse potential to the lowest portion of the ab initio potential [31]. The different theoretical treatments of the lowest excited states of the Li_2 have been reviewed and summarized [31]. A comparison between a reduced set of experimental molecular constants and the recent values of Konowalow and Fish [31] are given in Table 3. As with the ground state of the Li_2^+ ion, the agreement is excellent for the ground state of Li_2.

2.2 The "Double Minimum" $3^1\Sigma_g^+$ State of Li_2

In previous OODR experiments on the $3^1\Sigma_g^+$ state (also referred to as the $E^1\Sigma_g^+$ state), a Birge-Sponer extrapolation gave a dissociation energy that was 5000 cm^{-1} below the energy of the nearest pair of correlated atom states [15]. It was suggested that a possible source for this behavior was the existence of a double minimum potential such as exists for the H_2 molecule [35]. Ab initio calculations by Konowalow and Fish [36] showed that, indeed, the state had a double minimum potential with a shallow outer minimum. The origin of this behavior is the large mixing with the ion-pair configuration in this region.

In an effort to understand this interesting and unusual behavior

Table 3. Comparison between experiment (upper value) and *ab initio* theory (lower value) for the molecular constants of the lower electronic states of Li_2.

State	R_e (Å)	ω_e (cm^{-1})	$\omega_e x_e$ (cm^{-1})	D_e (cm^{-1})	Reference
$X^1\Sigma_g^+$	2.673 2.672	351.4 350.6	2.61 3.62	8516.9 8501	29,30 31
$1^3\Sigma_u^+$	Determined from accidental predissociation *Ab initio* curves				32 31,33-35
$1^3\Pi_u$	2.58 2.591	344.63 346.5	1.82 2.51	12182[a] 11955	32 31
$1^1\Sigma_u^+$	3.108 3.105	255.4 257.6	1.6 1.75	9354 9470	7 31
$1^1\Pi_u$	2.934 2.921	270.7 273.1	2.95 7.19	2984 2595	8,29 31
$3^1\Sigma_g^+$	3.12 3.066	245.9 238.9	2.83 1.73	8313 8263	15[b] 31[b]

[a] From the value of $T_e(1^3\Pi_u) - T_e(1^1\Sigma_u^+) = -2828.17\ cm^{-1}$ from Ref. [32] and $D_e(1^1\Sigma_u^+)$.

[b] Inner minimum of double minimum potential.

better, spectral data were collected in the region of the top of the "barrier" between the two minima and for an extended region above the barrier. The results are shown in Fig. 1, where the theoretical potential and vibrational (J = 0) levels are shown as dashed lines. The solid lines are the experimental values. The solid potential curve is an RKR fit to the spectral data which is valid only below the barrier since it cannot handle a potential with more than two turning points. An inverted perturbation approach to derive a potential for the entire experimental data field is in progress [37]. The comparison between theory and experiment is shown graphically in Fig. 2 where the vibrational interval ΔG_v is shown for both theory and experiment. The agreement is quite good except in the immediate region of the barrier maximum where theory predicts an extra level. At the present time there is no evidence for the existence of levels in an outer minimum, but the state at least has a distinct "shelf" region.

Experiments on such unusual electronic states puts a high demand on the theory, and the extent of the agreement in this case is extremely heartening. Future examination will surely reveal the sources of the small discrepancies that presently exist.

There are several areas which remain to be examined by *ab initio* theory. A treatment of the Rydberg states, including a reasonable explanation of the size and variation of the quantum defects, would be extremely helpful for the assignment of the high members of the different series. Data is now available for states which correlate with the $(2^2S + n^2D)$ atomic states, and the corresponding potentials would be useful. Areas that would benefit by experimental investigations include the lower states in Table 2 for which the theory has already been carried out. In particular the $1^1\Pi_g$ and $2^1\Sigma_g^+$ states, the double minimum $2^1\Sigma_u^+$ state, and the triplet states should receive immediate attention.

We wish to acknowledge the National Science Foundation and the donors of the Petroleum Research Fund of the American Chemical Society for support.

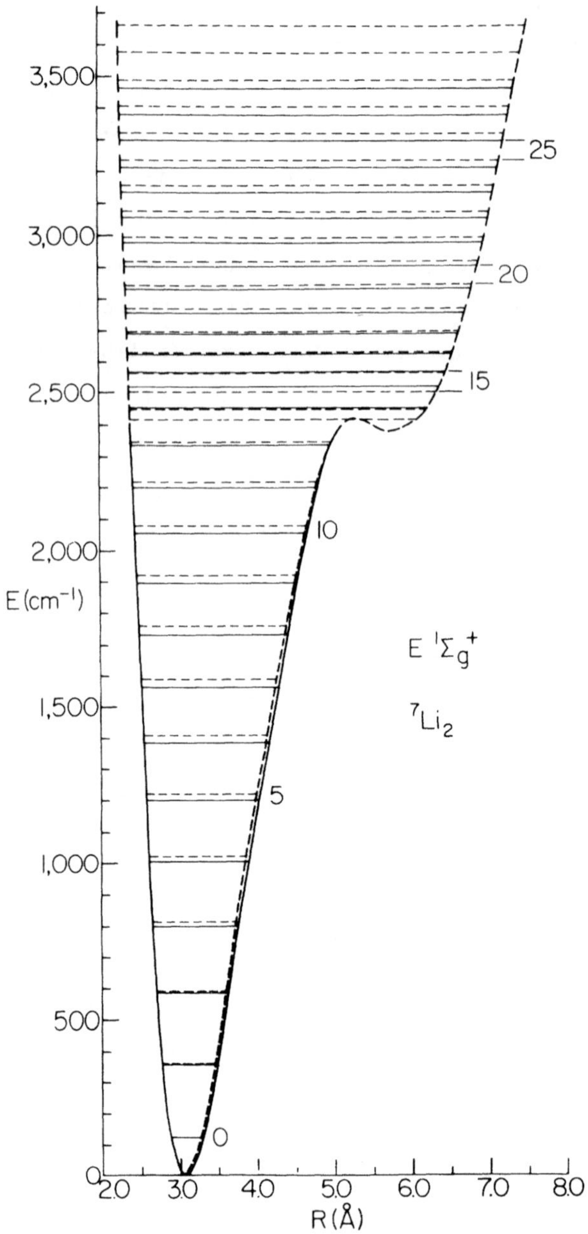

Figure 1. A comparision between experimental (solid) and *ab initio* (dashed) potential curves and vibrational states of the $3^1\Sigma_g^+$ state of Li_2.

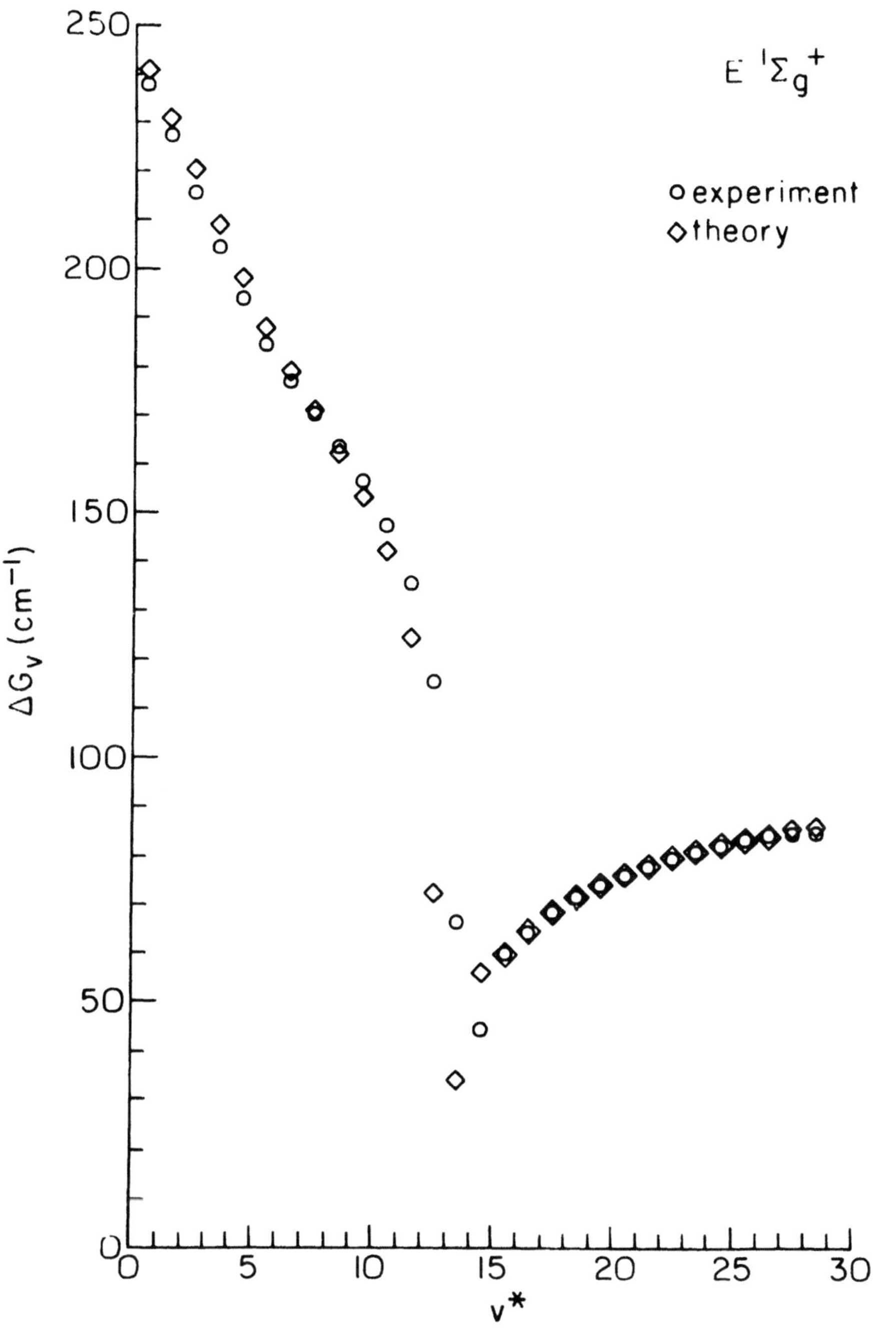

Figure 2. Comparison between experimental and *ab initio* Birge-Sponer plots of the $3^1\Sigma_g^+$ state of Li_2 between v = 0 - 29.

References:

1. H.M. James, J. Chem. Phys. 3, 9 (1935).
2. For discussion of this point see Y.T. Lee and B.H. Mahan, J. Chem. Phys. 42, 2893 (1965).
3. D. Wagman, W. Evans, R. Jacobson and T. Munson, J. Res. Natl. Bur. Std. 55, 83 (1955).
4. R.F. Barrow, N. Travis and C.V. Wright, Nature 187, 141 (1960).
5. K.P. Huber and G. Herzberg, Constants of Diatomic Molecules, Van Nostrand Reinhold Co., New York, 1979.
6. D.D. Konowalow and M.E. Rosenkrantz, Chem. Phys. Lett. 61, 489 (1979).
7. P. Kusch and M.M. Hessel, J. Chem. Phys. 67, 586 (1977).
8. M.M. Hessel and C.R. Vidal, J. Chem. Phys. 70, 4439 (1979).
9. F.R. Rico, Opt. Pura Apl. 2, 33 (1969).
10. J.L. Mercier, F.R. Rico and R. Velasco, Opt. Pura Apl. 2, 96 (1969).
11. R.A. Bernheim, L.P. Gold, P.B. Kelly, T. Tipton and D.K. Veirs, J. Chem. Phys. 74, 2749 (1981).
12. R.A. Bernheim, L.P. Gold, P.B. Kelly, C.A. Tomczyk and D.K. Veirs, J. Chem. Phys. 74, 3249 (1981).
13. J. Balz, R.A. Bernheim, L.P. Gold, P.B. Kelly and D.K. Veirs, J. Chem. Phys. 75, 5226 (1981).
14. R.A. Bernheim, L.P. Gold, P.B. Kelly, T. Tipton, C.A. Tomczyk and D.K. Veirs, in Laser Spectroscopy, A.R.W. McKellar, T. Oka and B.P. Stoicheff, ed., (Springer-Verlag, New York, 1981), p. 122.
15. R.A. Bernheim, L.P. Gold, P.B. Kelly, T. Tipton and D.K. Veirs, J. Chem. Phys. 76, 57 (1982).
16. R.A. Bernheim, L.P. Gold and T. Tipton, in Lasers '81, C.B. Collins, ed. (Soc. for Opt. and Quant. Electronics, 1982), p. 193.
17. R.A. Bernheim, L.P. Gold and T. Tipton, Chem. Phys. Lett. 92, 13 (1982).
18. R.A. Bernheim, L.P. Gold and T. Tipton, J. Chem. Phys. 78, 3635 (1983).
19. R.A. Bernheim, L.P. Gold, T. Tipton and D.D. Konowalow, Chem. Phys. Lett. 105, 201 (1984).
20. D. Eisel, W. Demtröder, W. Müller and P. Botschwine, Chem. Phys. 80, 329 (1983).
21. D.D. Konowalow and J.L. Fish, Chem. Phys. Lett. 104, 210 (1984).
22. G. Herzberg, Molecular Spectra and Molecular Structure, 2nd ed. (Van Nostrand, Princeton, 1950), Vol. I.
23. P.J. Foster, R.E. Leckenby and E.J. Robbins, J. Phys. B 2, 478 (1969).
24. A.M. Emel'yanov, V.A. Peredvigina and L.N. Goroknov, High Temp. (USSR) 9, 164 (1971).
25. C.H. Wu, J. Chem. Phys. 65, 3181 (1976).
26. B.P. Mathur, E.W. Rothe, G.P. Reck and A.J. Lightman, Chem. Phys. Lett. 56, 336 (1978).

27. M.W. McGeoch and R.E. Schlier, Chem. Phys. Lett. 99, 347 (1983).
28. W. Müller and W. Meyer, J. Chem. Phys. 80, 3311 (1984).
29. K.K. Verma, M.E. Koch and W.C. Stwalley, J. Chem. Phys. 78, 3614 (1983).
30. J. Verges, R. Bacis, B. Barakat, P. Carrot, S. Churassy and P. Crozet, Chem. Phys. Lett. 98, 203 (1983).
31. D.D. Konowalow and J.L. Fish, Chem. Phys. 84, 463 (1984).
32. W. Preuss and G. Baumgartner, Z. Physik. A (in press).
33. D.L. Cooper, J.M. Huston and T. Uzer, Chem. Phys. Lett. 86, 472 (1982).
34. Y. Uang, R.F. Ferrante and W.C. Stwalley, J. Chem. Phys. 74, 6267 (1981).
35. T.E. Sharp, At. Data 2, 119 (1971).
36. D.D. Konowalow and J.L. Fish, J. Chem. Phys. 76, 4571 (1982).
37. With C.R. Vidal.

THEORY AND CALCULATIONS ON SMALL MOLECULES USING PROPAGATOR METHODS WITH AN AGP REFERENCE*

Henry A. Kurtz, Department of Chemistry,
Memphis State University, Memphis, TN 38152
and
Brian Weiner and Yngve Öhrn, Department of Chemistry,
University of Florida, Gainesville, FL 32611

ABSTRACT: Propagator calculations with an Antisymmetrized Geminal Power (AGP) reference state are shown to give excellent results for excitation energies, state properties and transition moments for smal molecules. Applications to LiH and Li_2 are discussed.

1. INTRODUCTION

Antisymmetrized Geminal Power (AGP) [1] wavefunctions

$$|AGP\rangle = O_{AS} \prod_{i-1}^{n} g(2i-1, 2i), \qquad (1)$$

where O_{AS} is the antisymmetric projector of N = 2n electron indices, and the antisymmetric geminal

$$g(1,2) = \sum_{k=1}^{s} g_k \, |u_k(1), \; u_{k+s}(2)|$$

expressed in terms of two-electron determinants over the M = 2s spin orbitals $\{u_k\}$, have gained renewed attention because the (most) consistent reference state for the polarization propagator [2] in the random phase approximation (RPA) [2] necessarily takes this form.

*This work was supported in part through a grant from the national Science Foundation

R. J. Bartlett (ed.), Comparison of Ab Initio Quantum Chemistry with Experiment for Small Molecules, 339–355.

Consistency in this context refers to a reference state, $|REF\rangle$, such that

$$Q_K |REF\rangle = 0 \tag{3}$$

when the excited state $|K\rangle$ of the system are given by

$$Q_K^{\dagger} |REF\rangle = |K\rangle, \tag{4}$$

and the set of operators $\{Q_k^{\dagger}\}$ are obtained from the diagonalization of the propagator matrix [3]

$$\underline{P}(E) = \begin{vmatrix} \underline{\lambda} & \underline{0} \\ \underline{0} & \underline{\lambda} \end{vmatrix} \begin{vmatrix} E\underline{\lambda} - \underline{A} & \underline{B} \\ \underline{B}^T & -E\underline{\lambda} - \underline{A}^T \end{vmatrix}^{-1} \begin{vmatrix} \underline{\lambda} & \underline{0} \\ \underline{0} & \underline{\lambda} \end{vmatrix}$$

In this equation we have that

$$A_{\nu\nu'} = \langle[[q_\nu,H]_-,q_{\nu'}^{\dagger}]_-\rangle,$$

$$B_{\nu\nu'} = \langle[[q_\nu,H]_-,q_{\nu'}]_-\rangle, \tag{6}$$

$$\text{and} \quad \lambda_{\nu\nu'} = \langle[q_\nu,q_{\nu'}^{\dagger}]_-\rangle,$$

with the electronic hamiltonian H, and

$$\{q_\nu\} = \{q_{ki},q_{k\bar{i}},q_{\bar{k}i},q_{\bar{k}\bar{i}},q_{kk},q_{\bar{k}\bar{k}}\}$$

expressed more explicitly in terms of the electron field operators $\{a_i,\ 1 \leq i \leq s$ and $a_{\bar{i}} \equiv a_{i+s},\ 1 \leq i \leq s\}$ and their adjoints [4]:

$$q_{ki} = g_i a_i^\dagger a_k - g_k a_{\bar{k}}^\dagger a_{\bar{i}},$$

$$q_{k\bar{i}} = g_i a_{\bar{i}}^\dagger a_k + g_k a_{\bar{k}}^\dagger a_i,$$

$$q_{\bar{k}i} = g_i a_i^\dagger a_{\bar{k}} + g_k a_k^\dagger a_{\bar{i}}, \tag{7}$$

$$q_{\bar{k}\bar{i}} = g_i a_{\bar{i}}^\dagger a_{\bar{k}} - g_k a_k^\dagger a_i,$$

$$q_{kk} = (a_k^\dagger a_k - \langle a_k^\dagger a_k \rangle)|REF\rangle\langle REF|,$$

$$q_{\bar{k}\bar{k}} = (a_{\bar{k}}^\dagger a_{\bar{k}} - \langle a_{\bar{k}}^\dagger a_{\bar{k}} \rangle)|REF\rangle\langle REF|.$$

The average values in Eqns. (6) and (7) are taken over the reference state, which is chosen as an AGP energy optimized both with respect to the geminal coefficients $\{g_k\}$ and the orbitals $\{u_k\}$. In the applications made so far, the AGP is chosen as a spin singlet formed from a pure singlet geminal and expressed in natural form [5],

$$g(1,2) = \sum_k g_k |u_k(1)\ u_{\bar{k}}(2)| \tag{8}$$

with spin orbitals u_k and $u_{\bar{k}}$ having the same orbital and opposite spin parts. The natural spin orbitals (NSO's) $\{u_k\}$ of the geminal are also the NSO's of the AGP.

For the case of an odd number of electrons, we would choose a generalized AGP (GAGP):

$$|GAGP\rangle = O_{AS}|\nu_{2m+1}(2m+1)\ldots\nu_N(N)|\prod_{k=1}^{m} g(2i-1,2i) \tag{9}$$

with the spin orbitals ν_i of the determinational factor being kept orthogonal to the spin orbitals u_k of the geminal. This form of reference, which also satisfies the criteria of consistency, could also be used for the case of even N when for some reason one would choose an uncorrelated representation for part of the electrons. The AGP class of wavefunctions is very flexible [6] and contains as special cases [7] the single determinant, the single valence bond structure, and the alternant molecular orbital (AMO) wave function. When expanded as a superposition of determinants, the AGP reveals its makeup in configuration interaction (CI) terms as even excitations $(1\bar{1}) \rightarrow (k\bar{k})$, $(1\bar{1},2\bar{2}) \rightarrow (k\bar{k},\ell\bar{\ell})$, etc., through all orders out of a leading closed shell determinant of natural orbitals. The mixing coefficients of such a CI expansion are not all free to vary, but each is a product [7] of n geminal coefficients g_k, selected from the total number s.

The consistency condition, Eqn. (3), is crucial for the identification of the residues of the propagator with the transition probabilities between the reference state and the other states. It is also important for the orthogonality $\langle K|L\rangle = 0$ and noninteraction across the hamiltonian $\langle K|H|L\rangle = 0$ so essential for the identification of spectroscopic states. However, even when the reference is an AGP or a GAGP it can be shown [8] that operators Q_k which diagonalize the RPA propagator matrix do not satisfy Eqn. (3). If the $\underline{B}$ matrix of Eqn. (5) was set to zero then, of course, the consistency relation would be satisfied and we would have a generalized Tamm-Dancoff approximation (TDA) [4]. In the cases tested, the $\underline{B}$ matrix elements are small and do not contribute significantly (about 10^{-6} Hartrees) to the excitation energies, nor to individual transition moments or state properties. Thus, in these cases a generalized TDA is almost identical to a consistent RPA in these aspects. Generally, however, the inclusion of the $\underline{B}$ matrix improves the calculated sum rules and also leads to more accurate details of potential energy surfaces. The inclusion of $\underline{B}$ causes $Q_K|REF\rangle \neq 0$ but according to our experience, small in norm ($<10^{-6}$).

The AGP family of N-electron wavefunctions on the spin orbital basis of rank M has some interesting properties. If we form the propagator matrix in Eqn. (5) with the averages of Eqn. (6) calculated over an energy optimized AGP, and then neglect the $\underline{B}$ matrix, we obtain the same results as one would from monoexcited CI treatment with the AGP reference, i.e., the $\underline{A}$ matrix is identical to the CI singles hamiltonian matrix formed with the AGP reference. This fact is the rationale for the name generalized TDA for a resulting propagator approximation, since the standard TDA is obtained with a single determinant SCF reference, the $q^\dagger$ operators of Eqn. (7) in that case being simple particle-hole operators ($q^\dagger_{ph} = q^\dagger_p q_h$), and $\underline{B} = \underline{0}$. In this case the $\underline{A}$ matrix also equals the hamiltonian matrix for the singly excited CI. This equality between the propagator results and the CI results does not hold for other references as, say, a general MCSCF wavefunction.

In the next section, we review some of the characteristics of the propagator theory based on an AGP or a GAGP reference, and in the third section we give selected results for the application of the theory to small atomic and molecular systems. Comparisons with experiment and with other theoretical treatments are also shown.

2. REVIEW OF SOME THEORETICAL RESULTS

The total energy for the AGP with a pure spin singlet geminal in real form can be expressed as

$$E(g) = \sum_{j=1}^{s} b_{jj}(h_{jj} + 1/2 \langle jj|jj\rangle) + \sum_{1\leq j<k\leq s} t_{jk;jk} (4 \langle kj|kj\rangle - 2 \langle kj|jk\rangle) + 2\sum_{1\leq j<k\leq s} b_{jk}\langle kk|jj\rangle \quad (10)$$

where the coefficients of the one- and two-electron integrals are elements of the two-electron reduced density matrix of the AGP. We have that

$$b_{jj} = S_n^{-1}\, n_j S_{n-1}(j)$$

$$b_{jk} = S_n^{-1}\, g_j g_k S_{n-1}(jk) \tag{11}$$

and $$t_{jk;jk} = S_n^{-1}\, n_j n_k S_{n-2}(jk),$$

where $$n_j = |g_j|^2, \text{ and}$$

$$S_n = \sum_{1 \leq j_1 < j_2 \cdots j_n \leq s} n_{j_1} n_{j_2} \cdots n_{j_n} \tag{12}$$

is the "symmetric function" of order $n=N/2$, with $S_{n-1}(j) = \partial S_n/\partial n_j$, and $S_{n-2}(jk) = \partial^2 S_n/\partial n_j \partial n_k$. The two-matrix elements are thus readily computed from the normalization S_n of the AGP and efficient algorithms for this have been coded.

The total energy expression is minimized with respect to orbital variations and with respect to variations of the geminal coefficients. At each molecular geometry where an optimized AGP or GAGP reference is obtained we can then perform a propagator calculation to obtain excitation (and de-excitation) energies, and transition amplitudes.

It holds [3], in general, that if $E(g) = \langle REF|H\, REF\rangle$ is stationary with respect to spin orbital variations, then

$$\langle REF|[a_i^\dagger a_j, H]_-\, REF\rangle = 0, \tag{13}$$

for all i and j ($i \neq j$) where the field operators refer to an arbitrary spin orbital basis. When the reference state is a single determinantal (SCF) state, then Eqn. (13) implies the Brillouin condition

$$\langle REF|H a_p^\dagger a_h\, REF\rangle = 0 \quad 1 \leq h \leq N;\ N+1 \leq p \leq M \tag{14}$$

where $a_p^\dagger a_h|REF\rangle$ is a singly excited determinant produced from the reference determinant by the particle (p) -hole (h) excitation operator $a_p^\dagger a_h$, whose adjoint $a_h^\dagger a_p$ annihilates the reference determinant $|REF\rangle$. For GAGP and AGP states one can introduce linear combinations $\{q_{ij}^\dagger\}$ and their adjoints of the operators $\{a_i^\dagger a_j\}$ as shown in Eqn. (7), that have the property

$$q_{ij}^\dagger|AGP\rangle = |ij\rangle$$

$$\text{for } 1 \leqslant i \leqslant j \leqslant M \tag{15}$$

$$q_{ij}|AGP\rangle = 0$$

where $\{|AGP\rangle, |ij\rangle\}$ form an orthogonal set of N-electron states. We can assume that the transformation T given by

$$\{\underline{q}^\dagger, \underline{q}\} = \{\underline{a^\dagger a}\}T \tag{16}$$

is non-singular. Possible singularities associated with the "diagonal" operators $q_{ii}^\dagger$ or q_{ii} can be easily removed [4,10]. The stationarity conditions [3] of $E(g) = \langle AGP|H\ AGP\rangle$ then take the form

$$\langle AGP|[q_{ij}^\dagger, H]\_AGP\rangle = \langle AGP|[q_{ij}, H]\_AGP\rangle = 0 \tag{17}$$

for $1 \leqslant i \leqslant j \leqslant M$. These relations for i=j assures stationarity with respect to the variations of the geminal coefficients. It is now straightforward to deduce [4,9] the non-interaction condition

$$\langle REF|H\ q_{ij}^\dagger\ REF\rangle = 0 \tag{18}$$

for the $|REF\rangle$ being either an AGP or a GAGP. The states $q_{ij}^\dagger|REF\rangle$ are direct analogues of the singly excited states in the single determinantal (SCF) case, and are themselves GAGP states.

One can construct a "mono-excited" CI matrix over the manifold generated by the states $\{|AGP\rangle, q_{ij}^\dagger|AGP\rangle\}$ and diagonalize it to give the states

$$|K\rangle = \sum_{1\leq i\leq j\leq M} C^K_{ij} q^\dagger_{ij} |REF\rangle = Q^\dagger_K |REF\rangle. \tag{19}$$

Thus Eqn. (19) represents a generalized TDA approximation [4]. Transition amplitudes for perturbing fields

$$F = \sum_{i,j} f_{ij} a^\dagger_i a_j \tag{20}$$

can be expressed as

$$\begin{aligned}\langle K|\ F\ L\rangle &= \langle REF|Q_K F Q^\dagger_L\ REF\rangle = \\ &= \langle REF|[Q_K,[F,Q^\dagger_L]\_]\_REF\rangle \\ &+ \langle REF|[Q_K,Q^\dagger_L]\_F\ REF\rangle,\end{aligned} \tag{21}$$

where the first term of the right hand side is obtained from the propagator A- block of Eqn. (5) and the second term involves matrix elements $\langle REF|\ a^\dagger_i a_j a^\dagger_k a_\ell\ REF\rangle$ i.e. the second order reduced density matrix of the reference state and integrals f_{ij}. The consistency relation of Eqn. (3) has been used in Eqn. (21) leading to these simplifications.

3. RESULTS

In this section we report results for some applications of the AGPTDA approach to various small atoms and diatomic molecules.

As a starting point, we should observe that for two-electron systems, our model is identical to a full CI treatment. This fact was illustrated by H. Kurtz, B. Weiner and H.J. Aa. Jensen [11] in calculations on the He atom. These calculations were also the first to demonstrate the effect of including the diagonal operators.

The LiH molecule is perhaps the simplest system showing sufficient complexity to illustrate the ability of the AGPTDA approach to account for the spectra and the properties of the individual electronic states. The early work on this system concentrated mainly on describing the ground state and its properties [12]. Docken and Hinze [13] have used an MC-SCF procedure to calculate accurate potential curves, term values, molecular properties, and transition moments for the $X^1\Sigma^+$, $A^1\Sigma^+$, $B^1\Pi$ and $^3\Pi$ states of LiH. They used a 23σ 8π 4δ Slater-type orbital (STO) basis and essentially a state-by-state optimization of orbitals and configuration expansion coefficients to obtain results in good agreement with experiment. The spectroscopy of the three lowest singlet states is presently well understood largely due to the work by Stwalley and co-workers [14]. Recently, Partridge and Langhoff [15] have reported results for the three lowest singlet states using a SCF+CI method with a 22σ 12π 7δ STO function basis. The SCF orbitals for all the CI calculations were determined for the $B^1\Pi$ state. From a reference configuration list for each state (consisting of those configurations with a coefficient equal to 0.05 or greater at any internuclear distance in the final CI wave functions) the CI wave functions were built as all single excitations from each configuration in the reference list. In addition, double excitations were included depending on whether they lowered any energy root considered by more than 0.1 μH. Estimated error in comparison to the full singles and doubles CI out of the reference list was given as 10-20 cm^{-1}.

We have calculated the $X^1\Sigma^+$ state as an AGP with a basis of CGTO's (17σ, 6π, 1δ) and used a AGPTDA propagator calculation with this reference to obtain all the other states. This means that no selection of configurations has to be made beyond this. All the orbitals (consisting of strongly and weakly occupied natural orbitals of the AGP) are optimized for the AGP reference as are the excitation operators producing the excited states. The potential energy curves obtained in this way for the $X^1\Sigma^+$, $A^1\Sigma^+$ and the $B^1\Pi$ states are shown in Figure 1.

The shape of the curves and their separations, over the range of interatomic distances shown, are indistinguishable from the results of Partridge and Langhoff. Closer examination shows that the shapes of the X- and the A-state curves are such that the vibrational level

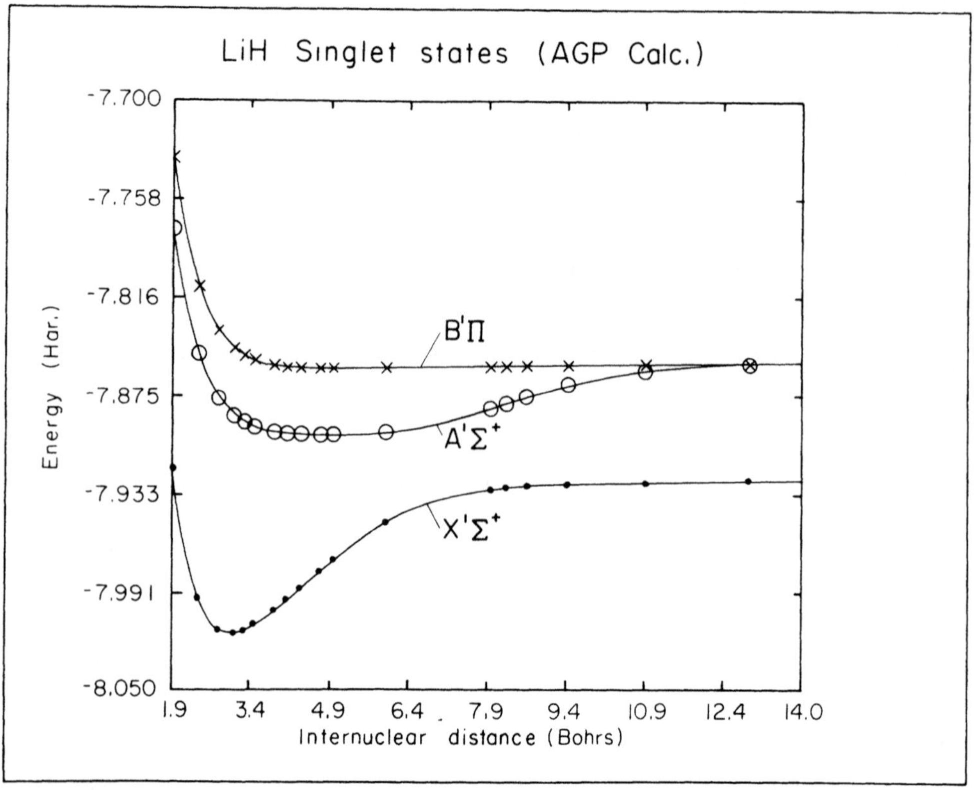

Figure 1. AGP potential energy curves for $X^1\Sigma^+$, $A^1\Sigma^+$ and $B^1\Pi$ states of LiH.

splittings differ from the best theoretical and the experimental results of Reference 15 by about 20 cm^{-1} or less. We do not obtain a bound $B^1\Pi$ state, which should be bound by 0.035 eV [14]. These discrepancies are likely due to basis set deficiencies.

The quality of the resulting wave functions is indicated by the electronic dipole moments given in Figure 2 together with the results of Partridge and Langhoff for the three lowest singlets. In Figure 3 we show the electric dipole moment for the four lowest triplet states and compare with the results of Docken and Hinze for the lowest $^3\Pi$ and $^3\Sigma^+$ (repulsive) states. The X ← A, X ← B and A ← B transition moments are very close to those calculated by Partridge and Langhoff as shown in Figure 4. Radiative lifetimes computed from the purely thoeretical Einstein coefficients are shown in Table 1 and compared with other theoretical results and with experiment.

Table 1. Comparison of calculated radiative and experimental (total) lifetimes τ (nsec) for vibrational levels v' of the $A^1\Sigma^+$ state of LiH; J' values are given in parenthesis.

v'	Theory PL[d]	ZCS[a]	KWO[e]	Experimental	
2	29.2(0)	30.2(0)	30.7(0)	29.4 ± 1.3(3)[b]	
5	31.0(0)	32.0(0)	32.5(0)	30.5 ± 1.3(3)[b]	33.0 ± 3.5(0)[c]
				32.6 ± 3.0(5)[c]	32.6 ± 3.0(10)[c]
				29.0 ± 3.2(15)[c]	
7	32.1(0)	33.1(0)	33.6(0)	36.9 ± 1.9(12)[b]	
8	32.6(0)	33.6(0)	34.9(0)	32.2 ± 5.9(15)[c]	

[a] W.T. Zemke, J.B. Crooks and W.C. Stwalley, J. Chem. Phys. 68, 4628 (1978).
[b] P.J. Dagdigian, J. Chem. Phys. 64, 2609 (1976).
[c] P.H. Wine and L.A. Melton, J. Chem. Phys. 64, 2692 (1976); L.A. Melton and P.H. Wine, ACS Symp. Ser. 56, 167 (1977).
[d] Reference [15]
[e] The present work.

The power of the AGPTDA method has also been demonstrated by the work of Sangfelt, et al. [16] on Li_2. Their calculations are similar to the above LiH calculations except the diagonal operators were not included. These diagonal operators will only affect the excited

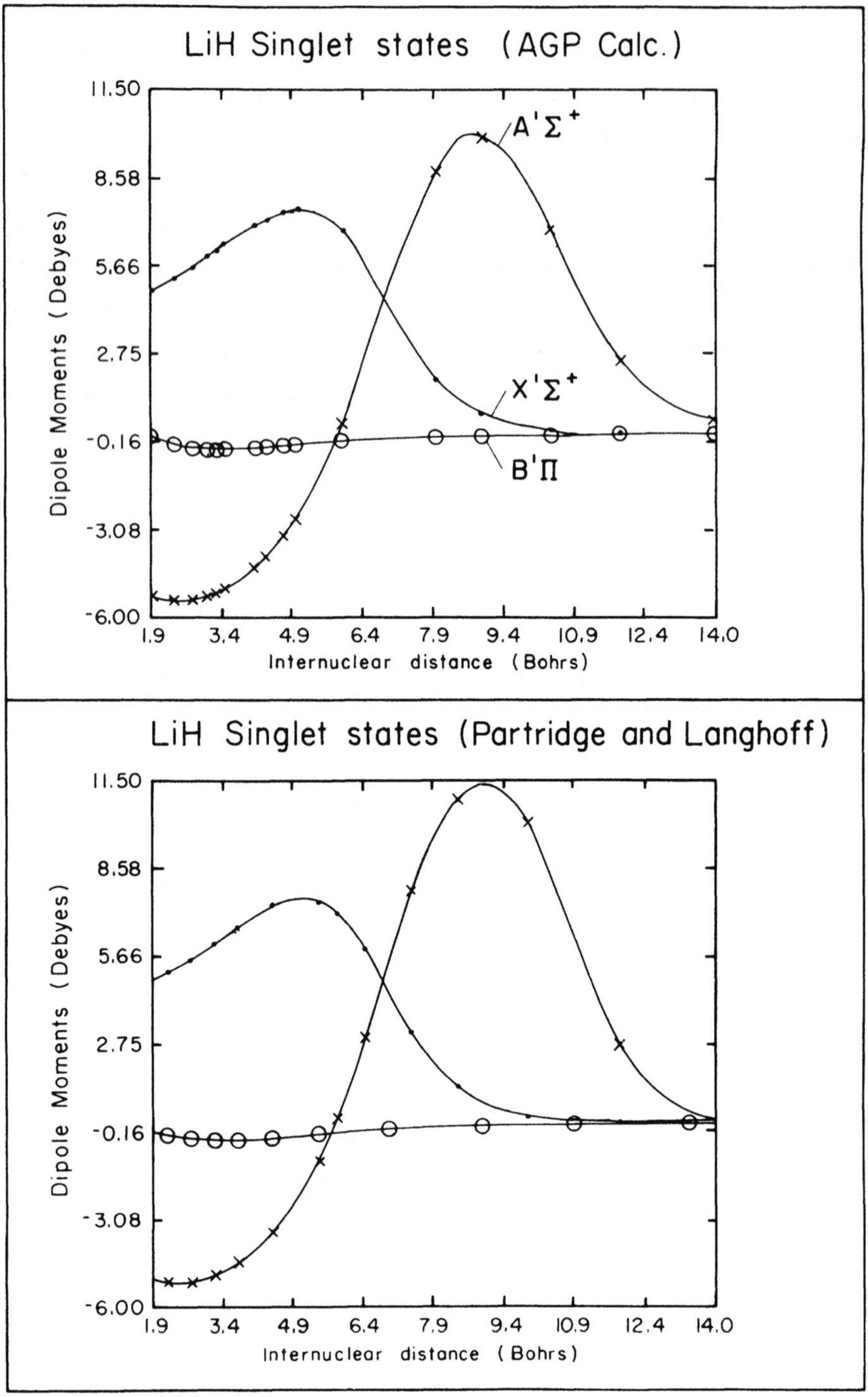

Figure 2. Dipole moment curves for lowest three singlet states of LiH.

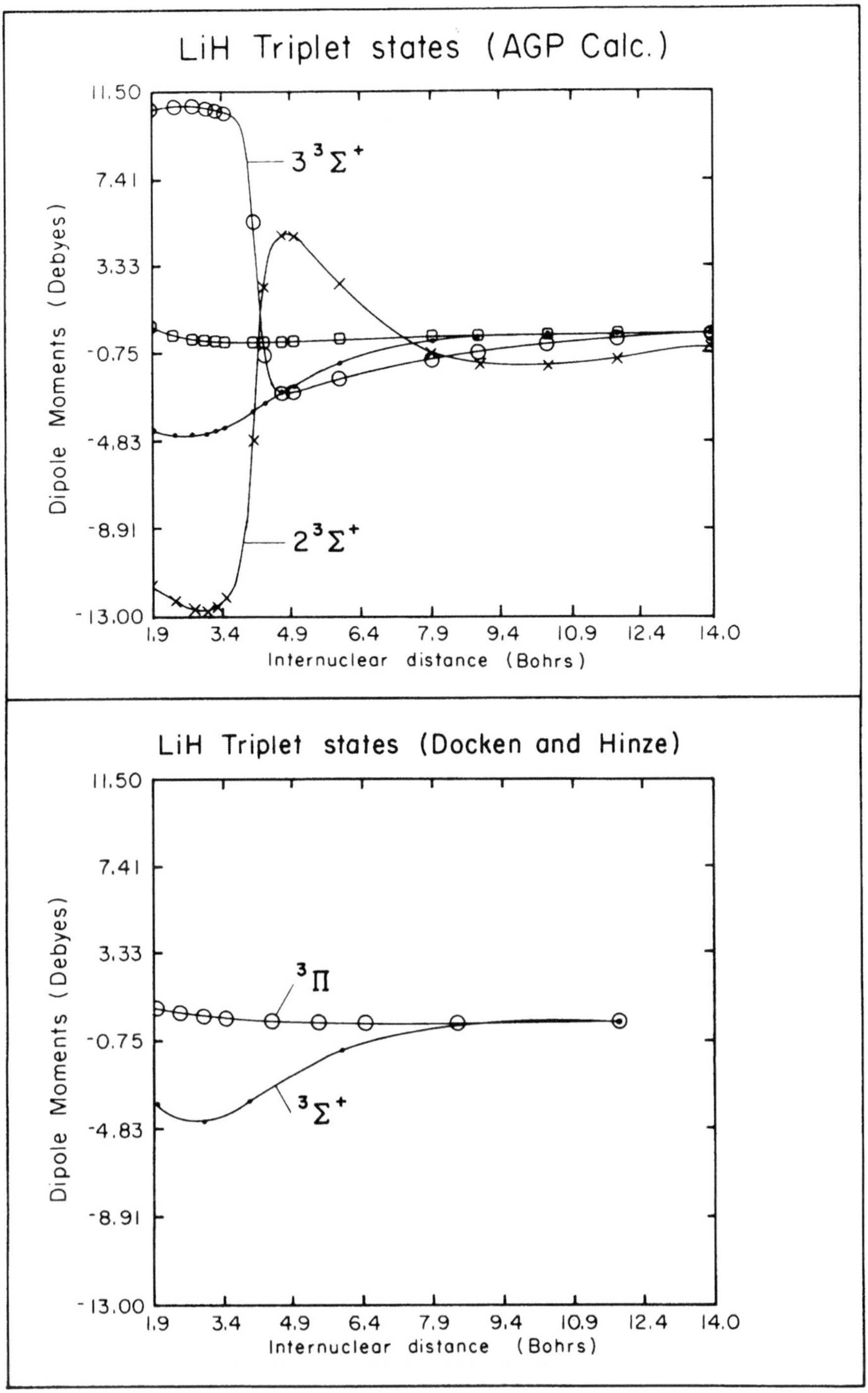

Figure 3. Dipole moment curves for lowest triple states of LiH.

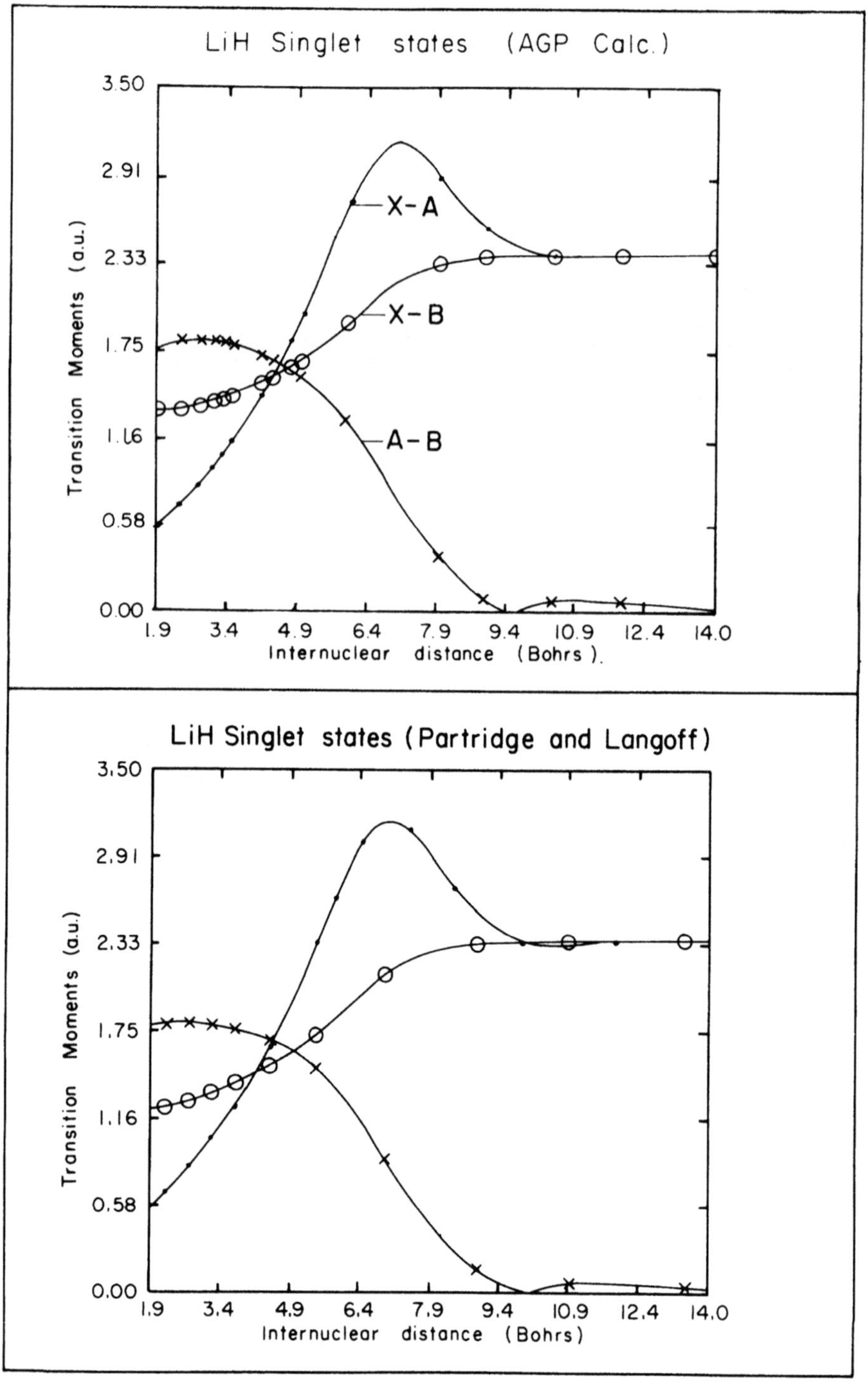

Figure 4. Transition moment curves for the X ← A, X ← B and A ← B transitions in LiH.

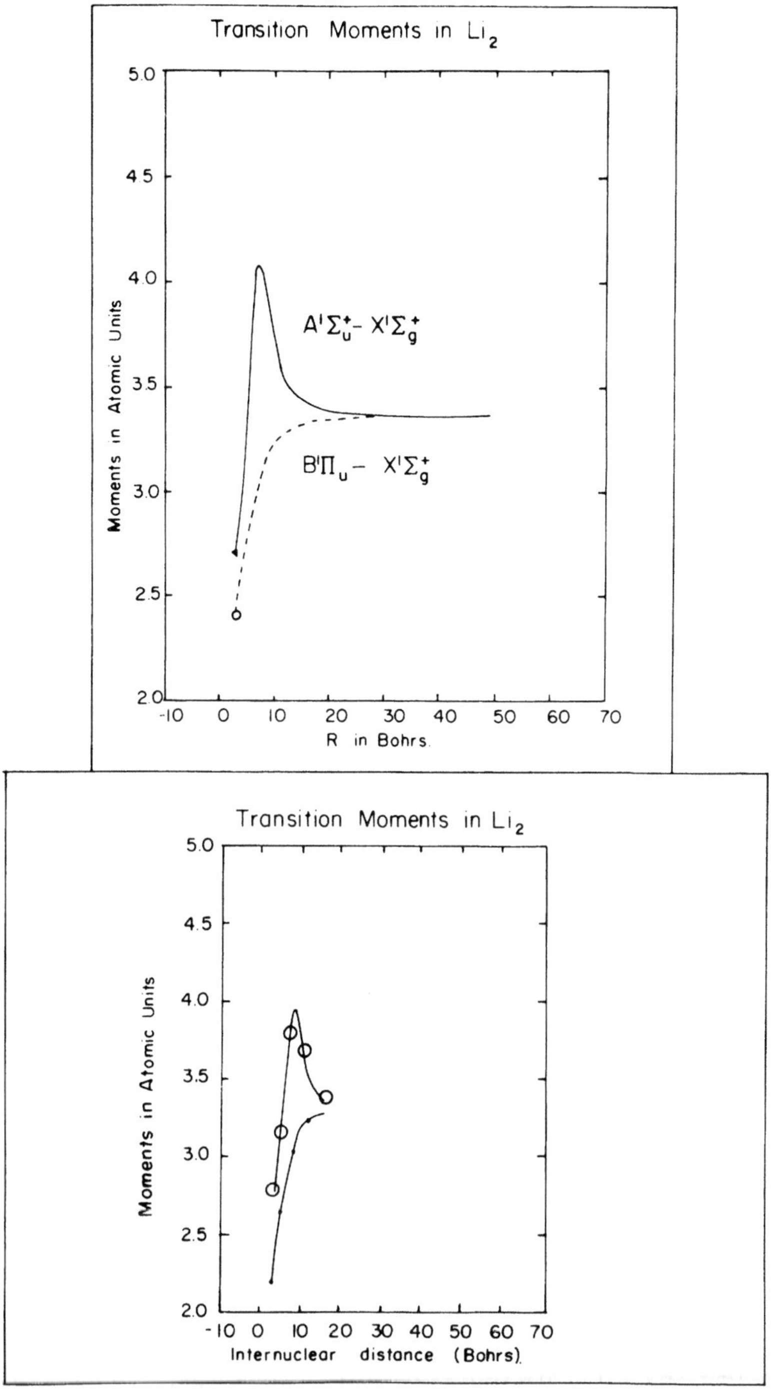

Figure 5. Transition moment curves for X ← A and X ← B transition in Li_2, Sangfelt, et al. (top); Konowalow and Olson (bottom).

$^1\Sigma_g^+$ states of Li_2. The results for the excitation energies and spectroscopic constants agree favorably with the extensive MCSCF results of Konowalow and Olson [17]. The transition moments (Figure 5) are also in excellent agreement and provide reasonable estimates for the radiative lifetimes.

References:

1. A.J. Coleman, Rev. Modern Phys. 35, 668 (1963); J. Math. Phys. 6, 1425 (1965). J. Linderberg, Israel J. Chem. 19, 93 (1980).
2. J. Linderberg and Y. Öhrn, Int. J. Quantum Chem. 12, 161 (1977). Y. Öhrn and J. Linderberg, Int. J. Quantum Chem. 15, 343 (1979).
3. J. Linderberg and Y. Öhrn, Propagators in Quantum Chemistry, Academic Press (London) 1973.
4. B. Weiner, H.J. Jensen and Y. Öhrn, J. Chem. Phys. 80, 2009 (1984).
5. P.O. Löwdin, Phys. Rev. 97, 1474 (1955).
6. O. Goscinski, in Horizons of Quantum Chemistry, D. Reidel Publishing Co., 1980, eds. K. Fukui and B. Pullman, p. 17.
7. J.V. Ortiz, B. Weiner and Y. Öhrn, Int. J. Quantum Chem. S15, 113 (1981).
8. J. Linderberg, Physica Scripta 21, 373 (1980); B. Weiner and O. Goscinski, Int. J. Quantum Chem. 18, 1109 (1980).
9. H.J. Jensen, B. Weiner, J.V. Ortiz and Y. Öhrn, Int. J. Quantum Chem. S16, 615 (1982).
10. B. Weiner and O. Goscinski, Phys. Rev. A27, 57 (1983).
11. H. Kurtz, B. Weiner and Y. Ohrn, Int. J. Quantum Chem. S17, 415 (1983).
12. A.M. Karo and A.R. Olson, J. Chem. Phys. 30, 1232 (1959); A.M. Karo, J. Chem. Phys. 30, 1241 (1959); H.S. Taylor, J. Chem. Phys. 39, 3382 (1963); S.L. Kahalas and R.K. Nesbet, J. Chem. Phys. 39, 529 (1963); J.C. Browne and F.A. Matsen, Phys. Rev. 135, A 1227 (1964); F.E. Harris and H.S. Taylor, Physica 30, 105 (1964); I.G. Cxizimadia, B.T. Sutcliffe and M.P. Barnett, Can. J. Chem. 42, 1645 (1964); D.D. Ebbing and R.C. Henderson, J. Chem. Phys. 42, 2225 (1965); C.F. Bender and E.R. Davidson, J. Chem. Phys. 75, 2675 (1966); P. Cade and W. Huo, J. Chem. Phys. 47, 614 (1967); R.E. Brown and H. Shull, Int. J. Quantum Chem. 2, 663 (1968); S.F. Boys and N.C. Handy, Proc. Roy. Soc. (London) A311, 309 (1969); W.E. Palke and W.A. Goddard, J. Chem. Phys. 50, 4524 (1969); J. Goodisman, J. Chem. Phys. 51, 3540 (1969); N.G. Mukherjee and R. McWeeny, Int. J. Quantum Chem. 4, 97 (1970).
13. K.K. Docken and J. Hinze, J. Chem. Phys. 57, 4928 (1972); and ibid. 57, 4936 (1972).
14. K.R. Way and W.C. Stwalley, J. Chem. Phys. 59, 5298 (1973); W.C. Stwalley, W.T. Zemke, K.R. Way, K.C. Li and T.R. Proctor, J. Chem. Phys. 66, 5412 (1977); W.T. Zemke and W.C. Stwalley, J. Chem. Phys. 68, 4619 (1978); and ibid. 69, 409 (1978); W.T.

Zemke, J.B. Crooks and W.C. Stwalley, J. Chem. Phys. 68, 4628 (1978); W.T. Zemke, K.R. Way and W.C. Stwalley, J. Chem. Phys. 69, 402 (1978); K.C. Li and W.C. Stwalley, J. Chem. Phys. 70, 1736 (1979); F.B. Orth and W.C. Stwalley, J. Mol. Spectroscopy 76, 17 (1979).

15. H. Partridge and S.R. Langhoff, J. Chem. Phys. 74, 2361 (1981).
16. E. Sangfelt, H.A. Kurtz, N. Elander and O. Goscinski, J. Chem. Phys., in press.
17. D.D. Konowalow and M.L. Olson, J. Chem. Phys. 71, 450 (1979); D.D. Konowalow and M.L. Olson, J. Chem. Phys. 67, 590 (1977); M.L. Olson and D.D. Konowalow, Chem. Phys. 22, 29 (1977). M.L. Olson and D.D. Konowalow, Chem. Phys. 21, 393 (1977); M.L. Olson and D.D. Konowalow, Chem. Phys. Lett. 39, 281 (1976).

THEORETICAL DISSOCIATION ENERGIES FOR IONIC MOLECULES

Stephen R. Langhoff and Charles W. Bauschlicher, Jr.,
NASA Ames Research Center, Moffet Field, California 94035
and
Harry Partridge, Research Institute for Advanced Computer Science, NASA Ames Research Center,
Moffet Field, California 94035

ABSTRACT. Ab initio calculations at the self-consistent-field and singles plus doubles configuration-interaction level are used to determine accurate spectroscopic parameters (D_e, r_e, ω_e) for most of the alkali and alkaline-earth fluorides, chlorides, oxides, sulfides, hydroxides and isocyanides. Numerical Hartree-Fock (NHF) calculations are performed on selected systems to ensure that the extended Slater basis sets employed for the diatomic systems are near the Hartree-Fock limit. Extended gaussian basis sets of at least triple-zeta plus double polarization quality are employed for the triatomic systems. By dissociating to the ionic limits, most of the differential correlation effects can be embedded in the accurate experimental electron affinities and ionization potentials. With this model, correlation effects are relatively small (0.0-0.3 eV), but invariably increase D_o. The importance of correlating the electrons on both the anion and the metal is discussed.

The theoretical dissociation energies (D_o) are critically compared with the literature to rule out disparate experimental values. The theoretical studies combined with the experimental literature allow us to recommend D_o values that are accurate to 0.1 eV for all systems considered. The systematic treatment of many different systems reveal many trends. For example, the dissociation energies of the alkali and alkaline-earth hydroxides are observed to be less than the corresponding fluorides by just

R. J. Bartlett (ed.), Comparison of Ab Initio Quantum Chemistry with Experiment for Small Molecules, 357–407.

slightly less than the difference in electron affinities of F and OH. In general, there is a strong correlation between the dissociation energy (to ions) and r_e, because the bonding is predominantly electrostatic in origin.

Theoretical $^2\Pi$-$^2\Sigma^+$ energy separations are presented for the alkali oxides and sulfides. The ground states of all the alkali sulfides are shown to be $X^2\Pi$. An extensive study of the $^2\Pi$-$^2\Sigma^+$ energy separation in KO reveals a $^2\Sigma^+$ ground state at all levels of theory. The separation is shown to be sensitive to basis set quality, and in the NHF limit the $^2\Sigma^+$ state is lower by about 250 cm^{-1}. The separation is almost unaffected when the 16 valence electrons are correlated at the singles plus doubles level using an extended Slater basis.

1. INTRODUCTION

The dissociation energies (D_o) of many of the ionic molecules containing the alkali and alkaline-earth atoms have been determined experimentally. The dissociation energies of the alkali halides reported by Brewer and Brackett [1] in 1961 are still accepted today. Their dissociation energies are limited in accuracy only by the quality of the initial vapor pressure data and by the necessary corrections for gaseous imperfection. In contrast, experimental dissociation energies for the alkaline-earth oxides show a wide variation. For MgO, the experimental D_o values [2-7] range from 3.5-4.3 eV, whereas a recent theoretical study [8] obtains 2.65±0.16 eV. For CaO, two chemiluminescence studies yielded disparate D_o values of ≥4.76±0.15 eV [9] and 4.11±0.07 eV [10]. Theoretical studies [11] gave excellent agreement with the lower value, supporting Dagdigian's conclusion that the higher value was incorrect as a result of interference with CaCl [10].

It is demonstrated in this paper that theoretical dissociation energies can be computed with 0.1 eV accuracy for large classes of ionic diatomic and triatomic systems. Considered herein are most of the alkali and alkaline-earth fluorides, chlorides, oxides, sulfides, hydroxides and isocyanides. The theoretical dissociation energies are capable of ruling out disparate experimental values and allow an overall assessment of the various experimental methods used to determine dissociation energies. Our theoretical methods described in detail in the next section reflect the fact that the charge distribution in ionic molecular systems much more closely resembles the constituent ions than the neutrals. Hence, by dissociation to the ionic limits and then correcting to the neutral limits using the accurate experimental ionization potentials and electron affinities, the problem can be formulated such that relativistic and correlation contributions to the dissociation energies are relatively small ($\leq$0.3 eV). This is documented for a large number of systems by comparing self-consistent-field (SCF) and singles plus doubles configuration-interaction, CI(SD), values for D_o.

For the alkali oxides and sulfides we have considered both the lowest $^2\Sigma^+$ and $^2\Pi$ states. This is of particular interest for the alkali oxides, where there is a change in the ground-state symmetry from LiO($^2\Pi$) to CsO($^2\Sigma^+$). For the KO molecule there is both conflicting experimental [12-13] and theoretical [14-15] evidence for the ground-state symmetry. This work provides rather strong support for KO($X^2\Sigma^+$). In contrast, the alkali sulfides Li-Rb are shown to have $X^2\Pi$ ground states, probably as a result of the increased bond lengths.

2. METHODS

The fraction (f) of ionic character in a bond is difficult to quantify. Pauling's criterion [16], $f=\mu/er_e$, where μ is the dipole moment and r_e is the equilibrium internuclear separation is not a good measure of ionicity in highly ionic systems owing to the large deformations caused by the electrostatic field of the ions. Although Mulliken populations can show ionic character, they are difficult to quantify and often vary significantly with the one-particle basis (e.g. diffuse functions increase the overlap population which is difficult to assign). The best measure of ionic character is probably the ratio r_x/r_e where r_x is the hypothetical crossing point where a purely ionic potential curve crosses the asymptote of the covalent curve. In Table 1 we have tabulated this ratio for selected alkali and alkaline-earth fluorides, chlorides, oxides and sulfides. Systems for which $r_x/r_e \leqslant 2.5$ potentially have some covalent character in the bond. By this criterion, all of the alkali and alkaline-earth oxides and sulfides as well as the alkaline-earth halides of Be and Mg could contain some covalent character.

For highly ionic systems, such as the alkali halides, the molecules can be considered to be composed of ions, each of which is polarized by the electrostatic field of the other. For these systems most of the binding energy arises from purely electrostatic interaction between ions: charge-charge interaction, charge-dipole interaction, dipole-dipole interaction, and quasi-elastic energy stored in the induced dipoles. According the the Rittner model [17], this portion of the binding (ϕ) can be expressed in terms of the atomic polarizabilities (α_1 and α_2) as

$$\phi = \frac{-e^2}{r} - \frac{e^2(\alpha_1+\alpha_2)}{2r^4} - \frac{2e^2\alpha_1\alpha_2}{r^7} \tag{1}$$

The Rittner model has yielded reliable binding energies for the alkali halides where accurate experimental r_e values are available. In addition, the model of Kim and Gordon [18] for describing forces between closed-shell atoms has been successfully applied to the alkali halides.

Probably the most rigorous theoretical treatment of the dissociation energies (D_e) of selected alkali halide molecules is given in a series of papers by Matcha [19]. He computed D_e at the SCF level with large Slater bases and then corrected for correlation and relativistic effects using atomic data. In his formalism the total

Table 1. Ratio r_x/r_e for selected alkali and alkaline earths

Molecule	r_e(Å)[a]	r_x(Å)[b]	Ratio
$LiF(^1\Sigma^+)$	1.564	7.232	4.62
$LiCl(^1\Sigma^+)$	2.021	8.113	4.01
$LiO(^2\Pi)$	1.695	3.666	2.16
$LiS(^2\Pi)$	(2.163)	4.347	2.01
$NaF(^1\Sigma^+)$	1.926	8.280	4.30
$NaCl(^1\Sigma^+)$	2.361	9.455	4.01
$NaO(^2\Pi)$	(2.040)	3.917	1.92
$NaS(^2\Pi)$	(2.487)	4.705	1.89
$KF(^1\Sigma^+)$	2.171	15.319	7.06
$KCl(^1\Sigma^+)$	2.667	19.889	7.46
$KO(^2\Sigma^+)$	(2.187)	5.005	2.29
$KS(^2\Pi)$	(2.843)	6.367	2.24
$RbF(^1\Sigma^+)$	2.270	18.533	8.16
$RbCl(^1\Sigma^+)$	2.787	25.668	9.21
$RbO(^2\Sigma^+)$	(2.288)	5.306	2.32
$RbS(^2\Pi)$	(2.983)	6.861	2.30
$BeF(^2\Sigma^+)$	1.361	2.432	1.79
$BeCl(^2\Sigma^+)$	1.797	2.524	1.41
$BeO(^1\Sigma^+)$	1.331	1.832	1.38
$BeS(^1\Sigma^+)$	1.742	1.988	1.14
$MgF(^2\Sigma^+)$	1.750	3.392	1.94
$MgCl(^2\Sigma^+)$	2.199	3.574	1.63
$MgO(^1\Sigma^+)$	1.749	2.329	1.33
$MgS(^1\Sigma^+)$	2.143	2.587	1.21
$CaF(^2\Sigma^+)$	1.967	5.310	2.70
$CaCl(^2\Sigma^+)$	2.439	5.769	2.37
$CaO(^1\Sigma^+)$	1.822	3.097	1.70
$CaS(^1\Sigma^+)$	2.318	3.570	1.54

Table 1 (cont.)

$SrF(^2\Sigma^+)$	2.075	6.280	3.03
$SrCl(^2\Sigma^+)$	2.576	6.933	2.69
$SrO(^1\Sigma^+)$	1.920	3.404	1.77
$SrS(^1\Sigma^+)$	2.441	3.984	1.63

[a] Bond distances in parentheses are theoretical values for cases where experimental values are unavailable.
[b] Hypothetical crossing point where a purely ionic potential curve (1/R) crosses the ground state asymptote.

dissociation energy of the molecule system MX is given by $D_e = D_e^{HF} + D_e^{corr} + D_e^{rel}$ where

$$D_e^{corr} \approx E_M^{corr} + E_X^{corr} - E_{M^+}^{corr} - E_{X^-}^{corr} \tag{2}$$

and

$$D_e^{rel} \approx E_M^{rel} + E_X^{rel} - E_{M^+}^{rel} - E_{X^-}^{rel} \tag{3}$$

Formulated in this way D_e^{corr} is positive and can be quite large. Matcha obtained rather good agreement with the available thermochemical D_o. His values are slightly low because his basis sets were still not at the HF limit, and he neglected the inter-fragment correlation energy (see later discussion).

The formalism that we employ herein is essentially equivalent to Matcha's, but avoids explicitly knowing the total correlation and relativistic energies of the atoms and corresponding ions. Also, our approach includes the interfragment correlation which, although small in total energy (e.g. 0.18 eV for NaF), is entirely differential since it vanishes in the asymptotic limit. The dissociation energy of MX can be written as

$$D_e = E(M^+) + E(X^-) - E(MX,r_e) - IP(M) + EA(X) \tag{4}$$

Hence, we explicitly dissociate to the ionic limits and then correct to the neutral ground state atomic limits using the accurate experimental ionization potentials (IP) available from Moore [20] and electron affinities (EA) available from Hotop and Lineberger [21]. Since most of the differential correlation and relativistic effects are absorbed into the IP and EA, the SCF dissociation energies, D_e^{SCF}, are potentially quite accurate. Note that it is necessary to apply this formalism to a molecular state that is well represented by the HF configuration. This is true of the ground states of all the systems considered herein except for the alkaline-earth oxides and sulfides whose $^1\Sigma^+$ ground states require a multireference description owing to the fact that they are a mixture of singly and doubly ionized structures. Hence, for these systems we have applied our formalism to either the excited $a^3\Pi$ or $A^1\Pi$ states, both of which are well described by a single reference configuration. The ground-state D_o can then be computed if the T_e is known for the excited state.

For the formalism in Eqn. 4 to be accurate, the basis sets must approach HF quality. Since the atoms approach the HF limit much more quickly than the molecular systems, the D_e^{SCF} values almost invariably increase monotonically with improvements in basis set quality. For all of the diatomics considered herein, we have employed extended Slater bases. They begin with the accurate sets of Clementi and Roetti [22] and were modified slightly where necessary to reduce problems with linear dependency. The basis sets are then further augmented with polarization functions to describe the considerable distortions that arise in the field of the ions. These basis sets are described briefly in Table 2. Additional details can be found in forthcoming publications [23-28]. To assess how close our SCF energies are to the HF limit, we have considered two approaches. First, for the lighter molecular systems (e.g. NaF, NaO, KO and BeCl) we have carried out numerical Hartree-Fock (NHF) calculations using an implementation of McCullough's numerical diatomic code [29] on the Cray XMP. The

Table 2. Summary of Slater and gaussian basis sets

A. Slater Basis Sets

Atom	Basis size	Basis set description[a]
O	7s5p3d1f	Clementi[b] (6s4p) O and O^- basis. Polarization functions optimized for O and O^-[c]
F	6s5p4d2f	Clementi (6s4p) F and F^- basis. Polarization functions optimized for F and F^-[d]
S	9s7p3d2f	Clementi (7s6p) S^- basis. Even tempered d and f functions optimized at the CI(SD) level.
Cl	9s7p3d2f	Clementi (7s6p) Cl^- basis. Even tempered d and f functions optimized at the CI(SD) level.
Li	6s6p4d2f	Konowalow[e] (5s5p3d) basis
Li(1s)	6s8p6d2f	Adds compact 2p (4.0, 8.0) and 3d (5.0, 8.0)
Na	8s6p4d2f	Clementi (7s,3p) basis
K	10s8p4d2f	Clementi (11s6p) basis with two most diffuse s functions deleted
Rb	11s9p6d2f	Clementi (11s7p3d) Rb basis minus most diffuse s
Cs	14s11p9d2f	McLean[f] (10s8p4d) Cs^+ basis reoptimized
Be	6s4p3d1f	Liu[g] Be_2 basis
Be(1s)	6s6p5d2f	Adds 2p (6.0, 3.5), 3d (8.0, 4.0) and 4f (2.8)
Mg	8s6p4d2f	8s6p4d1f basis used for MgO[h]
Ca(3d)	10s8p3d2f	Ca basis used for CaO[i] 3d (0.8007, 2.122, 4.77)
Ca(4d)	10s8p4d2f	Even tempered d basis 3d (0.75, 1.5, 3.0, 6.0)
Sr	13s10p7d3f	Clementi (11s7p3d) Sr^+ basis reoptimized. Valence exponents optimized for the Sr ^3P and ^3D states.

Table 2 (cont.)

Ba	15s13p9d3f	McLean (12s8p4d) Ba^+ basis reoptimized. Valence exponents optimized for the Ba ^3P and ^3D states.

B. Gaussian Basis Sets

O	(11s7p3d1f/6s4p3d1f)	van Duijneveldt[j] 11s6p.
H	(8s4p/5s3p)	van Duijneveldt 6s.
C	(11s7p2d/6s4p2d)	van Duijneveldt 11s6p.
	(11s7p3d1f/6s4p3d1f)	Expanded d and f basis.
N	(11s7p2d/6s4p2d)	van Duijneveldt 11s6p.
	(11s7p3d1f/6s4p3d1f)	Expanded d and f basis.
Li	(11s7p7d4f/7s7p7d3f)	van Duijneveldt 11s
Be	(11s5p2d/7s3p2d)	van Duijneveldt 11s.
	(11s6p7d4f/7s6p7d3f)	Tight p and expanded d and f basis.
Na	(12s9p6d3f/7s6p5d2f)	McLean and Chandler[k] 12s9p.
Mg	(12s9p5d/6s6p4d)	McLean and Chandler 12s9p.
	(12s9p6d4f/6s6p5d3f)	Expanded d and f basis.
K	(15s13p6d4f/9s9p5d3f)	Wachters[l] 14s9p.
Ca	(15s13p5d/9s9p3d)	Wachters 14s9p.
	(15s13p7d2f/9s9p6d2f)	Expanded d and f basis.
Rb	(18s14p11d4f/12s10p5d3f)	Huzinaga[m] 17s11p6d.
Sr	(17s14p11d4f/11s9p5d2f)	Huzinaga 17s11p6d.
Cs	(14s9p6d4f/7s7p5d3f)	RECP of Christiansen and Laskowski[n], modified valence basis.
Ba	(8s9p6d/6s7p4d)	RECP of Mascarello and Jaffe[o].
	(8s9p6d4f/7s7p4d3f)	Recontracted s basis and added f functions

[a] A full description of the basis sets is given in the complete works [23-28].

[b] E. Clementi and C. Roetti, At. Data Nucl. Data Table 14, 177 (1974).

[c] C.W. Bauschlicher, unpublished.

[d] B. Liu, private communication.

[e] D.D. Konowalow and M.L. Olson, J. Chem. Phys. 71, 450 (1980).

[f] A.D. McLean and R.S. McLean, IBM Research Report RJ 3187 (1981).

Table 2 (Cont.)

[g] B.H. Lengsfield, A.D. McLean, M. Yoshimine and B. Liu, J. Chem. Phys. 79, 1891 (1983).
[h] Ref. [8].
[i] Ref. [11].
[j] F.B. van Duijneveldt, IBM Research Report RJ 945 (1971).
[k] A.D. McLean and G.S. Chandler, J. Chem. Phys. 72, 5639 (1980).
[l] A.J.H. Wachters, J. Chem. Phys. 52, 1033 (1970).
[m] S. Huzinaga, J. Chem. Phys. 66, 4245 (1977).
[n] B.C. Laskowski, S.P. Walch and P.A. Christiansen, J. Chem. Phys. 78, 6824 (1983), modified valence basis tabulated in, S.R. Langhoff, C.W. Bauschlicher and H. Partridge, J. Phys. B, J. Phys. B: At. Mol. Phys. 18, 13 (1985).
[o] C.W. Bauschlicher, R.L. Jaffe, S.R. Langhoff, F.G. Mascarello and H. Partridge, J. Phys. B, in press; "Oscillator strengths of some Ba lines; a treatment including core-valence correlation and relativistic effects."

numerical calculations show that our Slater bases are very near the HF limit (error < 0.02 eV). For the heavier systems such as SrO, we have used basis set saturation studies to assess the degree of basis set incompleteness. Here the error is somewhat larger (0.20 to 0.25 eV), but most of the loss is in the core, so that our D_e^{SCF} is only low by about 0.07 eV. Also, for the triatomic systems considered herein, we have used extended gaussian basis sets, so again there is a tendency for our D_e^{SCF} values to be slightly low.

Since near HF quality D_e values can be produced using the formalism in Eqn. (4), the question naturally arises as to how well these D_e^{HF} reproduce the true values. In other words, the D_e^{HF} will be accurate if the two conditions

$$E_{MX}^{corr} \approx E_{M^+}^{corr} + E_{X^-}^{corr} \quad (5)$$

and

$$E_{MX}^{rel} \approx E_{M^+}^{rel} + E_{M^-}^{rel} \qquad (6)$$

are satisfied. The second condition is expected to be well satisfied for all systems considered herein, since only the valence electrons are substantially distorted by the electrostatic field. We have examined the approximation in Eqn. (5) by performing singles plus doubles configuration-interaction, CI(SD), calculations from the SCF reference configuration for all systems as well as more restricted CI calculations for selected systems. We find, invariably, that correlation increases the dissociation energies, making the D_e^{HF} lower bounds. Also, since E_{diff}^{corr} is always quite small (0.0-0.3 eV) the SCF model as used in Eqn. (4) is a good approximation to D_e. Apparently, SCF calculations using extended basis sets accurately portray the charge distribution in these ionic molecules near their equilibrium geometry.

In all CI calculations we correlate the n and (n-1) shells on the alkali and alkaline-earth cations, which include the 1s electrons for Li and Be. The outermost seven valence electrons are correlated for the oxide and sulfide anions, and the outermost eight valence electrons are correlated for the fluoride, chloride, hydroxide and isocyanide anions. Hence, for example, 17 electrons are correlated for the alkaline-earth halides, except for Be where 11 electrons are correlated. When this many valence electrons are correlated, one must potentially contend with size-consistency, basis set incompleteness, and basis set superposition errors, especially considering the small differential correlation contribution to D_o.

To facilitate the discussion of electron correlation let us divide the configuration-state functions (CSFs) into three classes. Classes 1 and 2 contain all single and double excitations from the orbitals that can be identified as X^- and M^+, respectively. Class 3 contains the important pair-pair terms, which are the double excitations where one electron is excited from an M^+ orbital and the other from an X^- orbital. This third class is size-consistent, since it contributes nothing at infinite separation, but increases in importance as the bond distance decreases.

If only class 1 CSFs are included, the calculation is size-consistent, but the energy is not invariant to a unitary transformation among the M^+ and X^- occupied orbitals. For example, accidental degeneracies between the M^+ and X^- orbitals result in arbitrary mixings that tend to reduce the correlation energy of both the alkali oxides and fluorides. Hence when the electrons on only one center are correlated, we use corresponding orbitals [30] to rotate the orbitals to give maximum overlap with the M^+ orbitals. If the orbitals are not rotated, then the effect of the class 1 CSFs, which is to decrease D_o and ω_e and to increase r_e, can be significantly accentuated. If only the class 1 CSFs are included, then the resulting spectroscopic constants (D_o, r_e, ω_e) are in worse agreement with experiment than are the HF values. These comments are illustrated by the CI results for the $^1\Sigma^+$ ground state of NaF in Table 3. If only the eight valence electrons on F^- are correlated, D_e is reduced by 0.08 eV and by an even larger amount, if the orbitals are not first localized. Also, r_e increases by 0.02Å, worsening the agreement with experiment. In contrast, correlating the eight valence electrons on Na has little effect on D_e or r_e, which is not surprising considering the compact nature of the Na^+ orbitals.

The class 3 excitations (pair-pair terms) have an opposite and somewhat larger differential effect than the class 1 excitations. Hence, including both classes (still size-consistent) tends to increase D_e and ω_e slightly and to decrease r_e, resulting in better agreement with experiment, particularly for the heavier alkali and alkaline-earths. The sum of the three disjoint classes, which contains all of the CSFs in the full CI(SD), should approximate the full calculation. In practice, the sum is a slight overestimation of the combined effect, because the importance of the pair-pair terms (class 3) is reduced somewhat in the full calculation. The fact that the effect of the three classes is nearly additive suggests that the calculations are nearly size consistent if compared with the supermolecule. One can also rationalize the absence of any significant size-consistency error based on the small differential correlation energy observed. This is further supported by the fact that corrections for higher excitations using either the Davidson's formula [31] or a coupled pair formalism (CPF) [32] have a negligible effect on the computed D_e. The energy of the supermolecule was based on a point computed at an internuclear separation of at least 20 Bohr. Note that there is a rather large size-consistency error (≈0.5 eV) if the calculations are referenced to the sum of the ions.

Table 3. Correlation effects on D_e and r_e in the $X^1\Sigma^+$ ground state of NaF

Description	D_e(eV)	r_e(A°)
HF	4.957	1.924
CI(SD) 8 electrons on F^- [a]	4.876 [b]	1.944
CI(SD) 8 electrons on Na^+ [a]	4.956	1.934
CI(SD) 8 electrons on F^- + pair-pair terms [a]	5.034	1.912
CI(SD) 16 electrons	5.023	1.922

[a] Orbitals were first localized using corresponding orbitals [30] by maximizing the overlap with the Na^+ orbitals.

[b] D_e is 4.83 eV if the orbitals are not localized.

Another potential problem that arises in the CI calculations is basis set superposition errors. At this level of correlation treatment, significant demands are put on the quality of the basis set. For example, it is necessary to have sufficiently tight polarization functions to correlate the contracted spatial distribution of the occupied orbitals on M^+. We have computed the basis set superposition errors in our large CI calculations using the counterpoise method. This approach is an upper bound to the error, since the counterpoise method tends to overestimate the superposition error, especially if the occupied space of the ghost atom is not excluded. The absence of a single basis function (such as a compact 3d function on M) can result in several tenths of an eV superposition error and artificially increase the calculated D_e. These statements are well illustrated by the following example for CaF. If we use three Slater 3d functions for Ca atom (α=4.76666, 2.1222, 0.8007) we find essentially no basis set superposition error at the SCF level, but a 0.129 eV error at the CI(SD) level when the fluorine ghost basis is brought up to 3.7 Bohr. When the Slater 3d basis is expanded to four Slater functions (α=6.0, 3.0, 1.5, 0.75) this error is reduced to 0.029 eV and further to 0.014 eV, when the occupied fluorine orbitals are excluded. Reducing the superposition error by 0.10 eV decreased the dissociation energy by 0.08 eV and increased r_e by 0.012Å. Using the counterpoise method, we observe a 0.035 eV superposition error for fluorine in the presence of the Ca basis (0.030 eV with the calcium occupied space deleted). Hence, with the larger CaF basis, the total superposition error is about 0.05 eV at the CI(SD) level. Based upon tests for several systems, we believe our basis set superposition errors are less than 0.1 eV and, to a large extent, are cancelled by basis set incompleteness. However, this example illustrates the necessity of using very complete basis sets at the CI(SD) level to keep basis set superposition errors below the small differential correlation effect on D_e.

In performing our CI(SD) calculations from the SCF reference configuration for open shells, we generally invoked the interacting space [33]; i.e., only those double excitations that have non-zero matrix elements with the SCF reference configuration are included.

This has a small effect on the total energy, but essentially no effect on the spectroscopic parameters. Although, in general, we included all virtual orbitals in the CI, our results indicate that eliminating the high-energy virtuals (core correlating orbitals) also has very little effect on the spectroscopic parameters. We also find that the spectroscopic parameters are insensitive to corrections for higher excitations [31], which is not surprising, considering the small total CI effect observed. We have performed CPF calculations [32] on several systems, but have not reported these results here since they are not significantly different from the CI(SD) values.

The initial phase of the study was done on the Cray-XMP using the DERIC-SWEDEN or MOLECULE-SWEDEN [34] molecular structure codes. When the Cyber-205 became available at NASA-Ames, we began calculations using the Karlsruhe adaptation [35] of the COLUMBUS codes [36-39]. The Slater integrals were again evaluated with the diatomics integral program DERIC [40]. Extensive tests were performed to ensure that the two independent sets of programs gave identical CI(SD) energies. More details of the program comparisons are given in the original papers [23-28].

3. RESULTS AND DISCUSSION

Before proceeding with a discussion of the dissociation energies of the individual molecular systems, it is useful to discuss the overall trends that have been observed. In Figure 1, we have plotted the dissociation energies, D_e, of the alkali and alkaline-earth fluorides, chlorides, oxides, sulfides and hydroxides with respect to ground state ions versus equilibrium bond length. The nearly linear relationship betwen bond length and D_e is striking. There is a tendency of the molecular systems containing Li and Be to fall above the curve, which probably arises from either the tendency of these systems to have a larger component of covalent character or the different core structure on the metal, i.e. s^2 versus s^2p^6. The similarity of the fluorides and

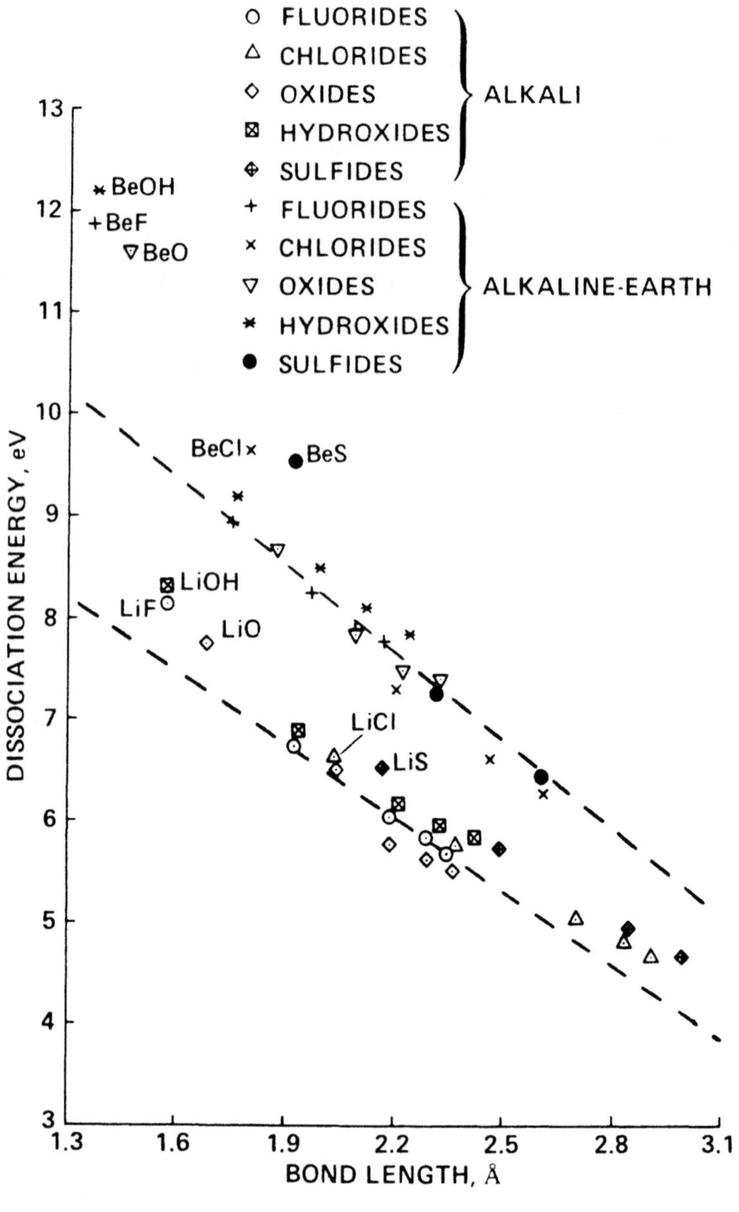

Figure 1. The dissociation energy (without zero point corrections) with respect to the ionic limits versus bond length.

hydroxides is also quite evident from Figure 1. Although the correlation diagram is not sufficiently accurate to predict D_e values, it is capable of ruling out widely divergent ones, such as all of the experimental values for MgO.

In the following sections where we discuss the different classes of molecular systems, we present D_o values at both the SCF and CI(SD) level. In almost every case correlation acts to slightly increase D_o and decrease r_e, generally improving agreement with experiment. As discussed previously, the D_o^{SCF} are probably lower bounds, and the CI(SD) D_o are more likely low than high, especially for the heavier metals where we do not fully account for the bond contraction. We feel that all of the CI(SD) D_o values reported herein for the diatomic systems (except those involving Cs and Ba) are accurate to 0.1 eV. Our results on triatomic systems (hydroxides, isocyanides) that employ extended gaussian basis sets may be slightly less accurate, but again are more likely too low than high. Our theoretical results in conjunction with the available experimental values allow us to recommend reliable D_o values for all systems considered in this study.

This study also reports values for the spectroscopic parameters r_e and ω_e. The CI(SD) results are generally in excellent agreement with experiment, although there is a tendency for theory to obtain slightly long bond lengths for the heavier systems. Note that there is several thousandths of an Angstrom uncertainty in the theoretical r_e values and a much larger ($\approx 30\ cm^{-1}$) uncertainty in the ω_e values, owing to the procedure of fitting the theoretical data to a parabola. In some instances we have done additional points on a tighter (0.05 Bohr) grid to stabilize the ω_e values.

3.1 Alkali fluorides

Two complete sets of experimental D_o values exist for the alkali fluorides. Brewer and Bracket [1] determined D_o^{298} values using a third-law treatment of vapor pressure data combined with the enthalpies of formation of the solid alkali halides and the gaseous monatomic elements. Their experimental numbers are limited in accuracy by the original vapor-pressure data and by the necessary corrections for gaseous imperfection. For an accurate comparison with our theoretical values given in Table 4, one should subtract 0.02-0.03 eV from their values to correct to absolute zero. The second set of experimental D_o values given in Table 4 are the flame photometric values of Bulewicz et al [41]. These values are in substantial agreement with the thermochemical values, but contain a large uncertainty due to the insensitivity of the flame photometric method for the fluorides. For NaF a chemiluminescent determination [42] led to a lower limit ($D_o^o \geqslant 5.33$ eV), in good agreement with the flame photometric result [41], but considerably larger than the generally accepted thermochemical value [1]. The bond energy of CsF is well established experimentally, since the bond energies determined by thermochemical [1], photoionization [43], flame photometry [41] and collision-induced dissociation [44] techniques give essential agreement.

Table 4. Dissociation energies (D_o) of the alkali and alkaline-earth fluorides

Molecule	Theoretical D_o(eV)[a]	Experimental D_o(eV)
LiF	6.05(6.06)	5.90±0.35[b], 5.96[c]
NaF	4.92(4.99)	5.25±0.30[b], 4.94[c], ⩾5.33±0.13[d]
KF	4.96(5.07)	5.07±0.35[b], 5.10[c]
RbF	4.90(5.04)	5.20±0.35[b], 5.04[c]
CsF	5.06(----)	5.33±0.35[b], 5.19[c], 5.27±0.06[e], 5.32±0.08[f]
BeF	5.88(5.94)	5.85±0.10[g], 6.25[h]
MgF	4.64(4.66)	4.79±0.05[i], 4.62±0.10[j]
CaF	5.43(5.51)	5.4±0.2[k], 5.85±0.30[l], 5.48±0.11[m], 5.51±0.09[n]
SrF	5.47(5.58)	5.58±0.07[m], 5.43±0.10[j], 5.72±0.30[l] 5.55±0.09[n]

Table 4 (cont.)

---

[a] SCF results given first with CI(SD) results in parentheses. Zero-point correction is one-half corresponding ω_e.
[b] Bulewicz, Phillips and Sugden (1961) [41], flame photometry.
[c] Brewer and Brackett (1961) [1], thermochemical (298K).
[d] Ham (1974) [42], chemiluminescent studies.
[e] Berkowitz (1971) [43], photoionization.
[f] Parks and Wexler (1984) [44], collision induced dissociation.
[g] Hildenbrand and Murad (1966) [46], mass spectrometry.
[h] Farber and Srivastava (1974) [47], mass spectrometry.
[i] Hildenbrand (1968) [49], mass spectrometry at 298K.
[j] Ehlert, Blue, Green and Margrave (1964) [48], mass spectrometry at 298K.
[k] Blue, Green, Bautista and Margrave (1963) [50], thermochemical at 298K.
[l] Ryabova and Gurvich (1964) [52], equilibria in flames.
[m] Kleinschmidt and Hildenbrand (1978) [49], mass spectrometry at 0°K.
[n] Karny and Zare (1978) [50], energy balance.

Comparing our SCF and CI(SD) D_o values in Table 4, we find that correlation uniformly increases D_o and that this effect is somewhat larger for the heavier alkali atoms. We have not yet performed the CI(SD) calculation for CsF, but predict a value of about 5.23 eV. This is in excellent agreement with experiment, except for the older thermochemical value of Scheer and Fine [45], which would appear to be about 0.2 eV too low. Overall, our CI(SD) D_o values are in excellent agreement with experiment, especially the thermochemical values of Brewer and Brackett [1].

The CI(SD) D_o values in Table 4 should be accurate to about 0.1 eV. Hence, we recommend adopting 5.0±0.1 eV for NaF, which supports the thermochemical value and is within the error bars of the flame experiments, but which rules out the chemiluminescent value of Ham [42]. We conclude, therefore, that there exists a small (0.3 eV) energy barrier in the formation of the emitting state in his reaction.

The theoretical r_e and ω_e values are compared with experiment in Table 5. The agreement with the experimental r_e and ω_e values is excellent. There is a tendency at the SCF level to overestimate the bond length for the heavier alkali atoms where the pair-pair correlation is more important. The fact that the CI(SD) calculations do not fully correct the deficiencies of the SCF may be due to basis set limitations or to the CI(SD) procedure itself, which may underestimate slightly the important differential effect of the pair-pair terms. As one can see from the strong correlation of D_e and r_e in Figure 1, it is likely that the theoretical D_o values are slightly small for the heavier systems where the CI(SD) r_e values are slightly larger than experiment.

Table 5. Theoretical spectroscopic parameters (r_e, ω_e) for the ground states of the alkali and alkaline-earth fluorides

Molecule[a]	r_e(Å)[b]			ω_e(cm^{-1})[b]		
	SCF	CI(SD)	Expt[c]	SCF	CI(SD)	Expt[c]
LiF	1.554	1.557	1.564	926	937	910
NaF	1.924	1.922	1.926	534	531	536
KF	2.205	2.185	2.171	402	437	428
RbF	2.311	2.287	2.270	367	377	376
CsF	2.400		2.345	327	352	
BeF	1.353	1.358	1.361	1220	1224	1247.36
MgF	1.740	1.746	1.750	731	729	711.69
CaF	1.980	1.966	1.967	611	595	581.1
SrF	2.104	2.086	2.075	478	524	502.4

[a] The ground state is $^1\Sigma^+$ for the alkali fluorides and $^2\Sigma^+$ for the alkaline-earth fluorides.

[b] Spectroscopic parameters are based on a parabolic fit to the lowest energy point on a 0.1 Bohr grid.

[c] Values taken from Huber and Herzberg [69].

3.2 Alkaline-earth fluorides

Our theoretical dissociation energies for the alkaline-earth fluorides (Be-Sr) are in excellent agreement with the available experimental values. For BeF we favor the older mass spectrometric determination of Hildenbrand and Murad [46] over the later determination by Farber and Srivastava [47]. Our D_o for BeF includes correlating the Be 1s electrons, which produces a D_o almost 0.1 eV greater than calculations in which the 1s is not correlated. For MgF our D_o of 4.66 eV lies between the two mass spectrometric determinations [48-49]. Since further basis set saturation is likely to increase our value slightly, we favor a value of 4.7±0.1 eV, which is consistent with all available values.

Our CI(SD) D_o for CaF is in excellent agreement with both the estimates of Karny and Zare [50] based on energy balance arguments and the mass spectrometric determination of Kleinschmidt and Hildenbrand [49]. Our value also lies within the error bars of the old thermochemical value of Blue et al. [51], but clearly suggests that the flame photometry D_o of Ryabova and Gurvich [52] is too large. Apparently, the D_o values for the fluorides determined by flame photometry are systematically large (except for LiF), but are usually correct within the rather large error bounds given. We would recommend adopting our theoretical value or a value slightly larger for CaF. For SrF, our CI(SD) D_o value agrees with all four experiments within their stated error bounds if one allows a 0.1 eV uncertainty in our value. However, our results again suggest that the flame photometry value [52] is too large and that the Ehlert et al. [48] mass spectrometric determination is too low. Again, we are in excellent agreement with both the mass spectrometric determination of Kleinschmidt and Hildenbrand (D_o=5.58±0.07) [49] and the value reported by Karny and Zare (D_o=5.55±0.09) [50] based on energy balance arguments. Work is presently in progress [23,53] on BaF. Preliminary results are in excellent agreement with experiment [48,49,51].

3.3 Alkali chlorides

For the alkali chlorides rather reliable experimental values are available for comparison. The flame photometry technique [41] is much more sensitive for the chlorides and produces values in good agreement with the thermochemical values [1]. Although our CI(SD) D_o values in Table 6 are in excellent agreement with experiment, especially for the lighter alkali chlorides (Li-K), there is a tendency for theory to underestimate D_o and overestimate r_e. Whereas the error for RbF is only 0.017Å, the error for RbCl is 0.043Å at the CI(SD) level. For the heavier alkali halides where the bonding is predominantly ionic, it is tempting to use the nearly linear relationship between D_e and r_e to "correct" the theoretical D_o values. Allowing for a 0.03 eV change in D_o per 0.01Å change in r produces "corrected" D_o^{SCF} values of 4.37 eV for KCl and RbCl and 4.54 eV for CsCl in excellent agreement with experiment. One produces similar results for KCl and RbCl by correcting the D_o^{CI} results by 0.02 eV per 0.01Å change in r, which is approximately the same correction that one would derive from the linear relationship among the alkali chlorides in Figure 1.

Table 6. Dissociation energies (D_o) of the alkali and alkaline-earth chlorides

Molecule	Theoretical D_o(eV)[a]	Experimental D_o(eV)
LiCl	4.79(4.89)	4.79±0.13[b], 4.85[c]
NaCl	4.08(4.21)	4.23±0.09[b], 4.23[c]
KCl	4.15(4.31)	4.32±0.09[b], 4.39[c]
RbCl	4.10(4.28)	4.40±0.09[b], 4.37[c]
CsCl	4.19(----)	4.65±0.13[b], 4.61[c], 4.58±0.07[d] 4.58±0.08[e]
BeCl	3.57(3.87)	3.98±0.10[f], 4.51[g]
MgCl	3.08(3.24)	3.26±0.13[h], 3.36±0.02[i]
CaCl	3.92(4.10)	4.09±0.13[h], 4.15±0.13[j], 4.08±0.07[k]
SrCl	4.04(4.17)	4.16±0.13[h], 4.15±0.08[k], ≥4.29[l]

[a] SCF results given first with CI(SD) results in parentheses.
[b] Bulewicz, Phillips and Sugden (1961) [41], flame photometry.
[c] Brewer and Brackett (1961) [1], thermochemical (298K).

Table 6 (cont.)

[d] Berkowitz (1969) [43], photoionization mass spectrometry.
[e] Parks and Wexler (1984) [44], collision-induced dissociation.
[f] Hildenbrand and Theard (1969) [54], mass spectrometry.
[g] Farber and Srivastava (1973) [55], mass spectrometry.
[h] Hildenbrand (1970) [56], mass spectrometry.
[i] Farber and Srivastava (1976) [57], mass spectrometry.
[j] Zmbov (1969) [58], mass spectrometry.
[k] Gurvich, Ryabova and Khitrov (1974) [59], flame photometry.
[l] Jonah and Zare (1971) [60], chemiluminescent.

The "corrected" theoretical D_o values provide strong support for the available experimental values. We prefer the thermochemical values of Brewer and Brackett [1] slightly over the flame photometry values of Bulewicz et al. [41]; since, our theoretical results suggest that the dissociation energy of RbCl is less than or equal to that for KCl. For CsCl we have not yet carried out the CI(SD) calculation, but we estimate a value of 4.4 eV, significantly below the experimental values [43-44] of 4.58 eV. Whether this is entirely a result of basis set incompleteness or due partially to a tendency of the CI(SD) procedure to underestimate the pair-pair correlation, we hope to elucidate in future work [23].

3.4 Alkaline-earth chlorides

The agreement between our CI(SD) D_o values and experiment [54,60] is excellent if we take the lower mass spectrometric determination of 3.98±0.10 eV [54] for BeCl. The very large positive correlation effect observed for BeCl arises from the large component of covalent character in the bonding. Note that BeCl has a r_x/r_e value of only 1.41 (see Table 1). When there is a significant differential correlation effect, there is a tendency to underestimate it; so, the mass spectrometric value of Hildenbrand [54] may be correct within their error bars. Thus, we recommend a value of 3.9±0.1 eV for BeCl. For MgCl, we also prefer the lower mass spectrometric determination of 3.26±0.13. The errors in the bond lengths for the heavier alkaline-earth chlorides are less than the corresponding alkali chlorides (see Table 7). Hence, the CI(SD) D_0 values for CaCl and SrCl are probably only slightly (0.05 eV) too low. Allowing for this, the lower bound of 4.29 for SrCl determined from chemiluminescent studies [60] would appear to be slightly high.

Table 7. Theoretical spectroscopic parameters (r_e,ω_e) for the ground states of the alkali and alkaline-earth chlorides

Molecule[a]	r_e(Å)[b]			ω_e(cm^{-1})[b]		
	SCF	CI(SD)	Expt	SCF	CI(SD)	Expt[c]
LiCl	2.036	2.018	2.021	675	668	643
NaCl	2.390	2.366	2.361	362	354	366
KCl	2.740	2.698	2.667	258	273	281
RbCl	2.877	2.830	2.787	223	235	228
CsCl	3.024	-----	2.906	196	---	214.17
BeCl	1.812	1.797	1.797	865	853	846.7
MgCl	2.212	2.199	2.199	478	469	462.12
CaCl	2.502	2.458	2.439	355	322	367.53
SrCl	2.645	2.607	2.576	287	304	302

[a] The ground state is $^1\Sigma^+$ for the alkali chlorides and $^2\Sigma^+$ for the alkaline-earth chlorides.

[b] Spectroscopic parameters are based on a parabolic fit to the lowest energy points on a 0.1 Bohr grid.

[c] Values taken from Huber and Herzberg [69]. For SrCl we used the value ascribed to T. Torring in Table A7.2 of the Ph.D. thesis of P. Bernath (MIT).

3.5 Alkali oxides

The alkali oxides are unique in that they undergo a change in ground state symmetry from $^2\Pi$(LiO, NaO) to $^2\Sigma^+$(KO, RbO, CsO). This has been explained [15] in terms of the competing effects of quadrupole interaction (favoring $^2\Pi$) and Pauli repulsion (favoring $^2\Sigma^+$). Both the theoretical [14-15] and experimental [12-13,64] evidence is that LiO and NaO have $^2\Pi$ ground states, whereas RbO and CsO have $^2\Sigma^+$ ground states. For KO there is conflicting experimental evidence; the ESR spectrum [13] suggests a $^2\Pi$ ground state, whereas a magnetic analysis of the K + NO_2 system [12] suggests a $^2\Sigma^+$ ground state. Previous theoretical results for KO are conflicting as well, with the SCF calculations of So and Richards [14] giving a $^2\Sigma^+$ ground state, and both the SCF and CI calculations of Allison et al. [15] giving a $^2\Pi$ ground state.

Our theoretical $^2\Pi$-$^2\Sigma^+$ energy separations are compared with previous theoretical and experimental evidence in Table 8.

Table 8. $^2\Pi$ - $^2\Sigma^+$ excitation energies of the alkali oxides

Molecule	$^2\Pi$ - $^2\Sigma^+$ excitation energies (cm^{-1})		
	This work[a]	Other Theory	Expt
LiO	2359(2391)	2894(2634)[b], 2342(2330)[c]	>0[g]
NaO	1429(1701)	2088(2177)[b], 1236[d], 1785[e]	>0[h]
KO	-205(-240)	233(831)[b], -347[d]	?[i]
RbO	-516(-650)	-138(-114)[b], -606[d]	< 0[j]
CsO	-798(----)	-735(-846)[b], -497(-726)[f]	< 0[h,j]

a HF results given first with the CI results in parentheses.
b (SCF/8-electron CI) results of Allison et al. [15].
c (SCF/CI) results of Yoshimine [61].
d SCF results of So and Richards [14].
e SCF results of O'Hare and Wahl [62].
f (SCF/CI) results of Laskowski et al. [63].
g Freund, et al. [64].
h Herm and Herschbach [12].
i Evidence has been presented separately for both a $^2\Sigma^+$ [12], and a $^2\Pi$ [13] ground state.
j Lindsay et al. [13].

Throughout we have correlated 15 valence electrons (11 for LiO), which we believe is essential for a balanced treatment of the two states. For all of the alkali oxides we find that the $^2\Pi-^2\Sigma^+$ energy separation is insensitive to electron correlation, because both states are equally well described by the HF configuration. This is also generally true of previous theoretical results except for the KO results of Allison et al. [15], where they observe a 600 cm^{-1} increase in the separation. However, this increase is probably an artifact of correlating only the oxygen valence electrons without first localizing the orbitals. In KO there is a substantial mixing of the oxygen 2s and potassium 3p orbitals, which is somewhat more pronounced in the $^2\Sigma^+$ state. The greater mixing in the $^2\Sigma^+$ state reduces its correlation energy with respect to the $^2\Pi$ state producing the larger separation. When both the oxygen and metal electrons are correlated with the CI(SD) procedure, the computed energies are invariant to these arbitrary mixings.

Apparently, basis set quality is a more important consideration than electron correlation for an accurate determination of the $^2\Pi-^2\Sigma^+$ energy separation. This statement is substantiated by the results in Table 9, where we have studied basis set effects on the energy separation in KO at 4.2 Bohr. Starting with the large Slater basis set in Table 9, and selectively deleting functions, we find that deleting the diffuse functions has little effect on the energy, but deleting the 4f functions and particularly the 4f functions on oxygen, has a substantial (250 cm^{-1}) effect on the separation. Apparently, f functions are more important for the $^2\Sigma^+$ state. Similarly, deleting all d- and f-functions results in a very large (960 cm^{-1}) effect on the $^2\Pi-^2\Sigma^+$ separation, since the d functions are more important for the $^2\Pi$ state. Finally, to assess how close our Slater basis is to the HF limit, we have carried out numerical HF calculations on both the $^2\Pi$ and $^2\Sigma^+$ states. The numerical HF calculations (see Table 9) show that our Slater basis is within 0.018 and 0.015 eV of the HF limit for the $^2\Sigma^+$ and $^2\Pi$ states at 4.20 Bohr, respectively. We obtain a $^2\Pi-^2\Sigma^+$ separation of -250 cm^{-1} in the HF limit, which lies between our Slater result and that of So and Richards [14]. Hence, we agree with So and Richards [14], and disagree with Allison et al. [15], who obtain a $^2\Pi$ ground state at the HF level, probably as a consequence of the relatively small basis set employed. Although our results are not definitive, they strongly support the assignment of a $^2\Sigma^+$ ground state for KO. However, the energy separation is probably sufficiently small to allow the population of both states even at room temperature.

Table 9. Basis set study of KO at the HF level[a]

Basis set Description	Energy(eV)[b] $^2\Pi$	$^2\Sigma^+$
NHF	-.015	-.018
[18s14p8d4f][b]	0.0	0.0
-diffuse functions(df)	0.0035	0.0030
-df and O(4f)	0.0042	0.029
-all f-functions	0.020	0.052
-all f-functions and all d-functions on K	0.359	0.320
-all d- and f-functions	0.608	0.489

[a] All calculations done at 4.20 Bohr.

[b] All energies are given relative to the [18s14p8d4f] Slater calculation. The basis set consists of the potassium and oxygen basis sets in Table 2 plus a set of diffuse s, p, d and f functions on oxygen. The numerical Hartree Fock (NHF) energies at 4.2 Bohr are -674.01582 a.u. for the $^2\Sigma^+$ state and -674.01249 a.u. for the $^2\Pi$ state.

The spectroscopic parameters (r_e, ω_e) for the $^2\Pi$ and $^2\Sigma^+$ states of the alkali oxides are summarized in Table 10. We observe almost identical bond contractions upon correlation in the $^2\Pi$ states of the alkali oxides as for the $^1\Sigma^+$ states of the alkali fluorides. Hence, a comparison of the CI(SD) and experimental r_e values for the alkali fluorides should provide a good measure of the remaining errors in the corresponding $^2\Pi$ states of the alkali oxides. However, we observe a significantly larger CI(SD) bond contraction in the $^2\Sigma^+$ states, probably because the hole in the $2p_\sigma$ orbital changes the character of the metal-oxygen repulsion. At the CI(SD) level we obtain nearly identical bond lengths for the $^2\Sigma^+$ states of KO and RbO as for the $^1\Sigma^+$ states of KF and RbF. This is significantly different from the assumption of $r_e(MO) = r_e(MF) + 0.05$Å made by Herm and Herschbach [12] for purposes of estimating MO dissociation energies based on the corresponding alkali fluorides.

Table 10. Theoretical spectroscopic parameters (r_e, ω_e) for the $^2\Pi$ and $^2\Sigma^+$ states of the alkali oxides[a]

Molecule	r_e(Å)[b]		ω_e(cm^{-1})[b]	
	$^2\Pi$	$^2\Sigma^+$	$^2\Pi$	$^2\Sigma^+$
LiO	1.674(1.677)	1.579(1.574)	816(805)	767(757)
NaO	2.046(2.040)	1.953(1.943)	519(511)	553(543)
KO	2.351(2.329)	2.225(2.187)	409(396)	417(439)
RbO	2.466(2.443)	2.334(2.288)	328(349)	366(380)
CsO	2.562(-----)	2.426(-----)	327(---)	331(---)

[a] Spectroscopic parameters are based on a parabolic fit to the lowest energy points on a 0.1 Bohr grid.
[b] The SCF values are given first with the CI(SD) values given in parentheses.

Our theoretical dissociation energies for the alkali oxides (with respect to ground state atoms) are compared with experiment in Table 11. Our results generally agree with the estimates of Herm and Herschbach [12], within their error bounds, illustrating again how well bond energies scale with bond length for these ionic systems. Our CI(SD) dissociation energies for LiO and NaO are significantly larger than the values derived from mass spectrometry [65-66]. In

Table 11. Dissociation energies (D_o) of the alkali and alkaline-earth oxides

Molecule	Theoretical D_o(eV)[a-c]	Experimental D_o(eV)
LiO($^2\Pi$)	3.79(3.78)	3.56±0.17[d], 3.49±0.06[e]
NaO($^2\Pi$)	2.75(2.83)	2.90±0.13[d], 2.61±0.20[f]
KO($^2\Sigma^+$)	2.76(2.86)	3.08±0.13[d], 2.86±0.13[g], 3.08±0.26[h]
RbO($^2\Sigma^+$)	2.74(2.88)	2.95±0.13[d]
CsO($^2\Sigma^+$)	2.90(----)	3.03±0.13[d]
BeO	4.61(4.69)	4.6±0.1[i], 4.53±0.13[j]
MgO	2.71(2.75)	3.71±0.13[j], 3.73±0.22[k]
CaO	4.07(4.14)	3.95±0.09[j], ≥4.76±0.15[l], 4.11±0.07[m] 4.03±0.22[k]
SrO	4.23(4.32)	4.27±0.09[j], ≥4.27±0.15[l,n]
BaO	5.61(----)	5.62±0.04[j], ≥5.79±0.15[l]

[a] SCF values are given first with CI(SD) values given in parentheses.
[b] For the alkali oxides we report D_o values for the designated ground state with respect to ground state atoms even though the $^2\Sigma^+$ states dissociate to M(^{2}P)+O(^{3}P). Zero-point corrections are made using the calculated ω_e values.
[c] For the alkaline-earth oxides, D_o values are reported for the $X^1\Sigma^+$ state with respect to ground state atoms. Zero-point corrections are made using the experimental ω_e values.
[d] Estimate of Herm and Herschbach, [12] based on the alkali halide molecules.
[e] Hildenbrand (1972) [65], electron-impact mass spectrometry.
[f] Hildenbrand and Murad (1970) [66], mass spectrometry.
[g] Ehlert (1977) [67], mass spectrometry at 298K.
[h] Gusarov and Gorokhov (1971) [68], third-law treatment of mass-spectrometric effusion data.
[i] Chupka, Berkowitz and Giese (1959) [70], mass spectrometry.
[j] Srivastava (1976) [7], recommended values- review of experimental literature through 1975. See also papers by Farber and Srivastava [47,55,57].
[k] Drowart, Exsteen and Verhaegen (1964) [3], mass spectrometry.
[l] Engelke, Sander and Zare (1976) [9], chemiluminescent studies.
[m] Irvin and Dagdigian (1980) [10], chemiluminescent studies.
[n] The value given is the reinterpreted one (see text). Original value was D_o≥4.67±0.15 eV.

fact, they are larger by about the $^2\Pi$-$^2\Sigma^+$ energy separation, which suggests the possibility that these molecules were prepared in the excited $A^2\Sigma^+$ state. Our D_o for LiO is also significantly larger than the theoretical value of Yoshimine [61], which was obtained by referencing directly to ground state atoms. This procedure underestimates the differential correlation energy, thereby underestimating D_o. We do, however, agree with the theoretical estimate of O'Hare and Wahl [62] of 2.72±0.3 eV for NaO. Apparently, all of the dissociation energies reported by Allison et al. [15] are significantly too low as a result of basis set incompleteness, accentuated by their procedure of correlating only eight valence electrons. For KO we are in excellent agreement with the Ehlert [67] mass spectrometric determination, and agree with the Gusarov and Gorokhov [68] value within their error bounds.

For RbO($^2\Sigma^+$) and CsO($^2\Sigma^+$), there are to our knowledge no experimental values available for comparison, but we are in good agreement with the estimates of Herm and Herschbach [12]. Since our theoretical values tend to be slightly low for the heavier alkali atoms, our recommended values are 2.9±0.1 eV for RbO and 3.1±0.1 for CsO. We arrive at the value for CsO by adding to our SCF value a 0.15 eV correction for electron correlation and a 0.05 eV correction for basis set incompleteness.

3.6 Alkaline-earth oxides

The $^1\Sigma^+$ ground states of the alkaline-earth oxides are not well represented by a single reference configuration reflecting the fact that the charge distribution is intermediate between M^+O^- and $M^{++}O^{--}$. Therefore, we have applied our method to the $a^3\Pi$ states, ($A^1\Pi$ for BeO) since they are well represented by the SCF reference and the T_e values are known experimentally [69]. For BeO we used the $A^1\Pi$ state since the T_e for the $a^3\Pi$ state has just recently been determined [69a]. The spectroscopic parameters (r_e, ω_e) are compared with the corresponding experimental values [69] in Table 12.

The CI(SD) values are uniformly better than the SCF values and show errors just slightly larger than for the fluorides.

Table 12. Theoretical spectroscopic parameters (r_e,ω_e) for the $A^1\Pi$ state of BeO and the $a^3\Pi$ states of MgO, CaO, SrO and BaO

Molecule/state	r_e(Å)[a]			ω_e(cm^{-1})[a]		
	SCF	CI(SD)	Expt[b]	SCF	CI(SD)	Expt[b]
BeO($^1\Pi$)	1.456	1.456	1.463	1180	1169	1144.24
MgO($^3\Pi$)	1.852	1.857	1.8_7	690	685	650
CaO($^3\Pi$)	2.105	2.087	2.09_9	561	542	556
SrO($^3\Pi$)	2.236	2.217	2.196	453	491	464
BaO($^3\Pi$)	2.347	-----	2.289	424	---	445.4

[a] Spectroscopic parameters are based on a parabolic fit to the lowest energy points on a 0.1 Bohr grid.

[b] Values taken from Huber and Herzberg [69].

The CI(SD) D_o values for the alkaline-earth oxides shown in Table 11 should be the most consistent set of numbers available and are thus capable of ruling out incorrect experimental values. Theoretical dissociation energies have been reported for both MgO [8] and CaO [11]. The values reported herein are just slightly larger owing to the larger basis sets employed and the treatment of electron correlation. Our CI(SD) value of D_o=4.69 eV for BeO is just slightly above the old mass spectrometric value of Chupka et al [70]. This value includes correlating the Be 1s electrons, and produces a bond length for the $A^1\Pi$ state that is 0.007Å shorter than experiment. If the Be 1s electrons are not correlated, the resulting D_o=4.60 eV and r_e=1.462Å are in almost perfect agreement with experiment [69]. However, we believe the higher value of D_o=4.69 eV is preferred, and that the experimental values are just slightly too low. Note that we arrive at our values using a $T_e(A^1\Pi-{}^1\Sigma^+)$ of 9405.6 cm^{-1} and correcting for the zero-point motion in the ground state using ω_e=1187.3 cm^{-1}.

All of the available experimental values [2-7] for the $^1\Sigma^+$ state of MgO are substantially too large. The value recommended by Huber and Herzberg [69] is 3.5 eV, which is a reinterpretation of the Srivastava value [7] in Table 11, taking into account the presence of the low-lying $a^3\Pi$ state. In contrast, the theoretical calculations by Bauschlicher et al. obtain D_o=2.65±0.16 eV. The CI(SD) value reported here of 2.75±0.1 eV is slightly larger as a result of the more extended basis set and CI treatment. Note that this D_o value is what would be expected based on the nearly linear relationship between bond distance and bond energy shown in Figure 1. In determining D_o, calculations were performed for the $a^3\Pi$ state, which is 0.326 eV above the ground state [71]. A recent chemiluminescence and laser fluorescence study [6] of several Mg oxidation reactions produces a dissociation energy for MgO that is consistent with theory, assuming there is a translational barrier of about 0.5 eV.

For CaO, recent chemiluminescence studies have yielded different values for the dissociation energy. The study of Engelke, Sander and Zare (ESZ) [9] yields a lower bound of ≥4.76±0.15 eV, while that of Irvin and Dagdigian (ID) [10] yields 4.11±0.07 eV. The work of ID is in excellent agreement with the older mass spectrometric values [3,7], but the ESZ result was recommended by Huber and Herzberg [69]. Recent SCF calculations by Bauschlicher and Partridge [11] gave strong support for the determination of ID. Our CI(SD) result of 4.14±0.1 is also in excellent agreement with the chemiluminescent value of ID. We thus recommend adopting this value for CaO, which again places the value on the linear plot of Figure 1. The higher value is now thought to be in error as a result of interference with CaCl.

For the SrO molecule there is again an apparent discrepancy between the recent chemiluminescence studies of ESZ [9] that gave a lower bound of D_o≥4.67±0.15 eV, and the older mass spectrometry and flame photometry values. Our value is in excellent agreement with the value recommended by Srivastava [7] in a review of the experimental literature prior to 1975. We expect our value is slightly (≈0.05 eV) too low judging from an overestimation of the bond length by 0.02Å.

Hence, we would recommend a value of 4.36±0.10 eV that rules out the determination by ESZ. The ESZ lower bound for SrO was based on the identification of a small spectral feature as the (18,0) band of the SrO $A^1\Pi$-$X^1\Sigma^+$ system (see Figure 3 of [9]). However, owing to the strong interference from SrCl, all that can be safely determined is that the (11,0) band is populated. Since the (18,0) band is at 17,420 cm^{-1} and the (11,0) band is at 14,237 cm^{-1}, 3203 cm^{-1} should be removed from the lower bound, giving a revised estimate [72] of $D_o(SrO) \geq 4.27 \pm 0.15$ eV.

For BaO we have not yet carried out the CI(SD) treatment, but have determined an SCF value of 5.61 eV. Since we anticipate that D_o will increase by at least 0.1 eV with the CI(SD) treatment as well as with further basis set saturation, we recommend a value of 5.75 ± 0.10 eV. Although the error bars given for BaO may be somewhat optimistic, our recommended value is in good agreement with the lower bound of ≥5.79±0.15 eV given by ESZ [9] determined from the chemiluminescence spectrum of Ba + ClO_2. Note also that their determination for BaO seems much more definitive since there is no interference from BaCl (see Figure 2 of [9]).

3.7 Alkali sulfides

We are unaware of any experimental work on the alkali sulfides (MS) which are difficult to observe as a result of their strong tendency to form M_2S. In Table 13 we have summarized our theoretical $MS(X^2\Pi)$ D_o values and $X^2\Pi$-$A^2\Sigma^+$ energy separations for the alkali sulfides LiS-RbS. The ground states for all of these systems are definitively determined to be $X^2\Pi$. Although calculations have not yet been done for CsS, we predict a $^2\Pi$ ground state as well. The $^2\Pi$ states would appear to be more stable with respect to the $^2\Sigma^+$ states in the sulfides than the oxides, because the considerably longer bond lengths ($r_e(MS) \approx 1.22 r_e(MO)$, M=Na,K,Rb) reduce the importance of the Pauli repulsion terms.

Table 13. Theoretical dissociation energies (D_o) and $^2\Pi-^2\Sigma^+$ energy separations for the alkali sulfides Li-Rb

Molecule	D_o(eV)[a,b]	T_e(cm^{-1}) for $A^2\Sigma^+$[b]
LiS	3.19(3.30)	5122(5088)
NaS	2.50(2.66)	3814(4033)
KS	2.51(2.68)	1717(1827)
RbS	2.39(2.58)	1240(1280)

[a] Dissociation energies are for the $X^2\Pi$ states with respect to ground state atoms. Zero-point corrections based on theoretical ω_e values in Table 14.

[b] The SCF values are given first with CI(SD) values given in parentheses.

The spectroscopic parameters (r_e, ω_e) for the $X^2\Pi$ and $A^2\Sigma^+$ states of the alkali sulfides (Li-Rb) are given in Table 14. For the $X^2\Pi$ state we observe a nearly identical bond contraction upon correlation as observed for the $^1\Sigma^+$ states of the corresponding alkali chlorides. For the $A^2\Sigma^+$ states we observe a larger CI bond contraction than in the $^2\Pi$ states, in analogy with the alkali oxides.

Since the $X^2\Pi$ states of the alkali sulfides have considerable analogy with the $X^1\Sigma^+$ states of the corresponding alkali chlorides, we expect the same deficiencies in the CI(SD) D_o values and bond lengths of the heavier alkali sulfides (KS, RbS) as for the chlorides. Thus, in determining our recommended D_o values in Table 19, we have added 0.06 and 0.08 eV, respectively, based on the known deficiencies in our D_o values of the corresponding alkali chlorides. Note that the alkali sulfides obey the nearly linear relationship between dissociation energy and bond length illustrated in Figure 1, even though by the r_x/r_e criterion they are far less ionic than the corresponding chlorides.

Table 14. Theoretical spectroscopic parameters (r_e, ω_e) for the $X^2\Pi$ and $A^2\Sigma^+$ states of the alkali sulfides Li-Rb[a]

Molecule	r_e(Å)[b]		ω_e(cm^{-1})[b]	
	$X^2\Pi$	$A^2\Sigma^+$	$^2\Pi$	$A^2\Sigma^+$
LiS	2.166(2.148)	2.100(2.074)	579(564)	538(527)
NaS	2.515(2.487)	2.454(2.424)	397(339)	326(319)
KS	2.890(2.843)	2.776(2.717)	223(239)	250(261)
RbS	3.036(2.983)	2.903(2.834)	208(214)	198(200)

[a] Spectroscopic parameters are based on a parabolic fit to the lowest energy points on a 0.1 Bohr grid.

[b] The SCF values are given first with the CI(SD) values given in parentheses.

3.8 Alkaline-earth sulfides

For the alkaline-earth sulfides (BeS-CaS) we have encountered the same problem as for the oxides, namely that the $X^1\Sigma^+$ ground states are not well described by a single reference configuration, and, therefore, are not amenable to our model. Therefore, we have considered instead the $A^1\Pi$ and $a^3\Pi$ states, which are both equally well described by a single reference configuration. Our theoretical spectroscopic parameters (r_e, ω_e, D_e) for the $A^1\Pi$ states of BeS, MgS and CaS as well as the $a^3\Pi$-$A^1\Pi$ energy separations are summarized in Table 15. The corresponding spectroscopic parameters (r_e, ω_e) for the $a^3\Pi$ state (not reported) are essentially identical to those for the $A^1\Pi$ state. The theoretical $a^3\Pi$-$A^1\Pi$ energy separations for the sulfides decrease down the column as the $K(\sigma,\pi)$ exchange integral decreases with increasing internuclear distance, since the σ open-shell orbital is localized on the metal, and the π electron is primarily a $3p\pi$ sulfur orbital. The singlet-triplet separations are comparable or slightly smaller for BeS and MgS than for the corresponding oxides [73-76].

To convert our dissociation energies for the $A^1\Pi$ states in Table 15 to ground state D_o, we must add $T_e(A^1\Pi)$ and subtract the $X^1\Sigma^+$ zero-point energy. Unfortunately, $T_e(A^1\Pi)$ is known only for BeS [77]. Using $T_e(A^1\Pi)$=7961.64 cm^{-1} and $\omega_e(X^1\Sigma^+)$=997.94, our $D_o(X^1\Sigma^+)$ for BeS is 3.21 eV. This theoretical D_o does not include correlating the Be 1s electrons, which is expected to increase D_o to about 3.30 eV. The theoretical value is consistent with the rather uncertain value of 3.47±0.65 eV reported in the JANAF tables [78]. A linear Birge-Sponer extrapolation of the ground state with corrections for ionic character [79] and excited state products yields a value of 3.30 eV [78] in excellent agreement with our theoretical value.

Table 15. Theoretical spectroscopic parameters (r_e,ω_e,D_e) for the $A^1\Pi$ states of BeS, MgS and CaS.[a]

Molecule	r_e(Å)[b]	ω_e(cm^{-1})[b]	D_e(eV)[b,c]	$a^3\Pi$-$A^1\Pi$ energy separation(cm^{-1})[b]
BeS[d]	1.926(1.921)[e]	796(784)[e]	2.04(2.29)	937(941)
MgS	2.335(2.310)	431(415)	1.49(1.70)	731(859)
CaS	2.640(2.601)	323(340)	2.25(2.41)	318(374)

[a] The $a^3\Pi$ r_e and ω_e values are not reported since they are essentially the same as those for the $A^1\Pi$ states.

[b] The SCF values are given first with the CI(SD) values in parentheses.

[c] Dissociation energies for the $A^1\Pi$ states (no zero-point correction) with respect to ground state atoms. To convert to ground state D_o, add $T_e(A^1\Pi)$ and subtract $X^1\Sigma^+$ zero-point energy.

[d] Calculations do not include correlating the 1s electrons on Be.

[e] Experimental values are r_e=1.9087 and ω_e=762.13 cm^{-1} from observation of the electronic spectrum by Pouilly et al [77].

Our theoretical dissociation (D_e) for the $A^1\Pi$ state of MgS is 1.70 eV. If we assume in analogy with MgO that the $A^1\Pi$ state is low-lying, assuming T_e=3000 cm^{-1} following JANAF, we obtain $D_o(X^1\Sigma^+)$=2.04 eV, which is considerably lower than the rather uncertain JANAF value

[78] of 2.86±0.69 eV. Hence, it is likely that like MgO, the JANAF value for MgS is significantly too large.

Our CI(SD) dissociation energy for the $A^1\Pi$ state of CaS is 2.41 eV. Adopting the rather uncertain JANAF estimate of 7200 cm^{-1} for $T_e(A^1\Pi)$ produces a $D_o(X^1\Sigma^+)$ of 3.27 eV in excellent agreement with the mass spectrometric studies of Colin et al. (D_o=3.20±0.20 eV) [80], but less than that of Marquart and Berkowitz (3.45) [81]. This suggests that the $a^3\Pi$ and $A^1\Pi$ excitation energies for the alkaline-earth sulfides are reasonably similar to the corresponding oxides. Theoretical work is in progress to accurately determine these excitation energies [26].

3.9 Alkali hydroxides

Our theoretical dissociation energies for the alkali and alkaline-earth hydroxides are compared to experiment [82-97] in Table 16. The most striking feature of the alkali hydroxides (especially the heavier ones) is their similarity to the alkali fluorides. The dissociation energies of the hydroxides are less than the fluorides by slightly less than the electron affinity difference (EA(F)-EA(OH)=1.57 eV) [21,98].

Comparison of the alkali hydroxides with the alkali oxides is also enlightening. Here the electron affinity difference (EA(OH) - EA(O)≈0.37) is only about half of the difference in dissociation energies (D_o(MOH)-D_o(MO)). The increased stability of the hydroxides relative to the oxides is a result of the ability of the hydrogen atom to pull charge density out of the metal-oxygen bond, thereby reducing the bond length and giving rise to greater electrostatic interaction [99]. The increased stability of the alkali hydroxides with respect to the oxides is also evident from the plot of D_e versus r_e in Figure 1.

The theoretical CI(SD) D_o values for the alkali hydroxides are in reasonably good agreement with the two sets of experimental values derived from the atomic absorption spectroscopy in flames. The theoretical values are probably more accurate than the flame

Table 16. Dissociation energies (D_o) of the alkali and alkaline-earth hydroxides

Molecule	Theoretical D_o(eV)[a]	Experimental D_o(eV)
LiOH	4.65(4.64)	4.55±0.09[b], 4.64±0.09[c], 4.54[d]
NaOH	3.48(3.51)	3.47±0.09[b], 3.38[d]
KOH	3.49(3.59)	3.69±0.09[b], 3.53[d]
RbOH	3.43(3.54)	3.77±0.09[b], 3.61[d]
CsOH	3.57(3.71)	3.90±0.09[b], 3.95[d]
BeOH	4.64(4.70)[e]	5.51[f], 4.94±0.43[g]
MgOH	3.33(3.31)	4.16±0.13[h], 2.43±0.22[i], 3.59±0.22[j] 3.19±0.22[k]
CaOH	4.10(4.13)	4.44±0.07[l], 4.51[m], 4.08±0.13[n] 4.00±0.17[o]
SrOH	4.11(4.16)	4.31±0.10[p], 4.47[m], 4.03±0.13[q] 4.00±0.17[o]
BaOH	4.30(4.38)	4.64[r], 4.94[m], 4.73±0.13[n] 4.55±0.17[o]

[a] SCF values are given first with CI(SD) values in parenthesis. We used the following zero-point corrections (eV) in converting our D_e values to D_o: LiOH(0.10), NaOH(0.08), KOH(0.07), RbOH(0.06), CsOH (0.06), BeOH(0.10), MgOH(0.07), CaOH(0.08), SrOH(0.08), and BaOH(0.08

[b] Cotton and Jenkins (1969) [82], atomic absorption spectroscopy in flames.

[c] Zeegers and Alkemade (1970) [83], flame photometry.

[d] Jensen (1970) [84], flame photometry.

[e] Calculation at linear geometry. Calculations with a small basis set show a bent geometry to be about 0.01 eV more stable.

[f] Ko, Greenbaum and Farber (1967) [85], molecular flow effusion.

[g] Inami and Ju (1968) [86], flame photometry, as reported in JANAF 1978

[h] Cotton and Jenkins (1969a) [88], flame photometry.

[i] Bulewicz and Sugden (1959) [89], flame photometry.

[j] Estimate from JANAF Tables (1978) [87].

[k] Murad (1980)[90], mass spectrometry.

[l] Kalff and Alkemade (1973) [96], flame photometry.

[m] Cotton and Jenkins (1968) [92], atomic absorption spectroscopy in flames.

[n] Ryabova, Khitrov and Gurvich (1972) [93], flame photometry.

[o] Murad (1981) [94], mass spectrometry.

[p] Hurk, Hollander and Alkemade (1974) [95], recalculation of work in footnote l.

[q] Gurvich, Ryabova, Khitrov and Starovoitov (1971) [91], flame photometry.

[r] Stafford and Berkowitz (1964) [97], mass spectrometry.

values, in part because the experimental values are based on equilibrium constants, which in turn require a knowledge of the rotational and vibrational partition functions of MOH. Our theoretical values for the hydroxides should be almost as accurate as for the fluorides, and show an amazing parallel. There is a tendency for theory to overestimate the bond length slightly for the heavier alkali hydroxides. Hence, based on the nearly linear relationship between bond length and D_e, our recommended values in Table 19 add 0.02, 0.04 and 0.06 to our CI(SD) D_o for KOH, RbOH and CsOH, respectively. These changes improve agreement with experiment, but our recommended value for CsOH of 3.77±0.10 eV is still considerably less than the experimental values [82,84]. Note that the difference in dissociation energy of CsOH and RbOH obtained from the two flame experiments was quite different (0.13 eV versus 0.34 eV). We obtain about 0.19 eV for this difference, which is essentially the same as the difference in dissociation energy of CsF and RbF.

Our theoretical r_e and ω_e values for the alkali hydroxides are compared with experiment [100-109] in Table 17. These results were obtained assuming a fixed OH bond of 0.947Å. If we correct our CI(SD) r_e values for the hydroxides by the known deficiencies in the CI(SD) values of the corresponding fluorides, then we are in satisfactory agreement with the rather uncertain experimental values, except for NaOH. For NaOH we estimate a Na-O bond length of 1.936Å, which is significantly shorter than the experimental estimate [101] of 1.95Å. Our theoretical ω_e values for the alkali hydroxides tend to be slightly larger than the corresponding fluorides, arising primarily from the difference in mass, i.e. $(M_O + M_H) < M_F$. The theoretical ω_e values are probably the most accurate available. For CsOH we agree with the rather uncertain gas phase value [106]. However, our results imply that all of the ω_e values derived from matrix isolation studies [102,104,107] are systematically low.

3.9 Alkaline-earth hydroxides

The similarity of the alkaline-earth hydroxides and fluorides is demonstrated by the recent high-resolution laser excitation spectra of CaOH [108] and SrOH [109]. The lowest electronic transitions for both the alkaline-earth fluorides and hydroxides involves the promotion of a nonbonding electron localized primarily in the metal nsσ orbital to a metal npπ or npσ orbital. Since these transitions are localized primarily on the metal, the potential curves for each state are similar, giving a spectrum consisting of very strong, badly overlapped $\Delta v=0$ sequences.

The theoretical dissociation energies of the alkaline-earth hydroxides are smaller than the corresponding fluorides by less than the difference of 1.57 in the electron affinities of F and OH. Again this demonstrates the increased stability of the hydroxides that arises from the ability of hydrogen to pull charge out of the metal-oxygen bond. Experimental studies of the alkaline-earth hydroxides in flames are complicated by the presence of the dihydroxide. In general, the flame-spectral data appear to give dissociation energies that are too high. The experimental values for BeOH are particularly uncertain. Our D_o value of 4.70 eV for BeOH is within the error bounds of the value of 4.94±0.43 reported in the JANAF Tables [87], which is the average of two values ascribed to Inami and Ju [86]. The considerably larger D_o value of Ko et al. [85] for BeOH was discarded based on a comparison of the trends in the alkaline-earth hydroxides and halides. Note that our value of D_o=4.70 eV was determined for a linear geometry and includes correlating the Be 1s electrons. However, calculations with a smaller gaussian basis set lead to a bent ($\theta=33^o$) equilibrium structure about 0.01 eV more stable than the linear one.

For MgOH, the two older experimental values of D_o=2.43±0.22 eV, derived by Bulewicz and Sugden [89] from flame spectra studies, and D_o=4.16±0.13 eV, derived by Cotton and Jenkins [88] using atomic absorption spectroscopy, appear to be in error. The JANAF Tables [87] have adopted an intermediate value of 3.59±0.22 eV, which is based

Table 17. Theoretical spectroscopic parameters (r_e,ω_e) for the ground states of the alkali and alkaline-earth hydroxides

Molecule[a]	r_e(Å)[b]			ω_e(cm^{-1})[b]		
	SCF	CI(SD)	Expt	SCF	CI(SD)	Expt
LiOH	1.576	1.573	1.58[c]	941	931	
NaOH	1.940	1.932	1.95[d]	588	579	431[e]
KOH	2.235	2.208	2.196[f]	455	467	408[g]
RbOH	2.349	2.323	2.301[h]	380	398	354[e]
CsOH	2.448	2.419	2.391[h]	350	378	400±80[i],336[j]
BeOH[k]	1.371	1.372	---	1340	1343	---
MgOH	1.757	1.758	---	810	802	---
CaOH	2.006	1.988	1.976[l]	612	629	623[l]
SrOH	2.134	2.117	2.10[m]	547	533	522[m]
BaOH	2.251	2.234	---	483	502	---

[a] The ground state is $^1\Sigma^+$ for the alkali hydroxides and $^2\Sigma^+$ for the alkaline-earth hydroxides.

[b] The metal-oxygen values assuming a rigid OH subunit with a fixed OH bond distance of 0.9472Å. The parameters are based on a parabolic fit to the lowest energy point on at least a 0.1 Bohr grid.

[c] Chase et al. (1974) [78], computed using moment of inertia reported by Freund et al., 25th Spectroscopy conference (Columbus, 1970) paper E8.

[d] Kuijpers, Torring and Dymanus (1976) [101], microwave.

[e] Acquista and Abramowitz (1969) [102], matrix isolation.

[f] Pearson, Winnewisser and Trueblood (1976) [103].

[g] Belyaeva, Dvorkin and Sheherba (1966) [104].

[h] Lide and Matsumura (1969) [105].

[i] Lide and Kuczkowski (1967) [106], gas phase.

[j] Acquista, Abramowitz and Lide (1968) [107], matrix isolation.

[k] Computed at the linear geometry. Calculations with a smaller basis yield a bent structure at the SCF(CI) level of R(Be-O)=1.390Å (1.396Å), R(O-H)=0.933Å(0.948Å), and θ=34°(33°).

[l] Hilborn, Qingshi and Harris (1983) [108], gas phase.

[m] Nakagawa, Wormsbecher and Harris (1983) [109], gas phase.

primarily on the trends between the hydroxides and halides. There is also a more recent mass spectrometric determination [90] of 3.19±0.22 eV that is more reliable. Our theoretical value of D_o = 3.31 eV is somewhat larger, but agrees with the mass spectrometric determination within its error bars.

For the heavier alkaline-earth hydroxides, there is a larger body of reliable experimental D_o values. Our theoretical D_o for BaOH appear to be systematically low by ≈0.2 eV as a result of using a relativistic effective core potential. Hence for BaOH we are only able to recommend a value of D_o=4.6±0.2 eV, which includes all of the experimental values [90,93,97] except that of Cotton and Jenkins [92], which appears to be too high as are their values for CaOH and SrOH.

For CaOH and SrOH our CI(SD) D_o values agree with the mass spectrometric determinations [90] within their error bounds, but strongly suggest that the values are systematically 0.10-0.15 eV low for the hydroxides of Mg, Ca and Sr. The lower flame values of Gurvich et al. [96] for SrOH and Ryabova et al. [93] for CaOH agree quite well with our theoretical values. Overall, our theoretical values should be the most reliable and may be capable of shedding light on the relative equilibrium constants for the formation of the mono- and di-hydroxides.

The theoretical spectroscopic constants r_e and ω_e for the alkaline-earth hydroxides are summarized in Table 17. These were determined using a fixed OH bond length of 0.947Å. For BeOH we obtain a non-linear equilibrium structure with an equilibrium bond angle of about 33°. The heavier alkaline-earths all favor linear structures, but they are exceptionally flat in the bending potential. The ω_e values for the alkaline-earth hydroxides are about 10% greater than the corresponding fluorides. Our ω_e values for CaOH and SrOH are in very good agreement with the gas phase values [108, 109]. Little is known about the equilibrium bond lengths for the alkaline-earth hydroxides. Our CI(SD) values for CaOH and SrOH are slightly larger than the gas phase values [108,109], as expected. However, our r_e values for BeOH and MgOH should be quite accurate (see corresponding fluorides).

.10 Alkaline-earth isocyanides

We summarize here the theoretical calculations [110] that were undertaken to determine the lowest energy structures and dissociation energies for the $^2A'$ ground-state surfaces of MCN (M=Be, Mg, Ca and Ba). There has been considerable interest in the corresponding alkali cyanide molecules recently with the discovery that both NaCN and KCN have non-linear T-shape equilibrium structures [111-115]. The dominant features in the spectra of the alkaline-earth (iso)cyanides are broad "quasicontinua," that occur at nearly the same wavelengths as the spectra of the homologous alkaline-earth monohalides [116]. These molecules are also expected to be ionic, based on the successive change in electronegativities Cl<CN<F. Note, however, that the electron affinity [117] of CN, 3.82±0.02, is considerably greater than even that of fluorine (3.399 eV).

For all of the alkaline-earths studied, the linear isocyanide structure was found to be most stable. At the CI(SD) level, the cyanide structure was found to lie above the isocyanide structure by 0.26 eV for Be, 0.13 eV for Mg and 0.20 eV for Ca. One interesting change in going from BeCN to BaCN is the loss of the interconversion barrier between the isocyanide and cyanide structures. On the basis of several cuts through the potential surface, we found an interconversion barrier for BeCN and MgCN, but none for the heavier alkaline-earths, where the bonding becomes increasingly ionic.

Our theoretical dissociation energies for the cyanide, isocyanide and bond-midpoint (metal atom approaching CN bond midpoint) structures are compared with experiment in Table 18. The experimental dissociation energies [118] for the monocyanides have been determined using electrothermal atomic absorption spectrometry. These values contain considerable uncertainty, since little was known about the number of low-lying electronic states, geometry or force constants of these molecules. The experimental values lie consistently above our values, particularly for MgCN which we feel must contain some systematic error. On the other hand, our value for the BaCN structures are probably 0.2 eV too low as a result of using a relativistic effective core potential treatment of the core.

Table 18. Theoretical dissociation energies for some alkaline-earth (iso)cyanides

Molecule	Dissociation energies(eV)[a]			
	Cyanide	Bmp[b]	Isocyanide	Expt[c]
BeCN	3.62(3.83)	3.17(3.38)	4.10(4.09)	4.46±0.29
MgCN	3.07(3.22)	3.00(3.12)	3.36(3.35)	4.21±0.26
CaCN	3.78(3.89)	3.88(3.96)	4.12(4.09)	4.46±0.22
BaCN	4.02(----)	4.17(----)	4.34(4.33)	4.87±0.26

[a] Dissociation energies (without zero-point corrections) relative to $CN(^2\Sigma^+)$ + $M(^1S)$. SCF values are given first with CI(SD) values in parentheses.

[b] The metal atom is approaching the CN bond midpoint.

[c] L'vov and Pelieva (1980) [118], electrothermal atomic absorption spectrometry.

Our dissociation energies for the isocyanides are very similar to the corresponding chlorides. This correlation provides additional support for our theoretical value for MgCN. Apparently, the delocalization of charge in CN makes the D_e of the isocyanides more comparable to the chlorides, even though the M-NC bond lengths are intermediate between the corresponding alkaline-earth fluorides and chlorides.

4. CONCLUSIONS

The theoretical dissociation energies presented here for most of the alkali and alkaline-earth fluorides, chlorides, oxides,

Table 19. Recommended dissociation energies (D_o) for selected alkali and alkaline-earth fluorides, chlorides, oxides, sulfides, hydroxides and isocyanides.

Molecule	State	D_o(eV)	Molecule	State	D_o(eV)
LiF	$^1\Sigma^+$	6.06	BeF	$^2\Sigma^+$	5.94
NaF	$^1\Sigma^+$	5.00	MgF	$^2\Sigma^+$	4.68
KF	$^1\Sigma^+$	5.10	CaF	$^2\Sigma^+$	5.53
RbF	$^1\Sigma^+$	5.07	SrF	$^2\Sigma^+$	5.62
CsF	$^1\Sigma^+$	5.27			
LiCl	$^1\Sigma^+$	4.89	BeCl	$^2\Sigma^+$	3.90
NaCl	$^1\Sigma^+$	4.22	MgCl	$^2\Sigma^+$	3.26
KCl	$^1\Sigma^+$	4.37	CaCl	$^2\Sigma^+$	4.14
RbCl	$^1\Sigma^+$	4.36	SrCl	$^2\Sigma^+$	4.23
CsCl	$^1\Sigma^+$	4.58			
LiO	$^2\Pi$	3.84	BeO	$^1\Sigma^+$	4.69
NaO	$^2\Pi$	2.83	MgO	$^1\Sigma^+$	2.75
KO	$^2\Sigma^+$	2.86	CaO	$^1\Sigma^+$	4.14
RbO	$^2\Sigma^+$	2.90	SrO	$^1\Sigma^+$	4.36
CsO	$^2\Sigma^+$	3.10	BaO	$^1\Sigma^+$	5.75
LiS	$^2\Pi$	3.30	BeS[a]	$^1\Pi$	2.38
NaS	$^2\Pi$	2.67	MgS[a]	$^1\Pi$	1.70
KS	$^2\Pi$	2.74	CaS[a]	$^1\Pi$	2.41
RbS	$^2\Pi$	2.66			
LiOH	$^1\Sigma^+$	4.64	BeOH	$^2\Sigma^+$	4.70
NaOH	$^1\Sigma^+$	3.51	MgOH	$^2\Sigma^+$	3.31
KOH	$^1\Sigma^+$	3.61	CaOH	$^2\Sigma^+$	4.15
RbOH	$^1\Sigma^+$	3.58	SrOH	$^2\Sigma^+$	4.18
CsOH	$^1\Sigma^+$	3.77	BaOH[b]	$^2\Sigma^+$	4.60
			BeNC	$^2\Sigma^+$	4.15
			MgNC	$^2\Sigma^+$	3.37
			CaNC	$^2\Sigma^+$	4.13
			BaNC[b]	$^2\Sigma^+$	4.50

[a] The dissociation energy (D_e) reported is for the excited $A^1\Pi$ state and does not include a zero-point correction (see Table 15).

[b] The results are less accurate (≈0.2 eV uncertainty) because the core electrons are described by a relativistic effective core potential (see text).

sulfides, hydroxides and isocyanides are sufficiently accurate to rule out disparate experimental values. Overall, we find that the thermochemical and mass spectrometric determinations of D_o are quite accurate. The D_o values determined by flame photometry are often accurate, but tend to be systematically high (especially for the fluorides). Finally, we find that D_o values determined from chemiluminescent studies, although potentially very accurate, are often incorrect.

An advantage of treating several classes of systems in a systematic way, is that trends in the dissociation energies emerge more clearly. For example, the MF and MOH bond lengths are comparable, and the difference in dissociation energies is slightly less than the electron affinity difference of 1.57 eV. However, the isocyanides have distinctly smaller dissociation energies than the fluorides, even though the electron affinity of CN is substantially greater than that of F. The difference arises because the charge is spread out in CN, but localized on oxygen in OH.

The theoretical D_o values in conjunction with the available experimental values allow us to recommend reliable D_o values for all systems considered. These recommended values in Table 19 are thought to be accurate to 0.1 eV. The model we have applied seems to be satisfactory even for systems (e.g. BeCl) that are not fully ionic, as long as we account for differential correlation effects with the CI(SD) procedure. We are willing to calculate the dissociation energy of any other diatomic or triatomic system that fits our model, if it is of sufficient scientific interest.

References:

1. L. Brewer and E. Brackett, Chem. Rev. 61, 425 (1961).
2. I.V. Veits and L.V. Gurvich, Zh. Fiz. Khim. 31, 2306 (1957).
3. J. Drowart, G. Exsteen and G. Verhaegen, Trans. Farad. Soc. 60, 1920 (1964).
4. D.H. Cotton and D.R. Jenkins, Trans. Farad. Soc. 65, 376 (1969).
5. P.J.T. Zeegers, W.P. Townsend and J.D. Winefordner, Spectrochim. Acta 26B, 234 (1969).
6. J.W. Cox and P.J. Dagdigian, J. Phys. Chem. 88, 2455 (1984).

7. R.D. Srivastava, High Temp. Sci. 8, 225 (1976). Recommended values based on review of the experimental literature through 1975.
8. C.W. Bauschlicher, B.H. Lengsfield III and B. Liu, J. Chem. Phys. 77, 4084 (1982).
9. F. Engelke, R.K. Sander and R.N. Zare, J. Chem. Phys. 65, 1146 (1976).
10. J.A. Irvin and P.J. Dagdigian, J. Chem. Phys. 73, 176 (1980).
11. C.W. Bauschlicher and H. Partridge, Chem. Phys. Letters 94, 366 (1983).
12. R.R. Herm and D.R. Herschbach, J. Chem. Phys. 52, 5783 (1970).
13. D.M. Lindsay, D.R. Herschbach and A.L. Kwiram, J. Chem. Phys. 60, 315 (1974).
14. S.P. So and W.G. Richards, Chem. Phys. Letters 32, 227 (1975).
15. J.N. Allison and W.A. Goddard III, J. Chem. Phys. 77, 4259 (1982). See also J.N. Allison, R.J. Cave and W.A. Goddard III, J. Phys. Chem. 88, 1262 (1984).
16. L. Pauling, "The Nature of the Chemical Bond," (Cornell University Press, Ithaca, New York, 1944), Second edition, p. 46.
17. E.S. Rittner, J. Chem. Phys. 19, 1030 (1951). The Rittner model is expected to be a good approximation only when the condition $r^6 \gg 4\alpha^+\alpha^-$ is satisfied. Since this is not the case for the alkaline-earth monohalides owing to the much larger polarizabilities of the alkaline-earth ions, improved ionic models have recently been proposed. See e.g. S.F. Rice, H. Martin and R.W. Field (J. Phys. Chem. in press) and T. Törring, W.E. Ernst and S. Kindt, J. Chem. Phys. 81, 4614 (1984).
18. Y.S. Kim and R.G. Gordon, J. Chem. Phys. 60, 4332 (1974).
19. R.L. Matcha, J. Chem. Phys. 47, 4595 (1967); 47, 5295 (1967); 48, 335 (1968); 49, 1264 (1968); 53, 485 (1970).
20. C.E. Moore, Atomic Energy Levels, Natl. Bur. Stand. (US) Circ. 467 (1949).
21. H. Hotop and W.C. Lineberger, J. Phys. and Chem. Ref. Data 4, 530 (1975).
22. E. Clementi and C. Roetti, At. Data Nuc. Data Tables 14, 177 (1974).
23. S.R. Langhoff, C.W. Bauschlicher and H. Partridge, "Theoretical dissociation energies of the alkali and alkaline-earth fluorides and chlorides" (to be published).
24. S.R. Langhoff, C.W. Bauschlicher and H. Partridge, "Theoretical dissociation energies of the alkali and alkaline-earth oxides" (to be published).
25. H. Partridge, C.W. Bauschlicher and S.R. Langhoff, "The Dissociation energy of SrO" (to be published).
26. H. Partridge, C.W. Bauschlicher and S.R. Langhoff, "Theoretical dissociation energies of the alkali and alkaline-earth sulfides" (to be published).
27. C.W. Bauschlicher, S.R. Langhoff and H. Partridge, "Ab initio study of the alkali and alkaline-earth hydroxides" (to be published).
28. H. Partridge, C.W. Bauschlicher and S.R. Langhoff, "Ab initio study of the positive ions of alkaline-earth hydroxides" (to be published).

29. E.A. McCullough Jr., J. Chem. Phys. 62, 3991 (1975); L. Adamowicz and E.A. McCullough Jr., J. Chem. Phys. 75, 2475 (1981); E.A. McCullough Jr., J. Phys. Chem. 86, 2178 (1982).
30. A.T. Amos and G.G. Hall, Proc. R. Soc. London Ser. A 263, 483 (1961).
31. S.R. Langhoff and E.R. Davidson, Int. J. Quant. Chem. 8, 61 (1974). See also E.R. Davidson in "The World of Quantum Chemistry," edited by R. Daudel and B. Pullman (Reidel, Dordrecht, 1974).
32. W. Meyer, Int. J. Quant. Chem. 55, 341 (1971). R. Ahlrichs, H. Lischka, V. Staemmler and W. Kutzelnigg, J. Chem. Phys. 62, 1225 (1975).
33. A. Bunge, J. Chem. Phys. 53, 20 (1970); A.D. McLean and B. Liu, ibid. 58, 1066 (1973); C.F. Bender and H.F. Schaefer, ibid. 55, 7498 (1971).
34. MOLECULE is a gaussian integral program written by J. Almlof. SWEDEN is a vectorized SCF-MCSCF-CI program written by P.E.M. Siegbahn, C.W. Bauschlicher, B.O. Roos, P.R. Taylor, A. Heiberg and J. Almlof.
35. The codes have been modified and vectorized for the Cyber 205 by R. Ahlrichs and coworkers. See also R. Ahlrichs, H.-J. Bohm, C. Ehrhardt, P. Scharf, H. Schiffer, H. Lischka and M. Schindler, J. Comp. Chem., in press.
36. The Columbus codes include the Gaussian integral and SCF programs of R. Pitzer and the unitary group CI codes of I. Shavitt, F. Brown, H. Lischka and R. Shepard.
37. R.M. Pitzer, J. Chem. Phys. 58, 3111 (1973).
38. H. Lischka, R. Shepard, F.B. Brown and I. Shavitt, Int. J. Quant. Chem. Symp. 15, 91 (1981).
39. R. Shepard, I. Shavitt and J. Simons, J. Chem. Phys. 76, 543 (1982).
40. S. Hagstrom, QCPE 10, 252 (1975); S. Hagstrom and H. Partridge (unpublished).
41. E.M. Bulewicz, L.F. Phillips and T.M. Sugden, Trans. Faraday Soc. 57, 921 (1961).
42. D.O. Ham, J. Chem. Phys. 60, 1802 (1974).
43. J. Berkowitz, J. Chem. Phys. 50, 3503 (1969); Adv. High Temp. Chem. 3, 158 (1971).
44. E.K. Parks and S. Wexler, J. Phys. Chem. 88, 4492 (1984).
45. M.D. Scheer and J. Fine, J. Chem. Phys. 36, 1647 (1961).
46. D.L. Hildenbrand and E. Murad, J. Chem. Phys. 44, 1524 (1966).
47. M. Farber and R.D. Srivastava, J. Chem. Soc. Faraday Trans. I. 70, 1581 (1974). See also M. Farber and R.D. Srivastava, High Temp. Sci. 8, 195 (1976).
48. T.C. Ehlert, G.D. Blue, J.W. Green and J.L. Margrave, J. Chem. Phys. 41, 2250 (1964).
49. P.D. Kleinschmidt and D.L. Hildenbrand, J. Chem. Phys. 68, 2823 (1978). See also D.L. Hildenbrand, J. Chem. Phys. 48, 3657 (1968).
50. Z. Karny and R.N. Zare, J. Chem. Phys. 68, 3360 (1978).
51. G.D. Blue, J.W. Green, R.G. Bautista and J.L. Margrave, J. Chem.

Phys. 67, 877 (1963). See also G.D. Blue, J.W. Green, T.C. Ehlert and J.L. Margrave, Nature 199, 804 (1963).

2. V.G. Ryabova and L.V. Gurvich, High Temperature 2, 749 (1964).

3. R.L. Jaffe (private communication).

4. D.L. Hildenbrand and L.P. Theard, J. Chem. Phys. 50, 5350 (1969).

5. M. Farber and R.D. Srivastava, J. Chem. Soc. Faraday I 69, 390 (1973).

6. D.L. Hildenbrand, J. Chem. Phys. 52, 5751 (1970).

7. M. Farber and R.D. Srivastava, Chem. Phys. Letters 42, 567 (1976).

8. K.F. Zmbov, Chem. Phys. Letters 4, 191 (1969).

9. L.V. Gurvich, V.G. Ryabova and A.N. Khitrov, Faraday Symp. Chem. Soc. 8, 83 (1973).

0. C.D. Jonah and R.N. Zare, Chem. Phys. Letters 9, 65 (1971).

1. M. Yoshimine, J. Chem. Phys. 57, 1108 (1972).

2. P.A.G. O'Hare and A.C. Wahl, J. Chem. Phys. 56, 4516 (1972).

3. B.C. Laskowski, S.R. Langhoff and P.E.M. Siegbahn, Int. J. Quant. Chem. 23, 483 (1983).

4. S.M. Freund, E. Herbst, R.P. Mariella and W. Klemperer, J. Chem. Phys. 56, 1467 (1972); R.A. Berg, L. Wharton, W. Klemperer, A. Buchler and J.L. Stauffer, ibid 43, 2416 (1965).

5. D.L. Hildenbrand, J. Chem. Phys. 57, 4556 (1972).

6. D.L. Hildenbrand and E. Murad, J. Chem. Phys. 53, 3403 (1970).

7. T.C. Ehlert, High Temp. Science 9, 237 (1977).

8. A.V. Gusarov and L.N. Gorokhov, Teplofiz Vys. Temp. 9, 505 (1971).

9. K.P. Huber and G. Herzberg, "Molecular Spectra and Molecular Structure," (Van Nostrand Reinhold, New York, 1979).

a. H. Lavendy, B. Pouilly and J.M. Robbe, J. Mol. Spectrosc. 103, 379 (1984).

0. W.A. Chupka, J. Berkowitz and C.F. Giese, J. Chem. Phys. 30, 827 (1959).

1. T. Ikeda, N.B. Wong, D.O. Harris and R.W. Field, J. Mol. Spectrosc. 68, 452 (1977). See also R.W. Field, Air Force Goephysics Laboratory Report No. AFGL-TR-83-0021.

2. R.N. Zare (private communication).

3. C.W. Bauschlicher, B.H. Lengsfield III, D.M. Silver and D.R. Yarkony, J. Chem. Phys. 74, 2379 (1981).

4. C.W. Bauschlicher and D.R. Yarkony, J. Chem. Phys. 68, 3990 (1978).

5. G.A. Capelle, H.P. Broida and R.W. Field, J. Chem. Phys. 62, 3131 (1975).

6. R.W. Field, J. Chem. Phys. 60, 2400 (1974).

7. B. Pouilly, J.M. Robbe, J. Schamps, R.W. Field and L. Young, J. Mol. Spectrosc. 96, 1 (1982). See also C.J. Cheetham, W.J.M. Gissane and R.F. Barrow, Trans. Faraday Soc. 61, 1308 (1965).

8. M.W. Chase, Jr., J.L. Curnutt, J.R. Downey, Jr., R.A. McDonald and A.N. Syverud, J. Phys. and Chem. Ref. Data 11, 695 (1982). JANAF Thermochemical Tables, 1982 supplement and related volumes.

9. D.L. Hildenbrand, "Advances in High Temperature Chemistry," Vol. 1, L. Eyring (ed.), pp. 198-206, Academic Press.

0. R. Colin, P. Goldfinger and M. Jeunehommer, Trans. Faraday Soc. 60, 306 (1964).

1. J.R. Marquart and J. Berkowitz, J. Chem. Phys. 39, 283 (1963).

82. D.H. Cotton and D.R. Jenkins, Trans. Faraday Soc. 65, 1537 (1969).
83. P.J.Th. Zeegers and C.Th.J. Alkemade, Combustion and Flame 15, 193 (1970).
84. D.E. Jensen, J. Phys. Chem. 74, 207 (1970). See also D.E. Jensen and P.J. Padley, Trans. Faraday Soc. 62, 2132 (1966).
85. H.C. Ko, M.A. Greenbaum and M. Farber, J. Phys. Chem. 71, 1875 (1967).
86. Y.H. Inami and F. Ju, work reported in JANAF Thermochemical Tables (Ref. 87).
87. M.W. Chase, J.L. Curnutt, R.A. McDonald and A.N. Syverud, J. Phys. and Chem. Ref. Data 7, 793 (1978).
88. D.H. Cotton and D.R. Jenkins, Trans. Faraday Soc. 65, 376 (1969).
89. E.M. Bulewicz and T.M. Sugden, Trans. Faraday Soc. 55, 720 (1959).
90. E. Murad, Chem. Phys. Letters 72, 295 (1980).
91. P.J. Kalff and C.Th.J. Alkemade, J. Chem. Phys. 59, 2572 (1973).
92. D.H. Cotton and D.R. Jenkins, Trans. Faraday Soc. 64, 2988 (1968).
93. V.G. Ryabova, A.N. Khitrov and L.V. Gurvich, High Temp. 10, 669 (1972).
94. E. Murad, J. Chem. Phys. 75, 4080 (1981).
95. J. van der Hurk, Tj. Hollander and C.Th.J. Alkemade, J. Quant. Spectrosc. Radiat. Transfer 14, 1167 (1974).
96. L.V. Gurvich, V.G. Ryabova, A.N. Khitrov and E.M. Starovoitov, High Temp. 9, 261 (1971).
97. F.E. Stafford and J. Berkowitz, J. Chem. Phys. 40, 2963 (1964).
98. R.J. Celotta, R.A. Bennet and J.L. Hall, J. Chem. Phys. 60, 1740 (1974).
99. C.W. Bauschlicher and H. Partridge, Chem. Phys. Letters 106, 65 (1984).
100. S.M. Freund, P.D. Godfrey and W. Klemperer, 25th Symposium on Molecular Structure and Spectroscopy, Ohio State University, Columbus, Ohio, 1970, paper E-8- moment of inertia reported for LiOH.
101. P. Kuijpers, T. Torring and A. Dymanus, Chem. Phys. 15, 457 (1976).
102. N. Acquista and S. Abramowitz, J. Chem. Phys. 51, 2911 (1969).
103. E.F. Pearson, B.P. Winnewisser and M.B. Trueblood, Z. Naturfosch 31, 1259 (1976).
104. A.A. Belyaeva, M.I. Dvorkin and L.D. Sheherba, Opt. Spectrosc. 31, 210 (1966).
105. D.R. Lide and C. Matsumura, J. Chem. Phys. 50, 3080 (1969).
106. D.R. Lide and R.L. Kuczkowski, J. Chem. Phys. 46, 4768 (1967).
107. N. Aquista, S. Abramowitz and D.R. Lide, J. Chem. Phys. 49, 780 (1968).
108. R.C. Hilborn, Z. Qingshi and D.O. Harris, J. Mol. Spectrosc. 97, 73 (1983).
109. J. Nakagawa, R.F. Wormsbecher and D.O. Harris, J. Mol. Spectrosc. 97, 37 (1983).
110. C.W. Bauschlicher, S.R. Langhoff and H. Partridge, "Ab Initio Study of BeCN, MgCN, CaCN and BaCN," Chem. Phys. Letters (in press).
111. T. Torring, J.P. Bekooy, W.L. Meerts, J. Hoeft, E. Tiemann and A. Dymanus, J. Chem. Phys. 73, 4875 (1980).

112. J.J. Van Vaals, W.L. Meerts and A. Dymanus, Chem. Phys. 86, 147 (1984).
113. P.E.S. Wormer and J. Tennyson, J. Chem. Phys. 75, 1245 (1981).
114. M.L. Klein, J.D. Goddard and D.G. Bounds, J. Chem. Phys. 75, 3909 (1981).
115. C.J. Marsden, J. Chem. Phys. 76, 6451 (1982).
116. L. Pasternack and P.J. Dagdigian, J. Chem. Phys. 65, 1320 (1976).
117. J. Berkowitz, W.A. Chupka and T.A. Walter, J. Chem. Phys. 50, 1497 (1969).
118. B.V. L'vov and L.A. Pelieva, Prog. Analyt. Atomic Spectrosc. 3, 65 (1980).

METAL CHEMICAL SHIFTS IN NMR SPECTROSCOPY - AB INITIO CALCULATIONS AND PREDICTIVE MODELS

Hiroshi Nakatsuji
Division of the School of Engineering
Graduate School of Engineering
Kyoto University
Kyoto 606
Japan

ABSTRACT. The metal chemical shifts in NMR spectroscopy are studied theoretically. Electronic origins and mechanisms of the metal chemical shifts in the Cu, Ag, Zn, Cd, and Mn complexes are clarified based on ab initio molecular orbital calculations. The mechanisms accounting for the chemical shift are found to be very different for metals with a $d^{10}s^{1-2}p^0$ configuration (Cu, Ag, Zn, Cd) and those with a $d^5s^2p^0$ configuration (Mn), because of the filled and half-filled nature of the valence d subshell.

1. INTRODUCTION

The recent developments in organometallic and inorganic chemistry are largely based on knowledge of the electronic structures of bonding in metal complexes [1]. The chemistry is very different than organic chemistry because of the essential participation of d electrons. The symmetry, multiplicity and flexibility of d orbitals cause the chemistry to be quite varied in comparison with the chemistry of s and p electrons alone. A convenient way of looking at the role of the d electrons is to observe those properties which are sensitive to the angular momentum of electrons, since the d electrons have larger angular momentum than the p electrons and the s electrons have no angular momentum.

Nuclear magnetic resonance spectroscopy of metal species [2-4] gives valuable information about the role of the d electrons in the

R. J. Bartlett (ed.), Comparison of Ab Initio Quantum Chemistry with Experiment for Small Molecules, 409–437.

chemical bonding of metal complexes, since the magnetic properties observed reflect the angular momentum of the electrons involved. The chemical shift of the metal nucleus is especially valuable, because it samples the angular momentum of electrons near the metal nucleus. It amplifies the role of the d electrons of the metal relative to that of the p electrons. The role of the s electron is effectively eliminated due to its having no angular momentum. Thus, the NMR spectroscopy of metal nuclei is important not only as a tool of analytical chemistry but also for the physical and chemical information concerning the rôle of the d (and also f) electrons in the chemistry of metal complexes.

We have recently carried out theoretical studies of the metal chemical shift in NMR spectroscopy [5-7]. The molecules studied are the Cu, Zn, Ag and Cd complexes [5-7], for which the metal atoms are characterized by the electronic configurations $d^{10}s^{1-2}p^{0}$ in their ground state, and the Mn complexes [6], for which the Mn metal is characterized by an open d-subshell $d^{5}s^{2}$ in its ground state. The purpose of the study is two-fold. One is to understand the origin of the metal chemical shift by performing reliable ab initio calculations and by analyzing the calculated results. We are especially interested in the role of the d electrons in the chemical bonds of the transition metal complexes. The other purpose is to clarify the mechanism of the metal chemical shift. The elucidation of this mechanism would be useful, since there seems to be no guiding concept, so far, which is useful for experimental chemists. This situation is completely different from that of the ^{1}H and ^{13}C chemical shifts. To the best of our knowledge, these are the first ab initio studies of the metal chemical shifts of transition metal complexes. We sketch here only the most important results of these studies in order to show an example of the close cooperation between quantum chemistry and experiment, which is the purpose of this book.

The metal complexes studied here are the following 27 molecules:

Cu complex: $CuCl$, $CuCl_4^{3-}$, $Cu(CN)_4^{3-}$, $Cu(NH_3)_2^{+}$.

Zn complex: $Zn(H_2O)_6^{2+}$, $ZnCl_2$, $ZnCl_4^{2-}$, $Zn(CN)_4^{2-}$, $Zn(NH_3)_4^{2+}$.

Ag complex: $Ag(H_2O)_6^+$, AgF_4^{3-}, $AgCl_2^-$, $AgCl_4^{3-}$, $Ag(CN)_4^{3-}$, $Ag(NH_3)_2^+$.

Cd complex: $Cd(H_2O)_6^{2+}$, $CdCl_2$, $CdCl_4^{2-}$, $Cd(CN)_4^{2-}$, $Cd(CH_3)_2$, CdMeEt, $CdEt_2$, CdMe(OMe).

Mn complex: $Mn(CO)_5H$, $Mn(CO)_5CN$, $Mn(CO)_5CH_3$, $Mn(CO)_5Cl$.

The experimental data for the chemical shifts are due to Cardin et al. [8], Ackerman et al. [9], Kennedy at al. [10], and Mennitt et al. [11] for the Cd complexes; Maciel et al. [12] for the Zn complexes, Endo et al. for the Ag and Cu complexes [13]; and, Calderazzo et al. [14] for the Mn complexes.

2. THEORETICAL BACKGROUND

The hamiltonian of a molecule in a magnetic field may be written as [15-17]

$$H = H_0 + \sum_x B_x H_x^{(1,0)} + \sum_A \sum_x \mu_{Ax} H_{Ax}^{(0,1)} +$$

$$+ (1/2) \sum_A \sum_x \sum_y B_x H_{Axy}^{(1,1)} \mu_{Ay} + \ldots \tag{1}$$

where H_0 is the hamiltonian free from the magnetic field, and B_x and μ_{Ax} are the magnetic field and the nuclear magnetic moment of nucleus A in the x direction. The perturbation hamiltonians $H_x^{(1,0)}$, $H_{Ax}^{(0,1)}$, and $H_{Axy}^{(1,1)}$ are given by

$$H_x^{(1,0)} = -(i\hbar e/2mc) \sum_j \ell_{jx}, \tag{2a}$$

$$H_{Ax}^{(0,1)} = -(i\hbar e/2mc) \sum_j \ell_{jx}/r_{Aj}^3, \tag{2b}$$

$$H_{Axy}^{(1,1)} = (e^2/4mc^2) \sum_j (r_j r_{Aj} \delta_{xy} - r_{jx} r_{Ajy})/r_{Aj}^3. \qquad (2c)$$

ℓ_{jx} is an angular momentum operator for electron j in the x direction. The nuclear magnetic shielding constant is defined by

$$\sigma_{xy}(A) = \left[\frac{\partial^2 E(\beta,\mu)}{\partial\beta_x \partial\mu_{Ay}}\right]_{B=\mu_A=0}, \qquad (3)$$

which is written with the use of the Hellmann-Feynman theorem as

$$\sigma_{xy}(A) = \langle\psi(0)|H_{Axy}^{(1,1)}|\psi(0)\rangle + \frac{\partial}{\partial B_x}[\langle\psi(B_x)|H_{Ay}^{(0,1)}|\psi(B_x)\rangle]_{B=0}. \qquad (4)$$

$\Psi(0)$ and $\psi(B_x)$ are the wavefunctions in the absence and the existence, respectively, of the magnetic field. The first term of Eqn. (4) is the diamagnetic term σ^{dia} and the second term is the paramagnetic term σ^{para}. If we expand the perturbed wavefunction $\Psi(B_x)$ in the set of unperturbed wavefunction $\{|n\rangle\}$, we obtain the familiar equation

$$\sigma_{xy}^{para}(A) = \sum_n \frac{\langle 0|H_x^{(1,0)}|n\rangle\langle n|H_{Ay}^{(0,1)}|0\rangle}{E_0 - E_n} + \text{C.C.} \qquad (5)$$

We note that both of the operators $H_x^{(1,0)}$ and $H_{Ay}^{(0,1)}$ include the angular momentum operator ℓ.

We calculated the magnetic shielding constant using the coupled perturbed Hartree-Fock theory or the finite perturbation theory of Cohen, Roothaan, and Pople et al. [18,19]. This theory has been shown to be the best expression of the second-order perturbation theory based upon the Hartree-Fock wavefunction [20]. We calculated the perturbed, complex wavefunction $\Psi(B_x)$ in the Hartree-Fock approximation using a finite magnetic field. Then, the nuclear shielding constant was calculated with Eqn. (4). The origin of the gauge was chosen at the position of the metal nucleus. We limit ourselves here only to the isotropic term,

$$\sigma_A = (1/3)\,\{\sigma_{xx}(A) + \sigma_{yy}(A) + \sigma_{zz}(A)\} \qquad (6)$$

The chemical shift is a relative value which is calculated by subtracting the nuclear magnetic shielding constant σ_0 from that of a standard molecule, i.e.

$$\Delta\sigma = \sigma - \sigma_0. \qquad (7)$$

The basis set we used for calculations are the MIDI-1 basis of Tatewaki, Huzinaga, and Sakai [21] plus two outer p-type Gaussians for metal atoms [5,6]. This basis is of a double-zeta quality for valence electrons. For the ligands of the Mn complexes, we used the MINI-1 set [21] which is poorer than the MIDI-1 set. The geometries of the complexes are summarized in previous papers [57]. For the metal-ligand length, we assumed a sum of the tetrahedral covalent radii [22].

3. CORRELATION BETWEEN THEORETICAL AND EXPERIMENTAL VALUES

Table 1 shows the theoretical values of the magnetic shielding constants and the chemical shifts of the metal nuclei of the complexes studied here. The diamagnetic and paramagnetic contributions and their decomposition into core and valence MO contributions are also given. Figures 1 - 5 show the correlations between the theoretical values and experimental results for the Cd, Ag, Zn, Cu, and Mn complexes, respectively. The open circle on the line indicates that only the theoretical value exists. For the Ag complexes, we compared the theoretical results for AgF_4^{3-} and $AgCl_4^{3-}$ with the experimental values of $AgBr_4^{3-}$ and AgI_4^{3-}.

The largest number of examples are seen in Figure 1 for the cadmium complexes. In addition to the results of Table 1, it includes the results for CdMeEt, $CdEt_2$, and CdMe(OMe) [7]. The correlation between theory and experiment is good except for $Cd(CN)_4^{2-}$. For the experiment in aqueous solution [8], the geometry of the complex is unknown and the effect of the solvent should be large.

Table 1. Diamagnetic and paramagnetic contributions, σ_M^{dia} and σ_M^{para}, to the metal magnetic shielding constant σ_M and their analyses into core and valence MO contributions (in ppm.).

	σ^{dia}				σ^{para}				σ	
Molecule	Core	Valence	Total	Shift	Core	Valence	Total	Shift	Total	Shift
$CuCl_4^{3-}$	2292	375	2667	0	-63	-818	-881	0	1786	0
$Cu(CN)_4^{3-}$	2190	387	2577	-90	-103	-1588	-1691	-810	886	-900
$Cu(NH_3)_2^{+}$	2150	314	2464	-203	-73	-2582	-2655	-1774	-191	-1977
CuCl	2171	279	2450	-217	-79	-4617	-4696	-3815	-2246	-4032
$Zn(H_2O)_6^{2+}$	2266	486	2752	0	-44	-272	-316	0	2436	0
$ZnCl_4^{2-}$	2376	406	2782	30	-55	-447	-502	-186	2280	-156
$ZnCl_2$	2294	342	2636	-116	-42	-414	-456	-140	2179	-257
$Zn(CN)_4^{2-}$	2272	420	2692	-60	-74	-654	-728	-412	1964	-472
$Zn(NH_3)_4^{2+}$	2250	418	2668	-84	-75	-711	-786	-470	1882	-554
$Ag(H_2O)_6^{+}$	4514	361	4875	0	-41	-431	-472	0	4403	0
$Ag(NH_3)_2^{+}$	4484	239	4722	-153	-33	-628	-661	-189	4062	-341
$AgCl_2^{-}$	4549	241	4790	-85	-46	-1001	-1047	-575	3742	-661
AgF_4^{3-}	4504	328	4832	-43	-75	-1083	-1158	-686	3673	-730
$AgCl_4^{3-}$	4631	307	4938	63	-76	-1296	-1372	-900	3566	-837
$Ag(CN)_4^{3-}$	4521	308	4829	-46	-96	-1416	-1512	-1040	3318	-1085

Table I (cont.)

Molecule	σ^{dia} Core	σ^{dia} Valence	σ^{dia} Total	σ^{dia} Shift	σ^{para} Core	σ^{para} Valence	σ^{para} Total	σ^{para} Shift	σ Total	σ Shift
$Cd(H_2O)_6^{2+}$	4626	378	5004	0	-33	-442	-475	0	4529	0
$CdCl_2$	4655	253	4908	-96	-25	-643	-668	-193	4240	-286
$CdCl_4^{2-}$	4731	314	5045	41	-40	-875	-915	-440	4129	-400
$Cd(CH_3)_2$	4595	256	4851	-153	-43	-1047	-1090	-615	3761	-768
$Cd(CN)_4^{2-}$	4640	343	4983	-21	-95	-1509	-1604	-1129	3378	-1151
$Mn(CO)_5H$	1889	318	2207	0	-141	-12927	-13068	0	-10861	0
$Mn(CO)_5CN$	1905	345	2249	42	-123	-13059	-13182	-114	-10933	-72
$Mn(CO)_5CH_3$	1898	339	2237	30	-130	-14327	-14457	-1389	-12220	-1359
$Mn(CO)_5Cl$	1929	339	2267	60	-91	-16149	-16240	-3172	-13973	-3112

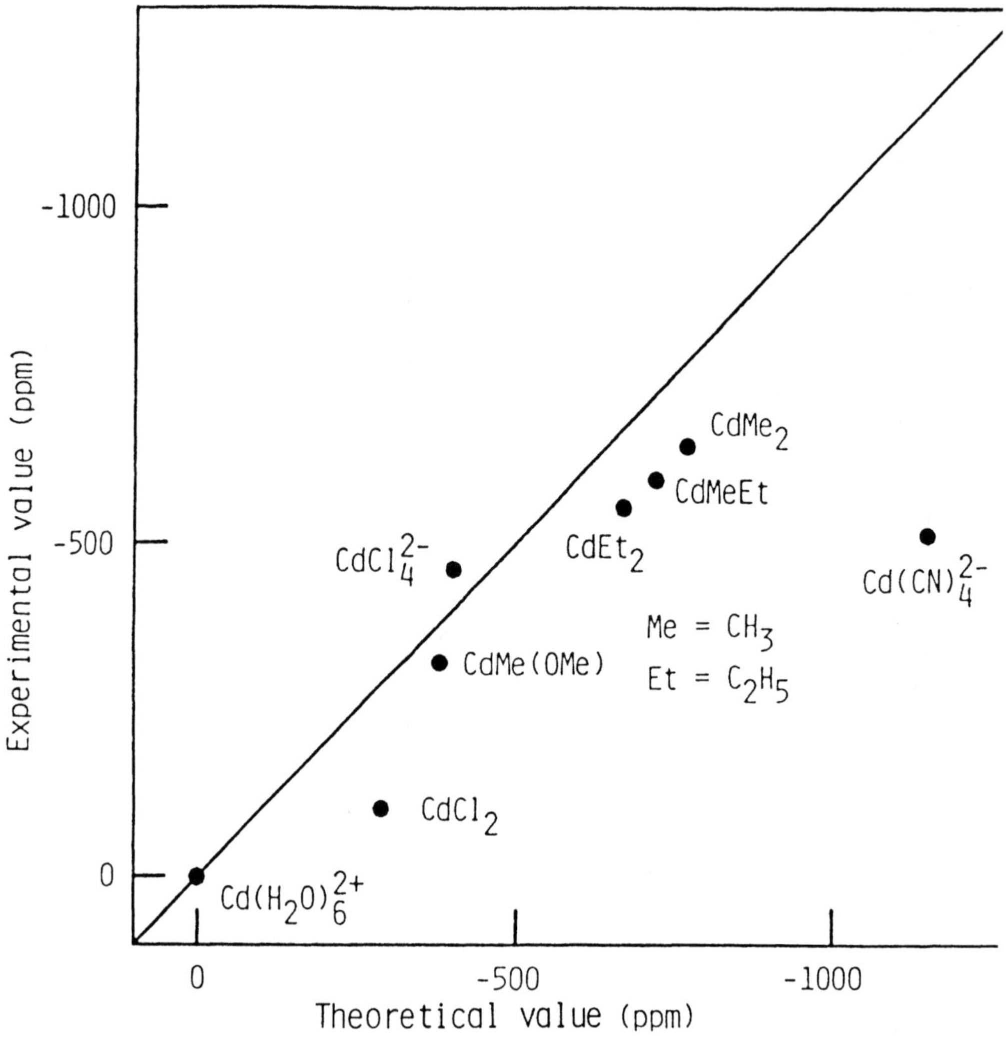

Figure 1. Comparison between theoretical and experimental values for the ^{113}Cd chemical shifts of the cadmium complexes.

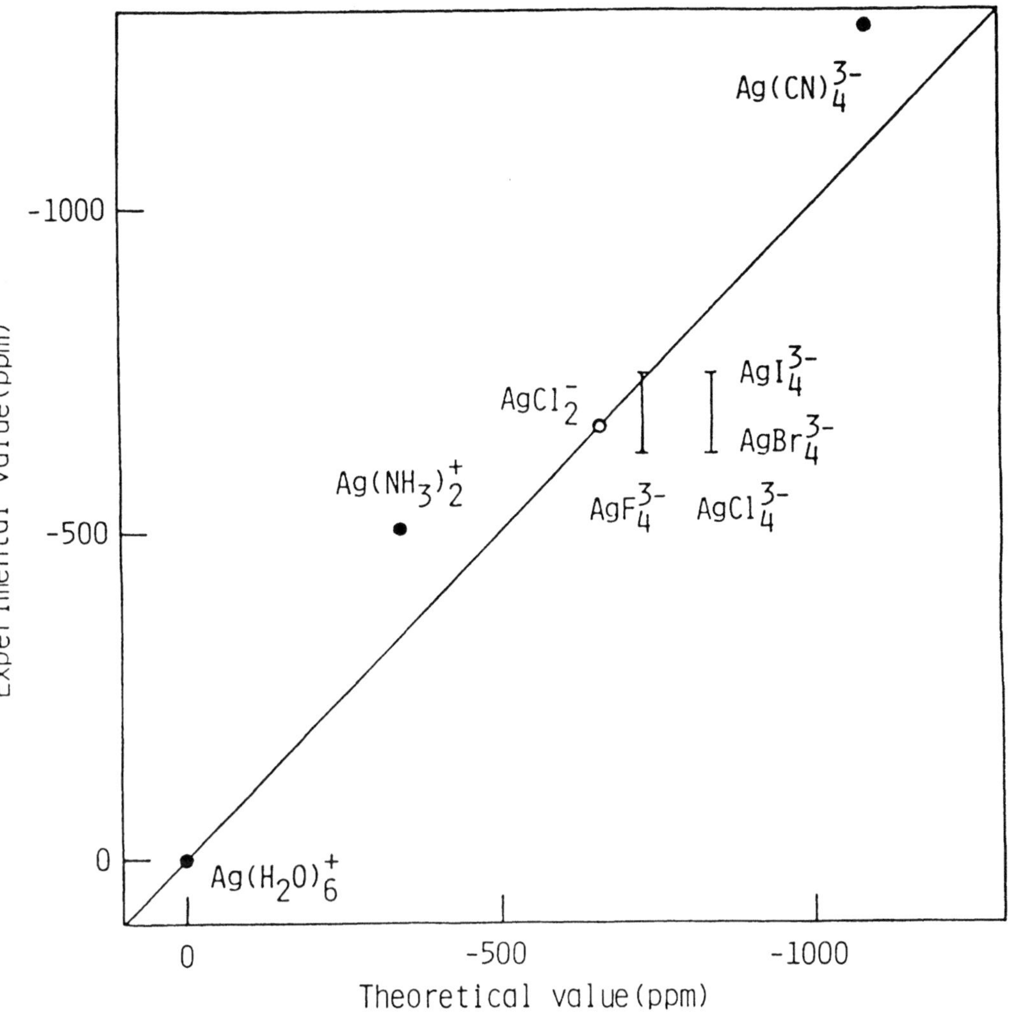

Figure 2. Comparison between theoretical and experimental vlaues for the ^{109}Ag chemical shifts of the silver complexes. For the halides see the text.

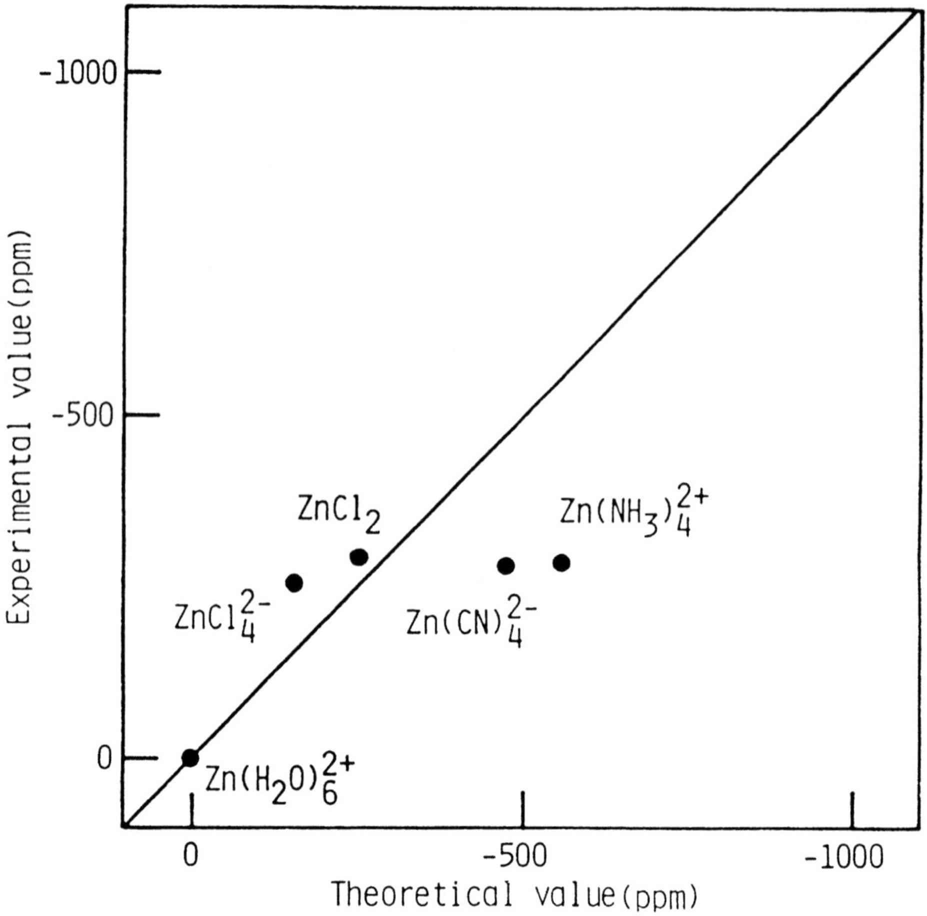

Figure 3. Comparison between theoretical and experimental values for the ^{67}Zn chemical shifts of the zinc complexes.

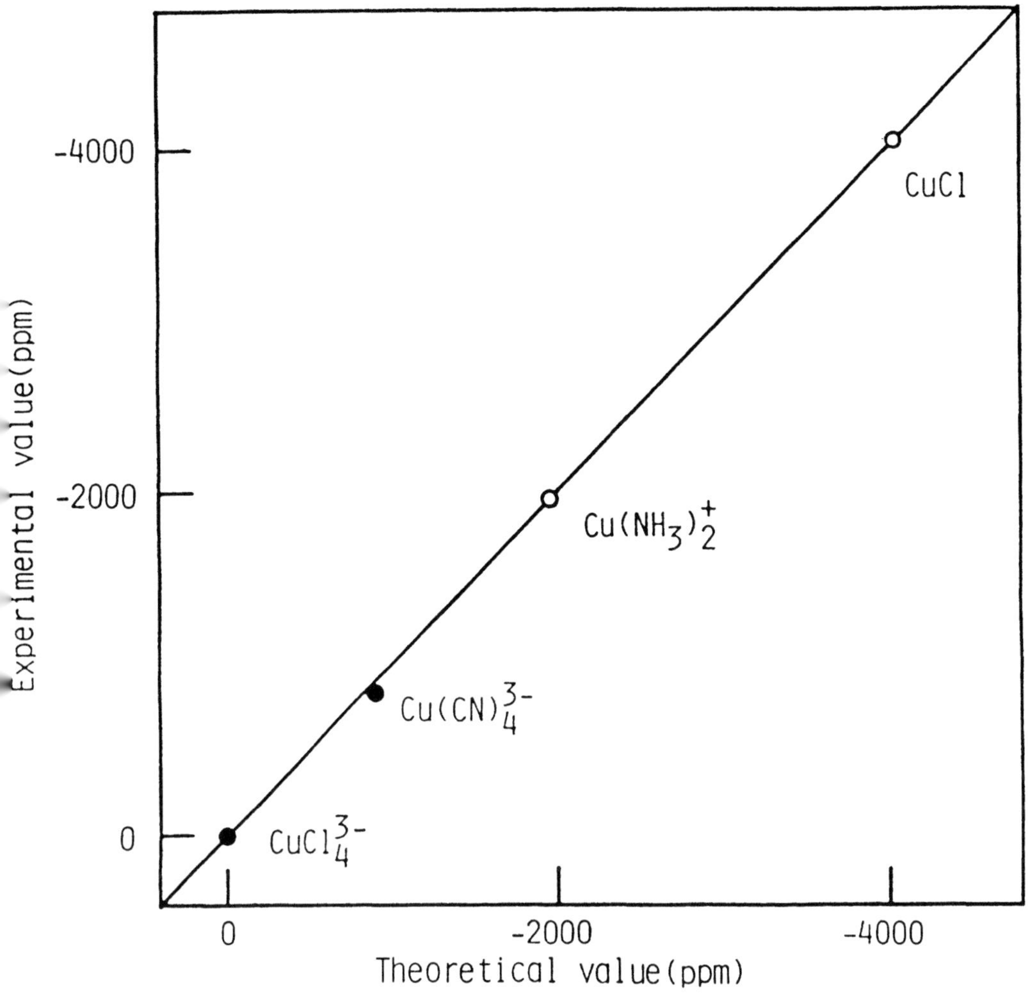

Figure 4. Comparison between theoretical and experimental values for the ^{63}Cu chemical shifts of the copper complexes.

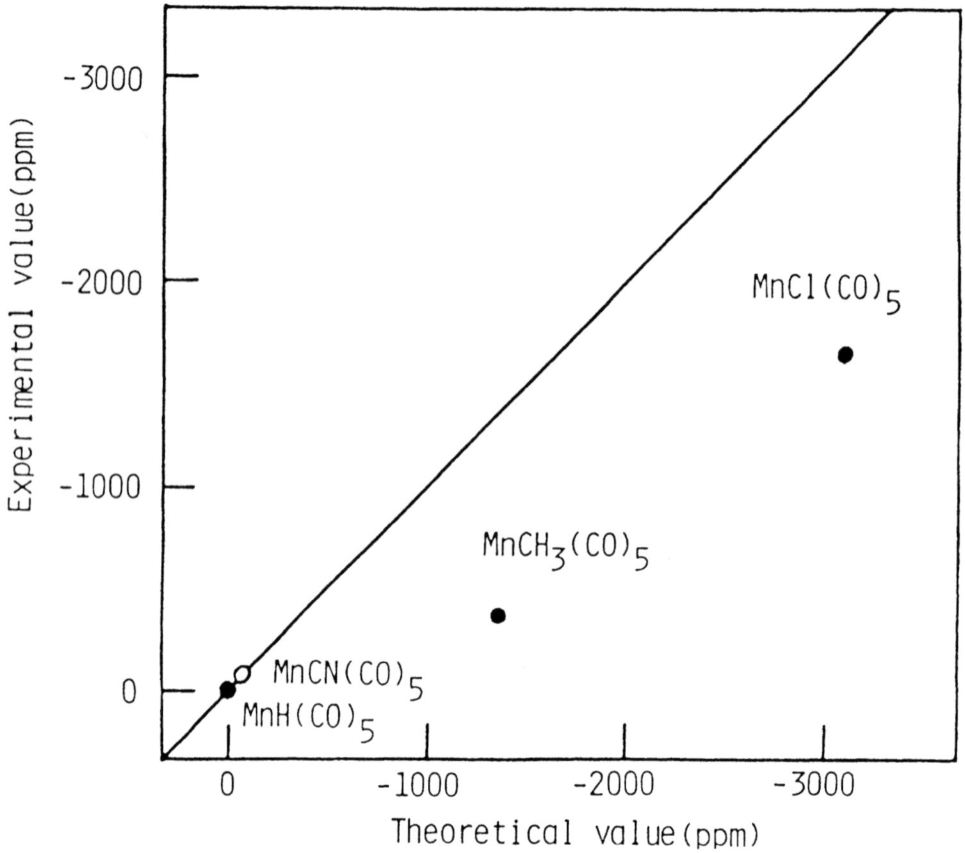

Figure 5. Comparison between theoretical and experimental values for the ^{55}Mn chemical shifts of the manganese complexes.

hese factors are not considered in the calculations. For the omplexes $CdMe_2$, CdMeEt, $CdEt_2$, and CdMe(OMe), the effects of the olvent are expected to be small for the experimental conditions 8,10]. The parallelism between theory and experiment is espeially excellent for these molecules. This fact is understood to how the reliability of the present theoretical results.

In Figures 2 and 3, we see that the correlation between theory nd experiment are good for the silver and zinc complexes. This is ather surprising considering the ambiguity in the structure and olvation of these complexes in aqueous solutions. For the silver omplexes, the experimental shift is larger for the iodide than for he bromide, and the calculated shift is larger for the chloride han for the fluoride. The ordering is natural and may be undertood to suggest a small relativistic effect due to the heavy ligand toms [5].

For the copper complexes, we have only two examples for which oth experimental and theoretical values exist. These two points ie almost on the diagonal line, showing the reliability of the calculated values. Recently, Kitagawa and Munakata reported a systematic xperimental study of the Cu chemical shift [23]. They reported that he copper chemical shift is larger for the ligand which is more π-lectron withdrawing. This observation is very important and is xplained clearly by the mechanism of the copper chemical shift, as hown in the next section.

For the Mn chemical shift, we have studied the complexes of he form $Mn(CO)_5L$ with L = H, CH_3, Cl, CN. The correlation betveen theory and experiment is only fair. The calculated shifts are arger than the experimental values. Referring to Table 1, we otice that for the Mn shielding constant, the paramagnetic term is n order of magnitude larger than those of the Ib and IIb metal omplexes. We will see later that this is due to the difference in he origin of the metal chemical shift for these complexes.

4. MECHANISM OF THE METAL CHEMICAL SHIFT OF THE Ib AND IIb METAL COMPLEXES

We now want to clarify the electronic origin of the metal chemical shift. Since the origin is very much different for complexes of the Ib and IIb metals and those of Mn, we will explain the origin separately. We want to identify an electronic mechanism that produces the metal chemical shift, which would be useful as a guiding principle for experimental chemists.

Referring to Table 1, we see that the chemical shifts are mostly due to the valence MO contributions to the paramagnetic term. We have shown in a previous paper [5] that the behavior of the valence electrons near the nucleus are observed as a chemical shift through NMR spectroscopy. The diamagnetic term itself is large but its contribution to the chemical shift is small. We therefore discuss the diamagnetic term separately in Section 6. In Tables 2 and 3, we have shown the metal AO contributions and the ligand contributions, respectively, to the paramagnetic term of the metal shielding constant. The definition of the AO contribution was given previously [5]. Comparing the numbers in Tables 2 and 3, we immediately understand that the ligand contributions are small and that the dominant term comes from the metal AO contributions. The metal s AO contribution is identically zero, because the s AO does not have angular momentum, so that the metal chemical shifts are primarily due to the metal p and d orbital contributions.

Looking at Table 2 in more detail, we notice that for the Cu chemical shift, the d orbital contribution is dominant, for the Zn and Cd chemical shifts; the p orbital contribution is more important than the d orbital contribution; and for the Ag chemical shift, the p and d orbital contributions are comparable. This observation is very important in understanding the trends in the chemical shift that are characteristic of metal species.

Now, how do these p and d orbital contributions arise? The Ib and IIb metals studied in this section have the electronic configurations $d^{10}s^{1-2}p^{0}$ in their ground state. As the perturbation heory shows [24], the closed p and d subshells give no contributions o the paramagnetic term. Table 1 shows that these p and d contribu- ions are due to the valence orbitals. Therefore, we conclude that hese paramagnetic terms are caused by the donation of electrons from he ligands to the metal p orbitals and by the back-donation of lectrons from the metal d orbitals to the ligands. In other words, he electrons in the valence p orbitals and the holes in the valence d rbitals produce angular momentum when the magnetic field is applied nd give paramagnetic contributions to the metal nuclear shielding onstants of the complexes. In Figure 6, we illustrated these two echanisms, namely the p mechanism and the d mechanism. Though this igure is drawn for the π interactions, these mechanisms involve oth σ and π interactions. For the p mechanism an electron donating igand produces a larger metal chemical shift. On the other hand, for he d mechanism an electron-withdrawing ligand produces a larger metal hemical shift. Thus, the effect of the ligand is entirely different or the p and d mechanisms.

From Table 3, we see that the d mechanism is much more important han the p mechanism for the Cu complexes. This suggests that the d rbitals are important in forming the bond between copper and igands. This further suggests that the Cu chemical shift would be arallel to the d-electron acceptability of the ligands, as long as he number of ligands is the same. This is supported by the recent ^{3}Cu NMR experiments of Kitagawa and Munakata [23]. They observed hat the Cu chemical shift increases with increasing π-electron ithdrawing ability of the ligand. For the Zn and Cd complexes, the mechanism is calculated to be more important than the d mechanism. or these IIb metal complexes, the p orbitals are more important han the d orbitals for the chemical bonds between the metal and the igands. For the CdL_2 complexes, the p mechanism is predominant. or the CdL_4 complexes, the d contribution increases though it is till smaller than the p contribution. This is in part due to the ybridization necessary for tetrahedral coordination. For the CdL_2 omplexes, sp hybridization is dominant, but for the CdL_4 complexes,

Table 2. Metal AO contributions to the paramagnetic term of the metal shielding constant (In ppm.)[a].

Ligand	Cu		Zn		Ag		Cd	
	p	d	p	d	p	d	p	d
$(H_2O)_6$	---	---	-195	-54	-204	-171	-291	-124
F_4	---	---	---	---	-633	-489	---	---
Cl	-210	-4486	---	---	---	---	---	---
Cl_2	---	---	-316	-128	-494	-540	-604	-52
Cl_4	-305	-553	-373	-101	-851	-491	-723	-166
$(CN)_4$	-290	-1360	-343	-335	-538	-925	-864	-687
$(NH_3)_2$	-165	-2475	---	---	-248	-390	---	---
$(NH_3)_4$	---	---	-420	-309	---	---	---	---
$(CH_3)_2$	---	---	---	---	---	---	-992	-68
Major contribution	d		p		p,d		p	

[a] The s orbital contribution is identically zero.

Table 3. Ligand contributions to the paramagnetic term of the metal shielding constant (In ppm.).

Ligand	Cu	Zn	Ag	Cd
$(H_2O)_6$	---	-67	-62	-60
F_4	---	---	-37	---
Cl	-1	---	---	---
Cl_2	---	-13	-13	-12
Cl_4	-24	-27	-30	-27
$(CN)_4$	-41	-50	-49	-53
$(NH_3)_2$	-15	---	-23	---
$(NH_3)_4$	---	-57	---	---
$(CH_3)_2$	---	---	---	-29

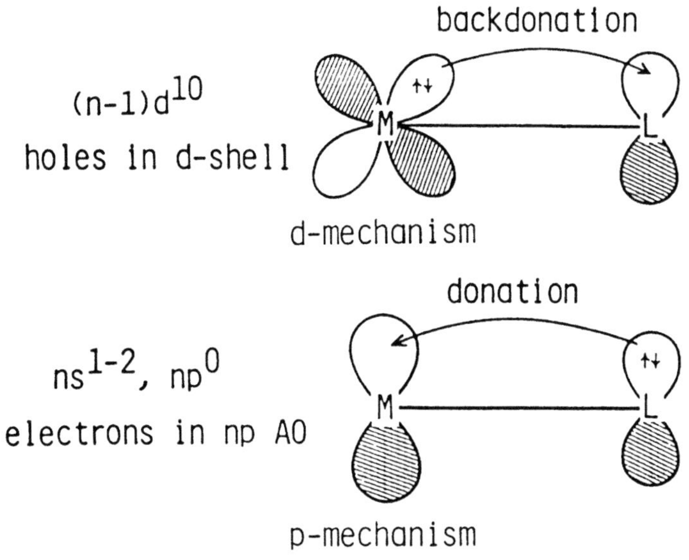

Figure 6. Illustration of the d and p mechanisms of the metal chemical shift. The holes in the d orbitals due to back-donation and the electrons in the p orbitals due to donation are the origin of the paramagnetic term of the metal chemical shift of the Ib and IIb metal complexes.

mixing of the sp^3 and sd^3 hybridizations occur. For the Zn complexes, we fail to see such a clear relation. Comparing the tetrahedral Cl and CN complexes, we see that the d mechanism is much facilitated by the CN ligand though the p contribution itself is larger than the d contribution. This is due to the π-withdrawing ability of the CN ligand through the d_π(metal) - p_π(ligand) conjugation. Further, the d mechanism is more effective than the p mechanism since the angular momentum is larger for the d orbital than for the p orbital. Lastly, for the Ag complexes, we see that the p and d mechanisms are competitive. The relative importance depends strongly on the number and the nature of the ligands. In Table 4, we summarize the major mechanisms for the chemical shifts of the Ib and IIb metal complexes and the roles expected of the ligands.

Now, why do these differences in the origin of the metal chemical shift occur? A possible answer is based on the atomic energy levels of the metal atoms. In Figure 7, we show the atomic energy levels of the Ib (Cu, Ag) and IIb (Zn, Cd) metal ions [25]. For these metals, the valence s AO is important for the metal-ligand bonds. For neutral species, the ns levels of these metals are close. Thus, we choose the ns levels as a standard in Figure 7. Note that the s AO does not contribute to the shielding constant, because it does not have any angular momentum. Looking at the differences between the (n-1)d - ns and ns - np level splittings, we see that the d - s splitting is smaller than the p - s splitting for Cu, the reverse relation for Zn and Cd, and the splittings are very close for Ag. In the Cu complex formation, the d orbitals of Cu would more easily mix with the metal-ligand bonds than the p orbitals. For the Zn and Cd complexes, the reverse relation is expected, and for the Ag complexes, the mixing tendency of the 4d and 5p AO's is expected to be similar. Thus, the relative importance of the p and d mechanisms in the metal chemical shift is attributed to the structures in the atomic energy levels of the metal atom itself.

Table 4. Summary of mechanisms which give major paramagnetic contributions to the metal chemical shifts.

Metal Complexes	p- or d- Mechanism	Role of ligand
Cu	holes in (n-1)d shell $\gg$ electrons in np AO	electron-acceptor $\gg$ electron-donor
Cd, Zn	electrons in np AO $>$ holes in (n-1)d shell	electron-donor $>$ (shift may be large)
Ag	electrons in np AO $\approx$ holes in (n-1)d shell	electron-donor $\approx$ electron-acceptor

Table 5. Analysis of the paramagnetic term, σ_{Mn}^{para}, of $Mn(CO)_5L$ complexes into metal AO and ligand contributions. (In ppm.)

Molecules	Mn		Ligand		
	p	d	CO_{ax}	$4xCO_{eq}$	L
$Mn(CO)_5H$	-129	-12983	10	30	4
$Mn(CO)_5CN$	-75	-13138	7	24	3
$Mn(CO)_5CH_3$	-106	-14385	9	27	-2
$Mn(CO)_5Cl$	1	-16276	8	24	3

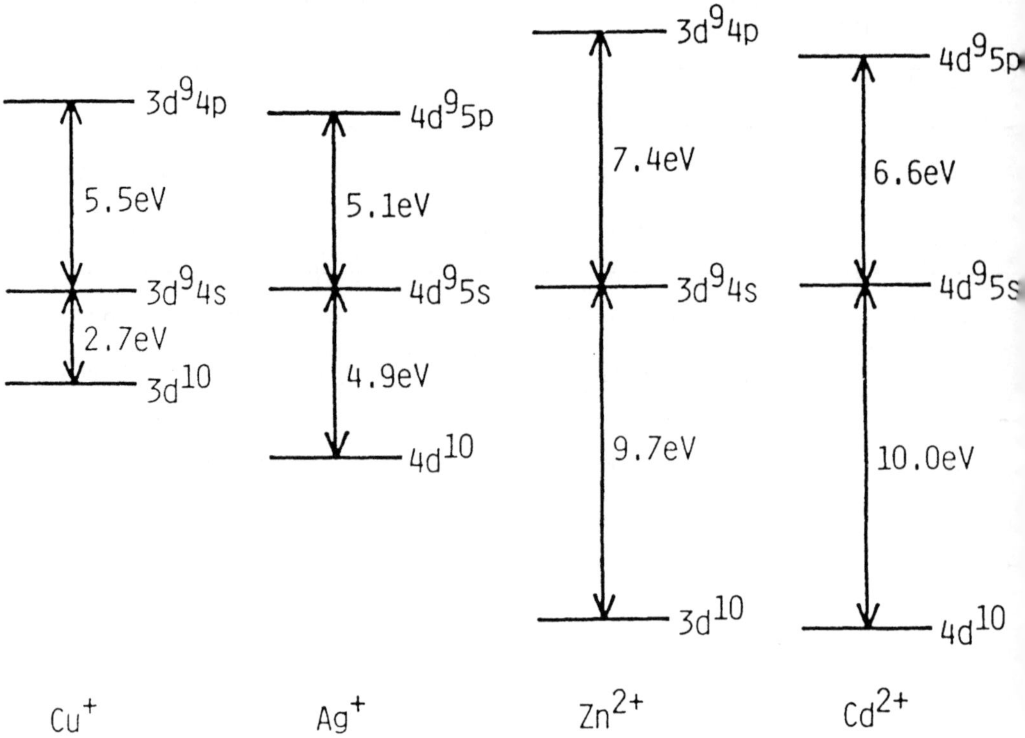

Figure 7. Atomic energy levels of the Ib (Cu, Ag) and IIb (Zn, Cd) metal ions. The energy levels of the d^9s^1 configurations are taken as a standard.

. MECHANISM OF THE METAL CHEMICAL SHIFT OF THE Mn COMPLEXES

The mechanisms of the metal chemical shift of the Mn complexes are entirely different from those of the complexes of the Ib and IIb metals. Referring to Table 1, we see that the Mn chemical shifts are entirely due to the valence MO contributions to the paramagnetic term. In Table 5, we analyzed the paramagnetic term into the metal AO and ligand contributions. We immediately understand that the Mn chemical shifts are almost entirely due to the d AO contribution of the Mn atom.

The electronic configuration of the Mn atom is d^5s^2 in its ground state. The d orbitals are half-filled so that they enter strongly into the bonds with the ligands. In Table 6, we analyzed further the d contributions into the d_σ, d_π, and d_δ AO contributions, where σ refers to an axis along the Mn-L bond in the $Mn(CO)_5L$ complexes. Figure 8 shows an illustration of these 3d orbitals and some orbitals of the ligands. From Table 6, we see that the $3d_\sigma$ and the $3d_\pi$ orbitals give almost fifty-fifty contributions to the chemical shift. This is very reasonable, as seen below.

From the MO analysis, we see that the $3d_\pi$ and $3d_{\delta 1}$ orbitals lie in the occupied MO's and the $3d_\sigma$ and $3d_{\delta 2}$ orbitals lie in the unoccupied MO's. The reason for this ordering is qualitatively given in the semiempirical treatment of Gray et. al. [26]. In the picture of second-order perturbation theory as given by Eqn. (5), the paramagnetic term is expressed as a sum of the contributions of various excitations. Among the d orbitals shown in Figure 8, there are four types of d-d transitions. They belong to the symmetry $A_2(3d_{\delta 1} \rightarrow 3d_{\delta 2})$, $B_2(3d_{\delta 1} \rightarrow 3d_\sigma)$, $E(3d_\pi \rightarrow 3d_\sigma)$, and $E(3d_\pi \rightarrow 3d_{\delta 2})$. Among these, the B_2 excitations do not contribute, because the angular momentum operator belongs to the $A_2(\ell z) + E(\ell_x, \ell_y)$ symmetry. Further, for the complexes studied here, no δ bond exists between Mn and the axial ligand L. The MO's of δ symmetry give only a secondary contribution to the chemical shift. Thus, the Mn chemical shift of the $Mn(CO)_5L$ complexes is due primarily to the $3d_\pi \rightarrow 3d_\sigma$ transitions. This is why the $3d_\pi$ and $3d_\sigma$ orbitals give almost fifty-fifty contributions to the chemical shift as analyzed in Table 6.

Table 6. Analysis of the Mn 3d contributions to the paramagnetic term[a] (In ppm.).

Molecules	$3d_\sigma$		$3d_\pi$		$3d_{\delta 1}$	
	Value	Shift	Value	Shift	Value	Shift
$Mn(CO)_5H$	-2669	0	-5313	0	-2453	0
$Mn(CO)_5CN$	-3009	-340	-5272	41	-2262	191
$Mn(CO)_5CH_3$	-3295	-626	-5988	-675	-2460	-7
$Mn(CO)_5Cl$	-4361	-1692	-6865	-1552	-2318	135

Molecules	$3d_{\delta 2}$		Total	
	Value	Shift	Value	Shift
$Mn(CO)_5H$	-2548	0	-12983	0
$Mn(CO)_5CN$	-2595	-47	-13138	-155
$Mn(CO)_5CH_3$	-2641	-93	-14384	-1401
$Mn(CO)_5Cl$	-2732	-184	-16276	-3293

[a] $3d_\sigma = 3d_{z^2}$; $3d_\pi = 3d_{xz}, 3d_{yz}$; $3d_{\delta 1} = 3d_{xy}$; $3d_{\delta 2} = 3d_{x^2-y^2}$, in the coordinate system shown in Fig. 8.

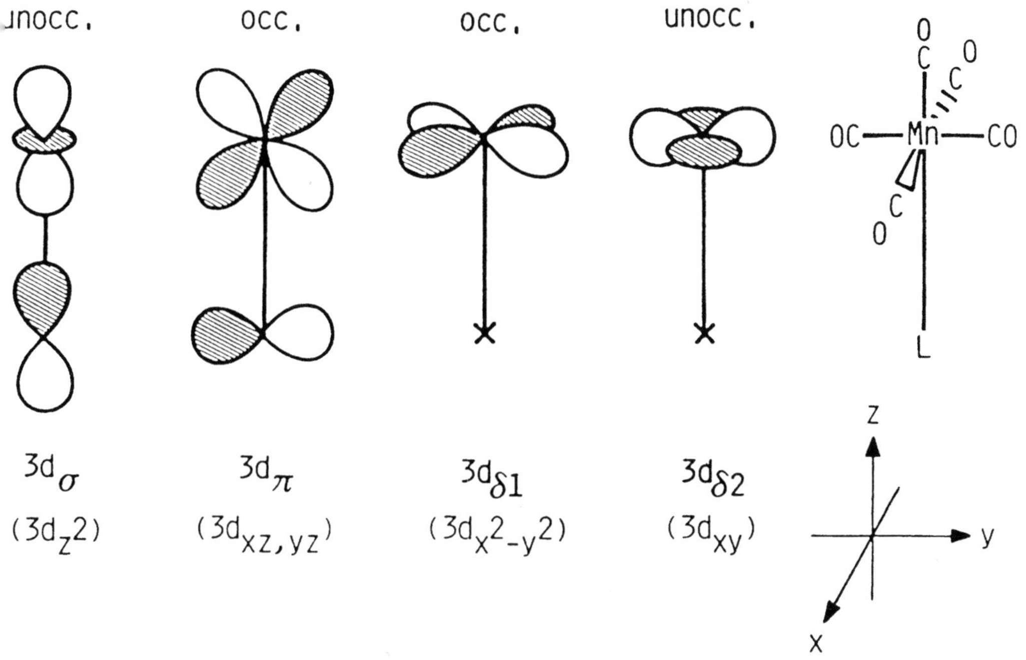

Figure 8. Illustration of the manganese 3d orbitals.

The $3d_\pi$ orbital of Mn lies in the HOMO region and its weight should increase as the ligand L becomes more π-donating. In Figure 9 we give an illustration in order to explain the mechanism of the Mn chemical shift in the $Mn(CO)_5L$ complexes. The $3d_\pi$ contribution would increase (become more negative) in the order of $CN<H<CH_3<Cl$, which is the empirical order of an π-donating ability. This agrees with the order given in Table 6. The $3d_\sigma$ orbital lies in the unoccupied MO region. Its weight would increase as the interaction between the $3d_\sigma$ AO and the ligand increases, because of the delocalization of the $3d_\sigma$ orbital. $Mn(CO)_5^+$ is a typical soft Lewis acid [27]. So, the interaction with the ligand L would increase as the softness of the ligand base increases. Then, the $3d_\sigma$ contribution to the chemical shift is expected to increase (become more negative) as the softness of the ligand L decreases, namely, in the order of $H^- \sim CN^- < CH_3^- < Cl^-$. In Table 6, the order of the $3d_\sigma$ contribution is $H < CN < CH_3 < Cl$, which reflects this ordering. For the unoccupied $3d_\sigma$ orbital, static properties are useless as a measure because only the occupied orbitals contribute to such properties. The "softness" or the "hardness" is a property which is a result of the chemical interaction [27] so that it is more appropriate for describing the property of the unoccupied orbitals. Calderazzo et. al. [14] explained the trend in the Mn chemical shift from the effects of the ligands on the excitation energies between the d orbitals.

The Mn atom in the $Mn(CO)_5L$ complexes has the electronic configuration $3d^{5.2}4s^{0}4p^{0.2}$. In comparison with the electronic structure of the free atom, $3d^54s^24p^0$, the 4s electrons are donated completely to the ligands and the 3d and 4p orbitals accept electrons from ligands. The net charge of Mn is $+1.4 \sim +1.8$ in comparison with the formal charge of 1.0.

6. DIAMAGNETIC TERM

The diamagnetic term itself is large but its contribution to the chemical shift is small as seen from Table 1. It increases with in-

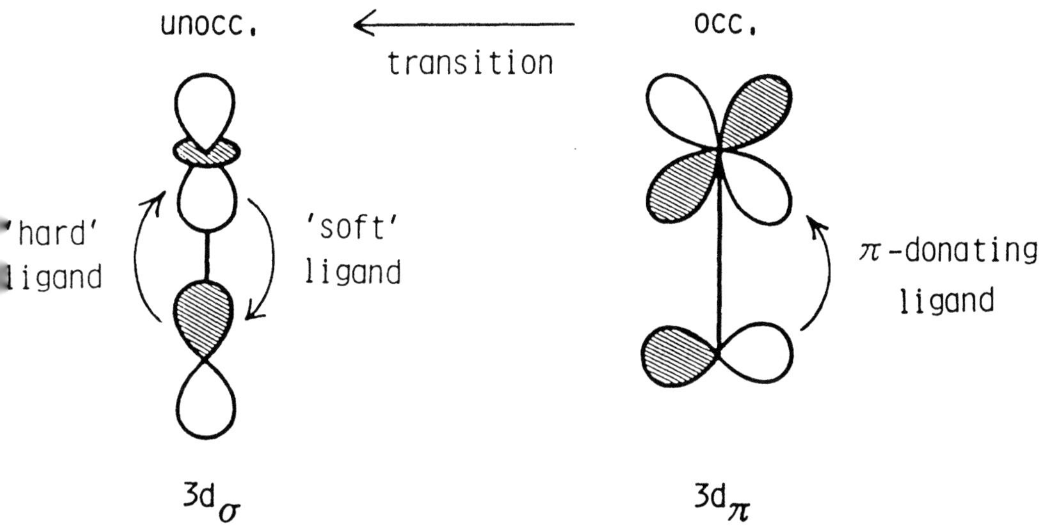

Figure 9. Illustration of the mechanism of the Mn chemical shift in the $Mn(CO)_5L$ complexes which shows that the chemical shift should be larger with increasing π-donating ability and the hardness of the ligand.

creasing nuclear charge [4] as expected from its operator given by Eqn. (2c). Flygare and Goodisman [28] have shown that the isotropic term of the diamagnetic shielding constant is written to a good approximation as

$$\sigma_A^{dia} = \sigma_A^{dia} \text{ (free atom)} + \frac{e^2}{3mc^2} \sum_{B(\neq A)} Z_B/R_{AB} \qquad (8)$$

where σ_A^{dia}(free atom) is the free-atom diamagnetic susceptibility, B denotes the nucleus other than A, Z_B its nuclear charge, and R_{AB} the internuclear distance. The free atom term σ_A^{dia}(free atom) is available from the literature [29,30]. This equation means that the diamagnetic contribution to the chemical shift is determined by the geometry and the ligand species of the complexes. A knowledge of the molecular electronic wavefunction is unnecessary.

In the previous two papers [5,6] on the metal chemical shift we reported that the diagmagnetic term is expressed in the following Pascal-rule-like formula

$$\sigma_A^{dia} = \sigma^{dia}(A) + \sum_L n_L \sigma^{dia}(L), \qquad (9)$$

where $\sigma^{dia}(A)$ is an atomic contribution, $\sigma^{dia}(L)$ the contribution from the single ligand L, and n_L the number of the ligands L. We calculated $\sigma^{dia}(A)$ and $\sigma^{dia}(L)$ by a least-squares fitting method from the diamagnetic terms shown in Table 1 and found that Eqn. (9) is accurate to within ±0.7%. It is evident that the formula given by Eqn. (9) is easily derivable from Eqn. (8), if we assume fixed metal-ligand length and fixed ligand geometry. We certainly adopted such an approximation in the previous calculations [5,6]. Thus, the validity of Eqn. (9) seems to support the validity of Eqn. (8), although we weren't aware of the relation given by Eqn. (8) when we wrote these papers. Thus, we conclude that the diamagnetic contribution to the chemical shift is determined by the geometry and the ligand species of the complexes. However, the diamagnetic contribution itself reflects ainly the variations in the paramagnetic term which is discussed in he previous sections.

. CONCLUDING REMARKS

In this article, we wanted to show that the metal chemical shift s an important property which includes valuable information about the lectronic structures and bonding in metal complexes. This property s especially suited to clarifying the role of the d electrons, relaive to those of the s and p electrons. Further, we wanted to show hat *ab initio* molecular orbital theory gives reliable and useful nformation not only in guessing the magnitude of the metal chemical hift but also in investigating the electronic origin of the chemical hift which reflects the nature of the bonds between the metal and the igands in the complexes. We think the latter role is very important ince our aim is to construct an intuitive concept which is useful as guide for experimental chemists.

The metal chemical shift is a sum of two terms, the diamagnetic erm and paramagnetic term. The diamagnetic term has only a minor ffect which is determined by the molecular geometry and the species f the ligands. The paramagnetic term is the major term and reflects he valence electronic structure near the metal atom.

For complexes of the Ib (Cu, Ag) and IIb (Zn, Cd) metals, the p nd d mechanisms are shown to be the two major mechanisms. For the Cu omplexes, the p mechanism is more important than the d mechanism, and or the Ag complexes, the two mechanisms are competitive. These diferences are attributed to the differences in the atomic energy evels of the free metal atoms. Further, these mechanisms suffer diferent substituent effects from the ligand. The d mechanism is faciitated by the electron-withdrawing substituent and the p mechanism is acilitated by electron-donating substituents.

For the Mn complexes, $Mn(CO)_5L$, the transitions from the d_π rbitals to the d_σ orbital are shown to be the main source of the Mn hemical shift from a perturbation theoretic point of view. The effects f the ligand L on the occupied d_π orbitals and on the unoccupied d_σ

orbitals are therefore important in an investigation of the trends in chemical shifts. We have summarized these effects in terms of the π-donating ability and hardness of the ligands.

For metal chemical shifts, many interesting subjects remain to be studied. There are many nuclear species for which the electronic origin is not yet clarified. Systematic studies, both theoretical and experimental, on the effects of the ligands on metal chemical shifts are necessary for thorough understanding. Relativity effects should be studied for complexes of heavy metals such as Pt, Hg, and Pb. For all of these studies, active collaborations between theory and experiment should be most fruitful.

References:

1. F.A. Cotton, G. Wilkinson, "Advanced Inorganic Chemistry," 4th ed., Wiley & Sons, New York, 1980.
2. R. Garth Kidd and R.J. Goodfellow, in "NMR and the Periodic Table," ed. R.K. Harris and B.E. Mann, Academic Press, New York, 1978, p. 195.
3. R. Garth Kidd, Ann. Rep. NMR Spectr. 10A, 1 (1980).
4. J. Mason, Adv. Inorg. Chem. Radiochem. 18, 197 (1976); 22, 199 (1979).
5. H. Nakatsuji, K. Kanda, K. Endo and T. Yonezawa, J. Am. Chem. Soc. 106, 4653 (1984).
6. K. Kanda, H. Nakatsuji and T. Yonezawa, J. Am. Chem. Soc. 106, 5888 (1984).
7. H. Nakatsuji and K. Kanda, to be published.
8. A.D. Cardin, P.D. Ellis, J.D. Odom and J.W. Howard Jr., J. Am. Chem. Soc. 97, 1672 (1975).
9. J.J.H. Ackerman, T.V. Orr, V.J. Bartuska and G.E. Maciel, J. Am. Chem. Soc. 101, 341 (1979).
10. J.D. Kennedy and W. McFarlane, J. Chem. Soc. Perkin Trans. II, 1187 (1977).
11. P.G. Mennitt, M.P. Shatlock, V.J. Bartuska and G.E. Maciel, J. Phys. Chem. 85, 2087 (1981).
12. G.E. Maciel, L. Simeral and J.J.H. Ackerman, J. Phys. Chem. 81, 263 (1977).
13. (a) K. Endo, K. Matsushita, K. Deguchi, K. Yamamoto, S. Suzuki and K. Futaki, Chem. Lett. 1497 (1982).
(b) K. Endo, K. Yamamoto, K. Matsushita, K. Deguchi, K. Kanda and H. Nakatsuji, to be published.
(c) K. Endo, K. Yamamoto, K. Deguchi, k. Matsushita, T. Fujito, to be published.
14. F. Calderazzo, E.A.C. Lucken and D.F. Williams, J. Chem. Soc. A154 (1967).

15. N.F. Ramsey, Phys. Rev. 77, 567 (1950); 78, 699 (1950); 83, 540 (1951); 86, 243 (1952).
16. H.F. Hameka, "Advanced Quantum Chemistry," Addison-Wesley, Massachusetts, 1965.
17. G.A. Webb, in "NMR and the Periodic Table," eds. R.K. Harris and B.E. Mann, Academic Press, New York, 1978, p. 49.
18. (a) H.D. Cohen and C.C.J. Roothaan, J. Chem. Phys. 43, 534 (1965). (b) H.D. Cohen, ibid., 43, 3558 (1965); 45, 10 (1966).
19. (a) J.A. Pople, J.W. McIver and N.S. Ostlund, Chem. Phys. Lett. 1, 465 (1967); J. Chem. Phys. 49, 2960 (1968). (b) R. Ditchfield, D.P. Miller and J.A. Pople, ibid., 53, 613 (1970).
20. H. Nakatsuji, J. Chem. Phys. 61, 3728 (1974).
21. (a) H. Tatewaki and S. Huzinaga, J. Chem. Phys. 71, 4339 (1979); J. Comput. Chem. 1, 205 (1980). (b) Y. Sakai, H. Tatewaki and S. Huzinaga, J. Comput. Phys. 2, 100 (1981); 3, 6 (1982).
22. L. Pauling, "The Nature of the Chemical Bond," Cornell University Press, Ithaca, N.Y. 1960.
23. S. Kitagawa and Munakata, Inorg. Chem., in press.
24. C.J. Jameson and H.S. Gutowsky, J. Chem. Phys. 40, 1714 (1964).
25. C.E. Moore, "Atomic Energy Levels," National Bureau of Standard, Washington, 1971, Vol. 2, 3.
26. H.B. Gray, E. Billig, A. Wojcicki and M. Farona, Can. J. Chem. 41, 1281 (1963)
27. R.G. Pearson, J. Am. Chem. Soc., 85, 3533 (1963).
28. W.H. Flygare and J. Goodisman, J. Chem. Phys. 49, 3122 (1968).
29. R.A. Bonham and T.G. Strand, J. Chem. Phys. 40, 3447 (1964).
30. G. Malli and S. Fraga, Theor. Chim. Acta. 5, 275 (1966). G. Malli and C. Froese, Int. J. Quantum Chem. 1S, 95 (1967).

SCATTERING OF X-RAYS AND HIGH-ENERGY ELECTRONS FROM MOLECULES: COMPARISON OF AB INITIO CALCULATIONS WITH EXPERIMENT

A.N. Tripathi* and Vedene H. Smith, Jr.
Department of Chemistry
Queen's University
Kingston, Ontario K7L 3N6
Canada

1. INTRODUCTION

Recent high-precision measurements of the elastic and inelastic scattering of electrons over a wide range of momentum transfer have renewed interest in the theoretical investigation of electron scattering from molecular systems. It is well known that high-energy electron scattering measurements offer a reliable way of examining the quality of molecular wavefunctions and for estimating the molecular binding energy. It is generally accepted [1] that high quality Hartree Fock (HF) wavefunctions are well suited to predict one-electron properties and therefore can be expected to be sufficient to account for most of the effects of bonding on the electron density (and the elastic scattering cross-section) in a closed shell molecule. However, the observable quantity is very often the total (elastic + inelastic) cross-section. Thus total cross-sections -- which are related to the two-electron distribution function (intracule) -- evaluated from HF wavefunctions are not expected to be reliable, inasmuch as two-electron expectation values are known to be rather sensitive to electron correlation effects which are not properly accounted for in the HF theory.

The experimental data resulting from high-energy electron scattering measurements have often been compared with theoretical

* Permanent address: Department of Physics, University of Roorkee, Roorkee, 247667, India.

R. J. Bartlett (ed.), Comparison of Ab Initio Quantum Chemistry with Experiment for Small Molecules, 439–462.

calculations using the hydrogen molecule as the prototype (model). Because of its simplicity, very accurate theoretical calculations both at HF and beyond HF levels have been reported. Among them the early work of Liu and Smith [2] and Bentley and Stewart [3] are worth mentioning. The calculations of these authors differ in some angular regions with the earlier measurements of Kohl and Bonham [4], Jaegle and Duguet [5] and Ulsh, Wellenstein and Bonham [6] and are outside the experimental (statistical) error bars. Liu and Smith [2] have noted that this disagreement between theory and experiment might be due to the following possibilities: (i) the inadequacy of the Born approximation in treating the scattering process, (ii) failure of the closure approximation for summing the final states of the molecule, (iii) breakdown of the Born-Oppenheimer separation of the electronic ground state wavefunction, (v) improper vibrational averaging or (vi) the need for improved experiments. The recent measurements of Ketkar and Fink [7] exhibited a still larger discrepancy between these calculations and their measured values. As a result, theoreticians have been prompted to have a fresh look at the problem and attempt to interpret the scattering data in the light of the observations made by Liu and Smith [2]. Recently in an attempt to resolve this discrepancy, Kolos, Monkhorst and Szalewicz [8-9] have carried out a detailed and accurate calculation of elastic, inelastic and total energy unresolved differential cross-sections for high-energy electron scattering by the H_2 and D_2 molecules in the first Born approximation.

Before we discuss *ab initio* calculations of electron scattering in comparison with the present state of the art in experiment, it is useful to begin by recalling a few well established relations within the framework of the first Born approximation.

2. THEORY

For X-Ray scattering, the elastic (I_{el}^{xr}) and total (I_t^{xr}), intensities in the Waller-Hartree approximation [10-12] may be written as

$$I_{el}^{xr}(\mu)/I_T = \langle|F(\vec{\mu})|^2\rangle \tag{1}$$

and

$$I_t^{xr}(\mu)/I_T = N + 2K(\mu) \tag{2}$$

where the angular brackets denote the spherical average

$$\langle G(\vec{\mu})\rangle = (4\pi)^{-1}\int G(\vec{\mu})\, d\Omega_{\vec{\mu}}, \tag{3}$$

μ is the momentum transfer, I_T the Thomson factor [13], N is the number of electrons in the target molecule and Hartree atomic units [14] are used here and throughout this paper. Moreover the elastic, $F(\mu)$, and total, $K(\mu)$, scattering factors are given by

$$F(\vec{\mu}) = \int \rho(\vec{r})\exp[i\vec{\mu}\cdot\vec{r}]d\vec{r} \tag{4}$$

and

$$K(\mu) = \int_0^\infty P(u)j_0(\mu u)du \tag{5}$$

in which $\rho(\vec{r})$ is the one-electron charge density, $j_0(x) = x^{-1}\sin x$, and $P(u)$ is the radial intracule density [15-16] defined in terms of the spinless electron pair density Γ [with $N(N-1))/2$ normalization] by

$$P(u) = \int\Gamma(\vec{r}_1,\vec{r}_2)\delta(u-|\vec{r}_1-\vec{r}_2|)d\vec{r}_1 d\vec{r}_2 \; . \tag{6}$$

The incoherent scattering factor $S(\mu)$ is defined by

$$S(\mu) = I_t^{xr}(\mu)/I_T - I_{el}^{xr}(\mu)/I_T \tag{7}$$

$$= N + 2K(\mu) - \langle|F(\vec{\mu})|^2\rangle \tag{8}$$

These X-ray intensities are closely related to the elastic (I_{el}^{ed}) and total (I_t^{ed}) electron scattering intensities in the first Born approximation [12]. Thus

$$\mu^4 I_{el}^{ed}(\mu)/I_R = I_{el}^{xr}(\mu)/I_T + \sigma_{ne}(\mu) + \sigma_{nn}(\mu) \tag{9}$$

and

$$\mu^4 I_t^{ed}(\mu)/I_R = I_t^{xr}(\mu)/I_T + \sigma_{ne}(\mu) + \sigma_{nn}(\mu) \tag{10}$$

where I_R is the characteristic Rutherford cross-section and the electron-nuclear (σ_{ne}) and nuclear-nuclear (σ_{nn}) interference terms are given by

$$\sigma_{ne}(\mu) = -2\Sigma\{Z_A \mathrm{Re}\langle F(\vec{\mu})\exp[-i\vec{\mu}\cdot\vec{R}_A]\rangle\} \tag{11}$$

and

$$\sigma_{nn}(\mu) = \Sigma\ Z_A Z_B j_o(\mu|\vec{R}_A-\vec{R}_B|) \tag{12}$$

in which the sums are over the nuclei with charges Z_A and position vectors $\vec{R}_A$, and Re< > denotes the real part of < >.

The above elastic intensities relate to the usual experiments where rotational energy differences are unresolved. The appropriate expression for the fully elastic intensity for electron [8] and X-ray [17] scattering from the J=0 state of a diatomic molecule are:

$$I_{fel}^{xr}(\mu)/I_T = |\langle F(\vec{\mu})\rangle|^2 \tag{13}$$

and

$$\mu^4 I_{fel}^{ed}(\mu)/I_R = |\langle F(\vec{\mu})\rangle - \sigma_n(\mu)|^2$$

$$= I_{fel}^{xr}(\mu)/I_T - 2\sigma_n(\mu)\mathrm{Re}\langle F(\vec{\mu})\rangle + [\sigma_n(\mu)]^2 \tag{14}$$

in which

$$\sigma_n(\mu) = \Sigma Z_A j_o(\mu R_A) \ . \tag{15}$$

A semiquantitative understanding of the electron molecule scattering cross-section in terms of the geometrical and vibrational parameters of the molecule is achieved with the help of the independent atom model (IAM) or promolecule [1] which assumes that the molecular density is the simple sum of the spherically averaged atomic densities centered at the equilibrium positions of the pertinent nuclei. Although it is an approximate model, nevertheless its simplicity compensates somewhat for what it lacks. Except in the forward direction, predictions of the IAM are always in good agreement with experiment at large scattering angles. This is not surprising as the large scattering angle region is dominated by the nuclear charge and core-electron distributions. Accurate electron scattering measurements can be used to deduce the deviations of the correct charge densities from the IAM model. The IAM expressions take very simple forms. For a homonuclear diatomic A_2 with the origin at the center of mass and the internuclear axis aligned with the polar axis, the IAM form factor is:

$$F_{IAM}(\vec{\mu}) = 2f_A(\mu)\cos((\mu R_e \cos\theta)/2) \qquad (16)$$

where $f_A(\mu)$ is the atomic form factor. Thus in the bond direction $\theta=0$, Eqn. (16) reduces to

$$F_{IAM}(\mu e_z) = 2f_A(\mu)\cos(\mu R_e/2) \qquad (17)$$

and in a direction orthogonal to the bond ($\theta=\pi/2$) it yields

$$F_{IAM}(\mu e_x) = 2f_A(\mu). \qquad (18)$$

During the last few years, high quality ab initio molecular wavefunctions at the Hartree-Fock level have become available for many small and medium-sized linear and polyatomic molecules and therefore much attention has been devoted to using them in the analysis of high-precision electron scattering experiments. Various groups are currently engaged in calculating the elastic and inelastic cross-sections particularly for linear molecules and in a few cases for elastic scattering by polyatomic molecules. In principle there is no restriction in carrying out the calculations for bigger molecules, however in practice the amount of computer time required is very large. This is basically due to the manifold increase in the number of integrals and then to the necessity of performing angular integrations over the angular variables. The spherical averaging procedure needed in Eqn. (1) has not received much attention in the past. There are several possible ways to do this averaging: (i) numerical integration using quadrature; (ii) expansion of the charge density in Eqn. (4) in an analytical function in order to facilitate the evaluation of Eqn. (1) in closed form; and (iii) an accurate analytic approach which does not use any type of numerical approximation.

The problem of spherical averaging is not of any great concern as long as one is doing calculations for linear molecules. However, the problem becomes somewhat alarming if the molecule involved is a polyatomic with a large number of atoms. The problem has been recently examined very elegantly by Pulay and co-workers [18,19]. They have developed a compact one-integral routine for all types of cases occurring in the scattering amplitude with Gaussian-based wavefunctions up to and including d functions. Furthermore, they also use the fact that the IAM amplitudes approximate the ab initio results very well, particularly at larger μ. Following this prescription, they define a difference function,

$$\Delta F(\mu) = F(\mu) - F_{IAM}(\mu) \tag{19}$$

and expand Eqn. (1) as

$$\langle |F(\vec{\mu})|^2\rangle = \int d\Omega\, |F_{IAM}(\mu)|^2 + 2\mathrm{Re}\int d\Omega\, F^*_{IAM}(\mu)\Delta F(\mu) + + \int d\Omega\, |\Delta F(\mu)|^2. \tag{20}$$

The advantage of Eqn. (20) is that the first term becomes very trivial and the orientational averaging for the second term is done analytically. Only the last term needs to be evaluated numerically. The integrand, being the square of an already small quantity, practically vanishes above $\mu = 5a_0^{-1}$, where the numerical integration becomes difficult.

3. RESULTS AND DISCUSSION

In the following section we shall briefly discuss a few illustrative examples to see how ab initio studies based on high-quality wavefunctions are able to resolve the discrepancy between experimental measurements and theoretical calculations. We shall divide our study into three main categories: H_2 and D_2 molecules, linear molecules, and polyatomic molecules.

3.1 H_2 and D_2 Molecules

These molecules provide a spawning ground for any theoretical model and, therefore, have been subjected to extensive studies, both experimental and theoretical. The most up-to-date calculations for these systems have been reported recently by Kołos and his collaborators [8,9]. They have used a correlated wavefunction consisting of 36 correlated Gaussian geminals which contain factors of the type $\exp(-\gamma r_{12}^2)$. This wavefunction yields a total energy which is only 4 X 10^{-5}h above the exact non-relativistic Born-Oppenheimer energy (see Kołos and Wolniewicz [20]). Figure 1 shows a comparison of their calculations [9] of the difference functions ($\Delta N_t = I_t - I_t^{IAM}$) for H_2 and D_2 along with the recent experimental measurements of Ketkar and Fink [7] referred to the same IAM functions with the same geometrical parameters as in the theoretical calculations. It is seen that the values computed using a highly-accurate correlated wavefunction and the experimental data points are almost indistinguishable from each other up to μ = 2.5 a.u. and show oscillations in the region of large momentum transfer although theory and experiment are very close to each other.

It is worth mentioning that at high impact energy (i.e. >25

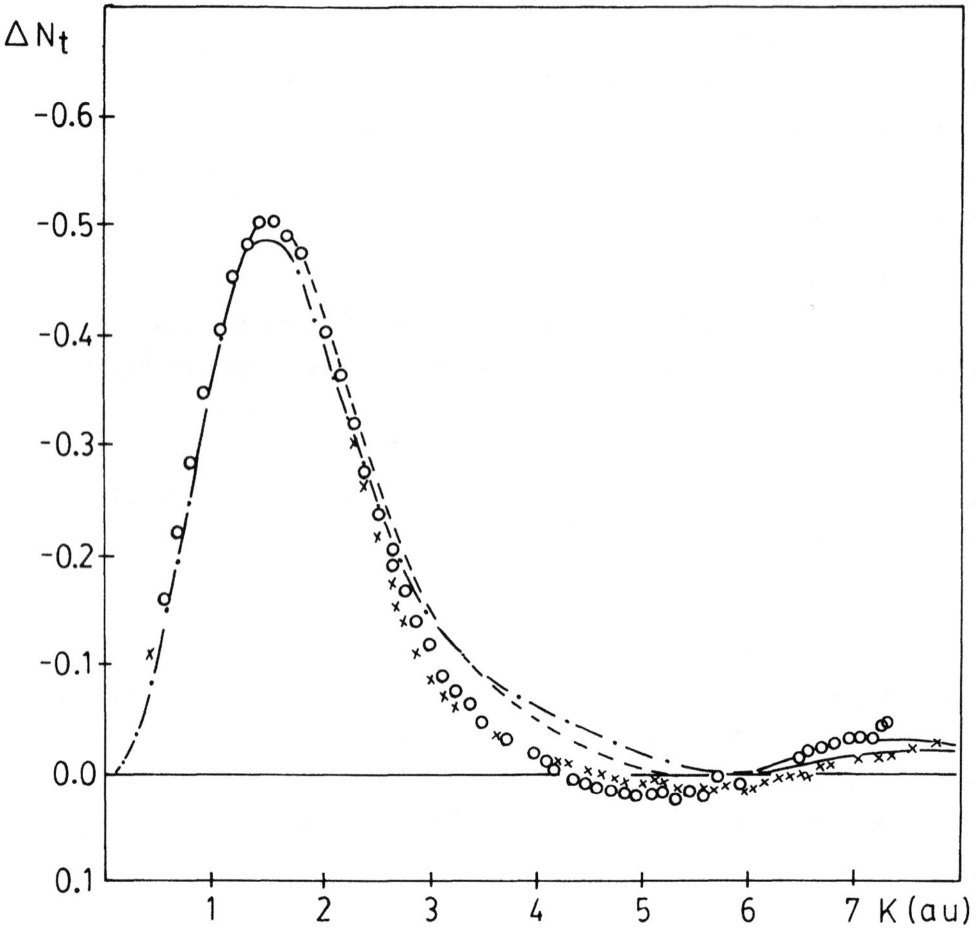

Figure 1. Comparison of ΔN_t curves for H_2 and D_2.

(a) -•- H_2 theoretical [8] with IAM, Rg=Re=0.763Å, l=0.098Å, a=2.0Å^{-1}

(b) --- D_2 [8] theoretical with IAM, Rg=Re=0.749Å, l=0.089Å, a=2.0Å^{-1}.

(c) Ketkar and Fink's results [7] crosses for H_2 and circles for D_2.

ev region), the vibrational damping caused by higher temperatures oes affect the calculated intensity. Therefore, to account proerly for the effect of molecular vibrations, the calculated ntensity should be rigorously averaged over the vibrational and otational wavefunctions computed by direct numerical solution of he radial Schrodinger equation using the most accurate potential 20] including adiabatic and relativistic corrections for the ground tate of H_2 and D_2 molecules. Kołos and coworkers have carried out uch a detailed vibrational averaging and observed that the averaging f ΔN_t over the zero-point vibrations (v=J=0) decreases ΔN_t by pproximately 5%. Further, in order to have a meaningful comparison f the theoretical difference function (ΔN_t) with experiment, they ave also averaged the IAM amplitudes with the same theoretical vibraional functions. These authors have further observed that the arlier differences in ΔN_t between computed and the measured values of etkar and Fink were largely an artifact due to an approximate vibraional averaging in their IAM model. After having obtained such a emarkably good agreement, it is believed that within the framework of he present theory there is no further scope of improvement for the umerical results. Thus from the theoretical point of view, the only ossibility of improvement that exists is to reexamine the first three f the possibilities noted by Liu and Smith [2], which we mentioned arlier in the Introduction.

.2 Linear Molecules

Among the small and medium-sized linear molecules, CO, N_2 and $_2$ have been studied the most. The early calculations [21] were ased mostly on molecular HF wavefunctions. However, it is enerally expected that better agreement between experimental and heoretical results will be attained with configuration interaction CI) wavefunctions instead of HF [1]. Epstein and Stewart [22] ave calculated the total and elastic X-ray scattering intensity or these molecules using near-HF-quality wavefunctions with xtensive basis sets at experimental R_e values. The disagreement between the experimental results of Fink et. al. [23] for N_2 and the theoretical results is pronounced. Later on, Epstein and Stewart carried out a multi-configuration self-consistent-field (MC-SCF) study on CO. However, this wavefunction was able to account for only 21% of the correlation energy.

In an attempt to see how important a role electron correlation plays in the computed total and elastic electron scattering intensities, Breitenstein and collaborators [24] recently carried out a series of moderately high-quality CI calculations which yielded about 60% of the correlation energy for a series of linear molecules (CO, N_2, C_2H_2, O_2 and F_2). Hirota et. al. [25] also carried out an electron correlation study on N_2, but examined the elastic scattering only. As an example, consider the total scattering difference function ΔN_t for N_2 (see Fig. 2). It is remarkable to see that as the energetic quality of the wavefunction improves, i.e. as one goes from a near HF to HF and then to a CI-type wavefunction, the total electron-scattering intensity functions move in the right direction, i.e. nearer to the experimental difference function [23]. The discrepancy between theory and experiment is still quite large indicating that additional correlation effects should be included in the representation of the ground state wavefunction and new experiments should be undertaken. It is not expected that vibrational effects play more than a minor role for N_2.

Apart from considering the role of electron correlation in the target wavefunction, Thakkar, Tripathi and Smith [17] examined some other aspects of scattering calculations including basis set effects and the difference between the usual elastic intensities for X-ray and electron scattering from non-vibrating but freely rotating diatomic molecules and the fully elastic intensities for scattering from the J=0 state.

The difference between the elastic and fully elastic X-ray and electron scattering intensities in the cases of H_2 and N_2 are shown in Figure 3. It may be shown [17] that

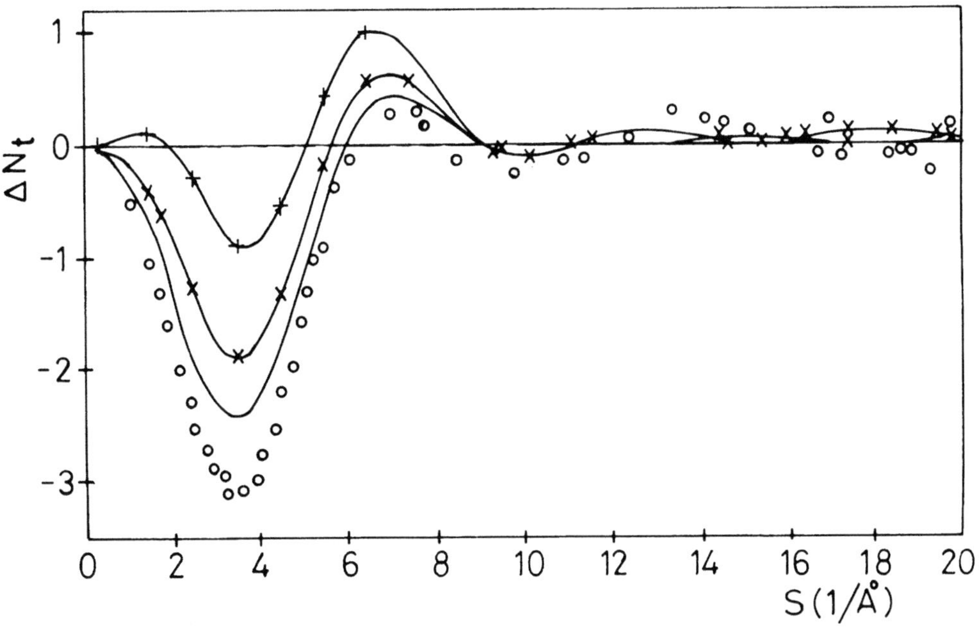

Figure 2. Difference function for the total intensity for N_2 molecules, obtained using the molecular wavefunction and the IAM [24]. +: approximate HF wavefunction (AHF); x: AHF plus calculated correlation correction; -: AHF plus full correlation; 0: experimental.

$$I_{el}^{xr}(\mu) \geqslant I_{fel}^{xr}(\mu), \tag{21}$$

and

$$I_{el}^{ed}(\mu) \geqslant I_{fel}^{ed}(\mu), \tag{22}$$

where equality holds in the case that $F(\mu)$ is spherically symmetric. Naturally, Figure 3 is in accordance with inequalities (21) and (22), with greater differences being seen for N_2 with its more anisotropic $F(\mu)$ than for H_2. The differences in the electron scattering case do not fall off with increasing μ in agreement with previous work on H_2 [8]. This behavior can be rationalized by noting that [8] large momentum transfer collisions are unlikely to occur without exciting rotation of the target molecule, and hence the fully elastic intensities should decrease with increasing μ unlike the elastic ones.

Since the effect of rotation is primarily controlled by the molecular geometry, which is properly accounted for by the IAM, one might expect [8] that these differences between elastic and fully elastic intensities would be correctly predicted by the IAM. The following IAM expressions that are applicable to homonuclear diatomics [17]

$$I_{el}^{xr}(\mu)/I_T = 2[f_A(\mu)]^2[1+j_o(\mu R_e)] \tag{23}$$

$$\langle F(\mu)\rangle = 2f_A(\mu)j_o(\mu R_e/2) \tag{24}$$

$$\sigma_{ne}(\mu) = -4Z_A f_A(\mu)[1+j_o(\mu R_e)], \tag{25}$$

were used to calculate the differences between the IAM elastic and

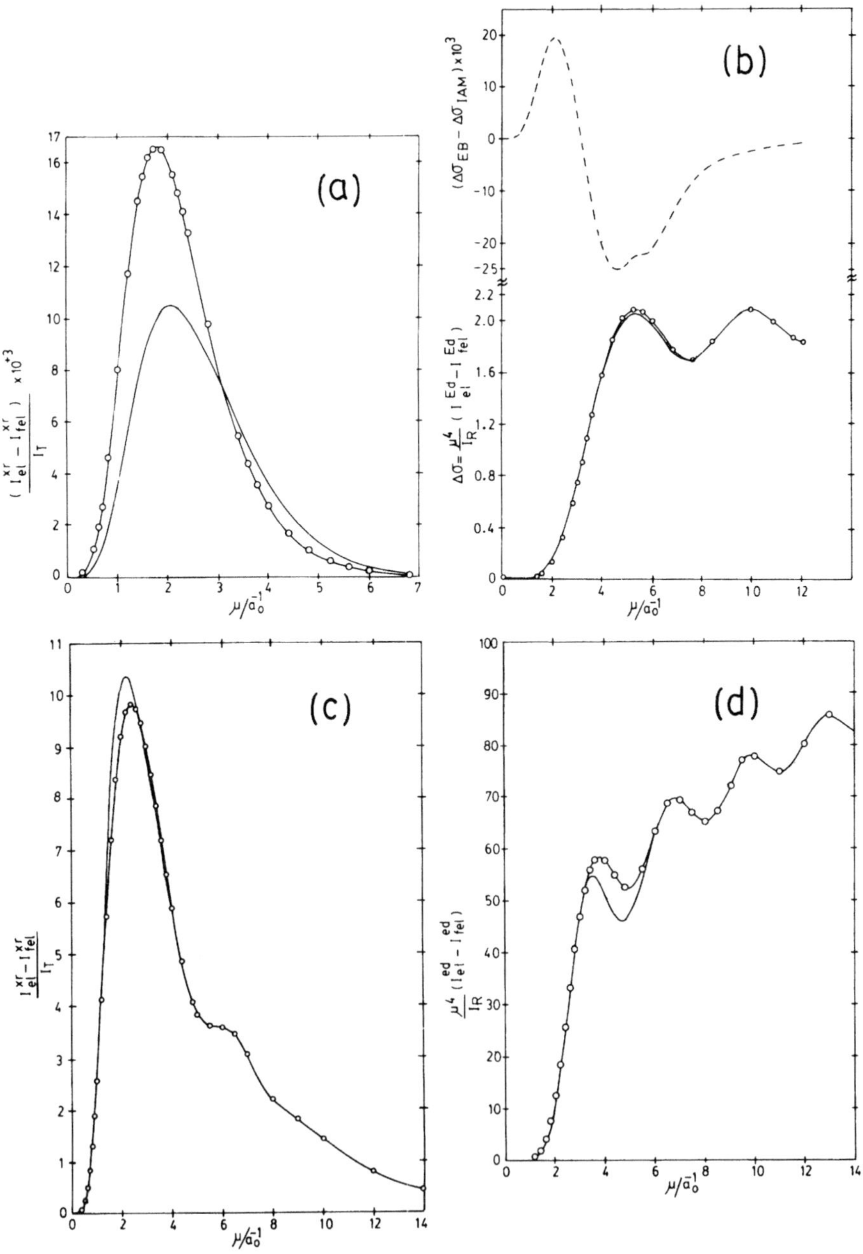

Figure 3. Differences between the elastic and fully elastic X-ray and electron-scattering intensities for H_2 and N_2 [17] (note the changes in scale). ———: calculated using molecular wavefunctions; -•-•-: calculated in the IAM; ----: represents the double differences ($\Delta\sigma_{EB}-\Delta\sigma_{IAM}$).

fully elastic intensities. The results are also shown in Figure 3 and it can be seen that they do mimic the true molecular differences.

Figure 4(a) shows the differences [17] between the $I_{el}^{xr}(\mu)/I_T$ values computed from extended basis (EB) and double zeta (DZ) functions for H_2. Clearly polarization functions in the EB function are needed to obtain accurate results at the restricted Hartree-Fock (RHF) level. This effect is most pronounced in the bond direction. The volume element sinθ damps this effect in the spherically averaged $I_{el}^{xr}(\mu)/I_T$ as illustrated in Figure 4(b). The latter also shows basis set effects on $I_t^{xr}(\mu)/I_T$ and $\sigma_{ne}(\mu)$. Of course, the sum of the basis set effects on σ_{ne} and the X-ray intensities constitute the basis set effects on the corresponding electron intensities (cf. Eqns. (9) - (10)). It is seen in Figure 4(b) that the basis set effects on the one-electron property I_{el}^{xr} in H_2 are about twice as large as those on the two-electron property I_t^{xr}. This result runs counter to one's intuition until one realizes that it is an artifact of the RHF approximation for the ground state of a two-electron system. In this very special case one has [17]

$$\Gamma_{HF}(\vec{r}_1,\vec{r}_2) = |\phi(\vec{r}_1)\phi(\vec{r}_2)|^2 = \rho_{HF}(\vec{r}_1)\rho_{HF}(\vec{r}_2)/4 \tag{26}$$

where $\rho(\vec{r}) = 2|\phi(\vec{r})|^2$ is the RHF charge density arising from the RHF orbital $\phi(\vec{r})$. It follows that, for a two-electron system described by a RHF wavefunction,

$$I_t^{xr}(\vec{\mu})/I_T = 2 + (1/2)I_{el}^{xr}(\vec{\mu})/I_T \tag{27}$$

and consequently [17]

$$I_{el}^{xr}(\vec{\mu};B_1) - I_{el}^{xr}(\vec{\mu};B_2) = 2[I_t^{xr}(\vec{\mu};B_1) - I_t^{xr}(\vec{\mu};B_2)] \tag{28}$$

where B_1 and B_2 denote two different basis sets. Figure 4(b) shows exactly the behavior predicted by Eqn. (28). Naturally this behavior does not persist for many-electron systems even in the RHF approximation. Figure 4(c) shows that, in line with one's intuition, basis set

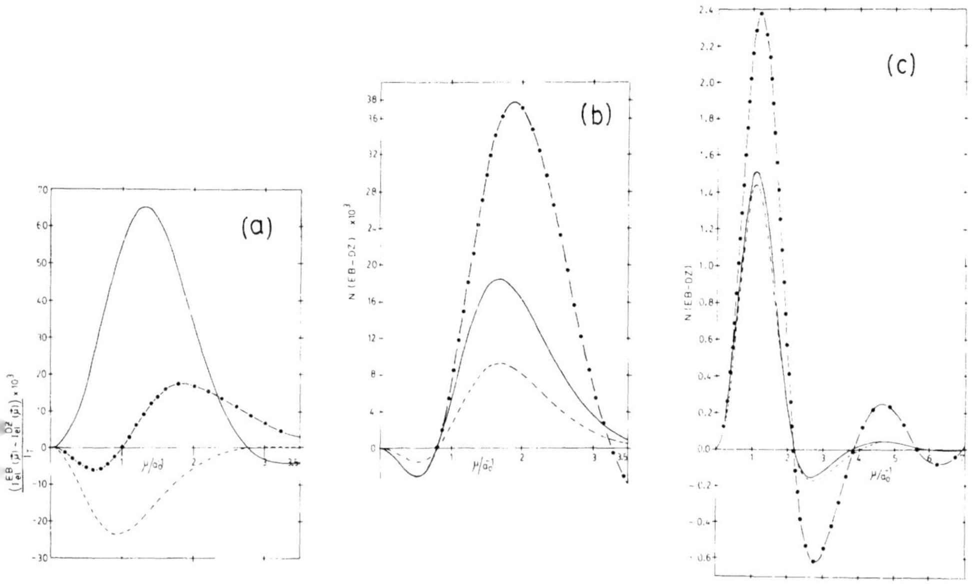

igure 4. Basis-set effects on various intensities [17] (note the changes in scale).

effects on I_t^{xr} are roughly equal to those on I_{el}^{xr} for N_2. Evidently polarization functions are needed to obtain a RHF-level description of these intensities for N_2. This is in agreement with the study of Hirota, Terada and Shibata [25] who found that polarization functions were necessary to obtain accurate RHF values of the elastic electron-scattering intensity for N_2 but not for H_2O.

The other linear molecule which has been studied somewhat extensively is CO_2. All the earlier experimental studies of CO_2 present very conflicting results [26]. A new measurement by Sasaki and collaborators [27] has been reported recently. They also performed theoretical calculations for the electron scattering intensities using a HF molecular wavefunction while the effect of correlation on the total intensity was estimated through the IAM model. There exists a wide disagreement between theory and experiment particularly in the region of small μ ($0 \leq \mu \leq 4 Å^{-1}$).

3.3 Polyatomic Molecules

Among the early calculations on the scattering of electrons by polyatomic molecules was one by Haberl and Hasse [28] on water using an <u>ab initio</u> wavefunction and an elastic study by Szabo and Ostlund [29] (H_2O, NH_3 and CH_4) using semi-empirical CNDO-type wavefunctions. These results showed improvement over the IAM model calculations but still differed significantly from the experimental data [30]. Smith and coworkers [31] calculated directional elastic scattering factors for C_6H_6 using a minimal basis set. It is only very recently that various groups have initiated elastic and total electron scattering calculations for medium and large-size polyatomic molecules. Hirota et al. [25] and Shibata et al. [32] reported elastic electron scattering cross-section calculations on water and diborane molecules employing several molecular wavefunctions such as double zeta, triple zeta with and without polarization functions and CI wavefunctions. Elastic scattering for electrons from H_2O, NH_3 and CH_4 and H_2O has also been carried out by Sharma and Tripathi [33] using the SCF-MO wavefunctions obtained in double-zeta quality basis of

Gaussian contracted orbitals by Snyder and Basch [34].

Pulay and coworkers [18] performed an ab initio Hartree-Fock calculation of the elastic cross-section for electron scattering from the SF_6 molecule. They used a triple-zeta basis set augmented with polarization functions and diffuse functions. Their results confirm the Bartell hypothesis [12] that the observed residuals from the IAM model intensities are due to electronic bonding effects. Therefore, for accurate electron diffraction work, the effect of the electron distribution must be incorporated via ab initio calculations rather than the IAM. Kohl et al. [19] and Shang de Xie et al. [35] also reported elastic electron scattering cross-sections for CH_4, CF_4 and C_2H_4 respectively, using ab initio HF wavefunctions. Shang de Xie et al. [35] made an extensive study of basis set effects on C_2H_4. They used six different basis functions ranging from 6-31G to $6\text{-}311G^{**}$. As there are no experimental measurements available, it is rather difficult to comment on the accuracy of the computed intensities. However, it is believed that the SCF calculations at the $6\text{-}311G^{**}$ level might be sufficient for the experimental calibration and related routine structure determination. Thakkar [36] recently reported inelastic scattering cross-sections for electrons from diborane, methane, ammonia and water. He pointed out that in the case of SCF wavefunctions, the calculation of the inelastic scattering cross-section can be carried out simultaneously, with a little additional effort, with the calculation of the elastic scattering cross-section. Results of inelastic scattering, even at the SCF level, can play a very useful role in analyzing experimental data, especially after correcting for electron correlation in the following manner [27, 36]:

$$S_{CORMO}(\mu) = S_{SCFMO}(\mu) + S_{IAMCI}(\mu) - S_{IAMHF}(\mu). \qquad (29)$$

Here $S(\mu)$ refers to the spherically averaged Waller-Hartree (WH) incoherent scattering factors defined by Eqns. (7) and (8). The subscript CORMO designates the corrected value, SCFMO the SCF value, IAMCI the IAM value using CI atomic data, and IAMHF, the IAM value using HF atomic data. By this procedure, the correlation contributions to the residual atoms are accounted for correctly and only the non-atomic portion of the electronic correlation for the molecule remains.

Thakkar demonstrated [36] in the case of diborane that except in the region of small momentum transfer the correlation corrected calculation of the inelastic scattering cross-section is in better agreement with experiment [37] than that of the usual SCF MO calculations (see Figure 5). The advantage of such a procedure is that one can calculate both the elastic and inelastic scattering cross-sections at the SCF level on the same footing which is better than using an IAM result for the inelastic contribution as these are strongly affected by the formation of the chemical bond. There have been no calculations reported yet for inelastic scattering cross-sections of polyatomic molecules at the same level of accuracy as has been done for linear molecules.

In this regard it is worth presenting a preliminary report on work [38] on total, elastic, and inelastic X-ray and high-energy electron scattering from CH_4 and C_2H_2 done in this group. Three different ab initio SCF wavefunctions with cartesian Gaussian type basis orbitals (CGTO) were constructed for CH_4 and C_2H_2. For identifying these wavefunctions, we use the number of primitive orbitals on the carbon atom. The 11.7.1 wavefunction means e.g. 11 s-type, 7 p-type and 1 d-type CGTO's. The 11.7.1 were contracted to 5.3.1. For hydrogen, the number of primitives/contracted orbitals are 5.1/3.1. One of these (SCF 11.7.1) was improved by a configuration interaction (CI) calculation with all single and double excitations from the SCF reference state. The best SCF wavefunction, with two contracted polarization functions on each atom, gave an energy of -40.2153 a.u. and -76.8514 a.u. respectively for CH_4 and C_2H_2 and is very close to the HF limit. As an illustration, we present briefly one of our results for basis set effects on the elastic intensity for methane molecules (see Fig. 6). Three sets of difference functions were constructed:

Set A: (13.8.2 - 11.7) SCF

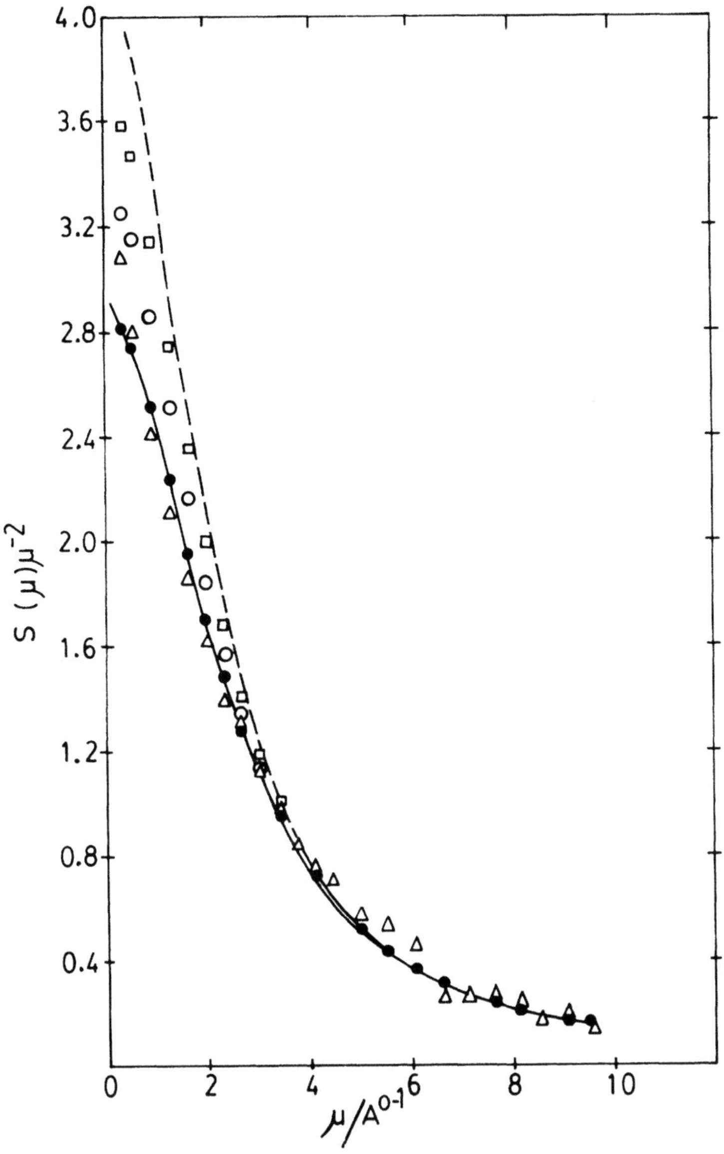

Figure 5. Inelastic cross-sections S(μ) for diborane [36]; ---: calculated in the IAM from the HF scattering factors; □: calculated in the IAM from the correlated scattering factors; O: calculated from the SCF wavefunction of [34]; ●: SCF results corrected for electron correlation; Δ: experimental results of Hirota et al [36].

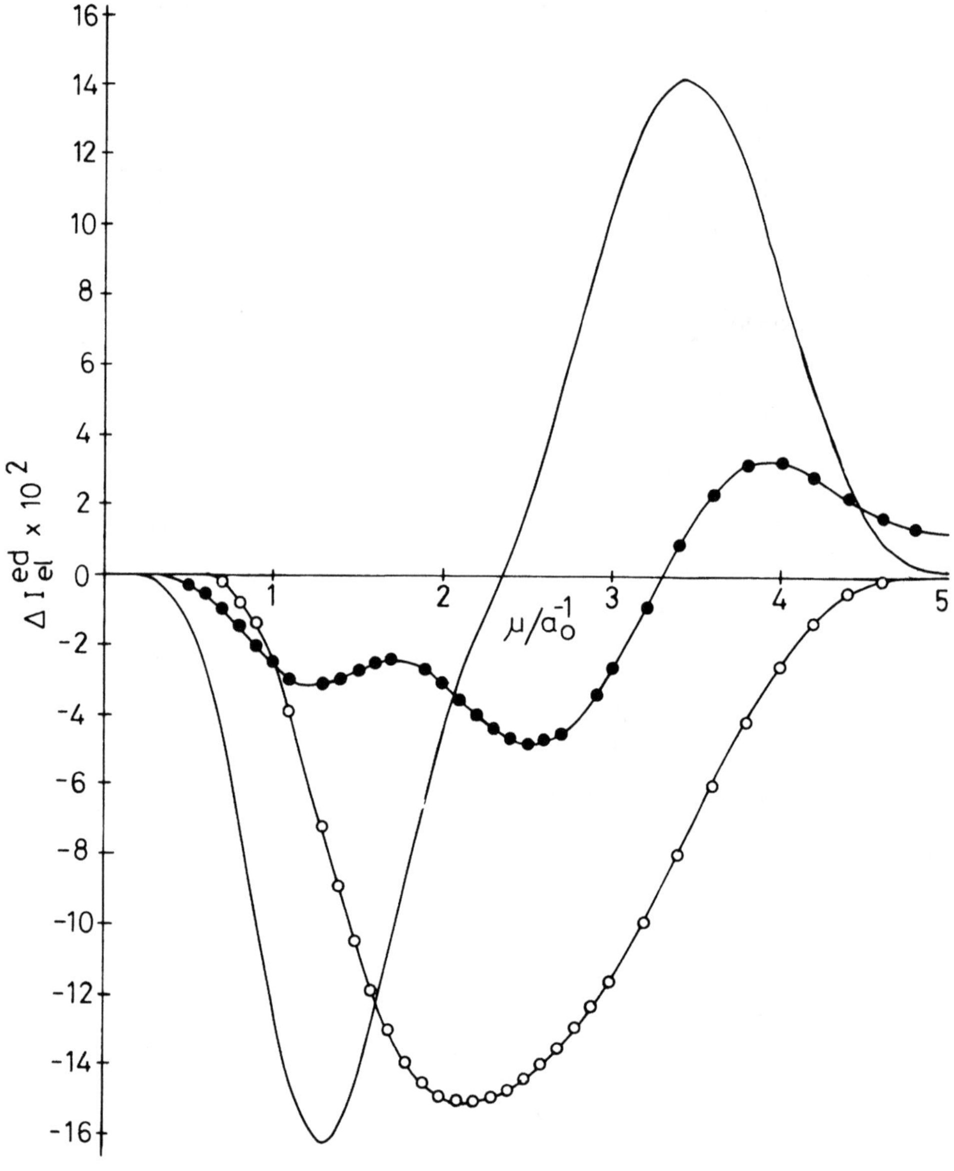

Figure 6. Basis set effects on elastic intensity for CH_4 molecules [38]: —— Set A; -●- Set B; -o- Set C.

Set B: (13.8.2 - 11.7.1) SCF

Set C: (11.7.1CI - 11.7.1 SCF)

We note that Set A shows a significant deviation in the ΔF curve both for small and large μ values. This reflects that an improved double zeta basis (11.7) can neither account for the correct charge density in the far-off region nor in the core region. However, this difference narrows down in the entire angular region for Set B, indicating a proper weighting of the charge density. Set C is not very much different qualitatively, except for having a large negative oscillation compared to Set B. This suggests that the core orbitals did not participate in the CI calculations. A detailed study of the basis-set dependence of different aspects of the electron scattering problem such as inelastic intensity, X-ray elastic intensity, and nuclear-electron interference terms apart from elastic intensity, is in progress and will be published elsewhere [38].

CONCLUSIONS

High quality ab initio calculations of X-ray and electron total, elastic and inelastic scattering cross-sections are now being done for molecules ranging from small to large size primarily for linear systems. Electron correlation effects have been found to be important for linear molecules. The basis set effect is also in no way less important than the effect of electron correlation. Basis-set effects on molecules already studied suggest quite strongly that a properly selected and optimized "set" is important to attain the measureable level of accuracy in the calculation of scattering cross-sections at the SCF level. On the experimental side a few high-precision measurements for total high-energy electron scattering on H_2, D_2, N_2, O_2, CO_2, B_2H_6, CH_4, H_2O and NH_3 have been reported [7,23,27,35,26,30,39]. The total electron scattering cross-sections for H_2 and D_2 in the entire angular region are now well understood. However, the situation is quite different for other molecular systems. For each molecule the calculated and experimental differences have similar nodal patterns but they always differ in the magnitude of oscillation.

In order to have a better understanding of electron correlation effects and also how well a HF-level calculation predicts the cross-section, new measurements made separately for both the elastic and inelastic components are needed.

In recent experiments of high-energy elastic scattering of electrons from the rare gases in the very forward direction by Geiger and Moron-Leon [40], there is a strong forward peak which two orders of magnitude larger than the first Born results. This is accompanied by rapid angular variations, i.e. essentially a Fraunhofer diffraction pattern. The observed deviations vary with atomic size and incident electron energy and therefore it seems unlikely they are an artifact of experiment. Geiger and Moron-Leon interpreted their data as being due to shadow scattering. Very recently Geltman and Nesbet [41] carried out a model calculation for atomic hydrogen and found that the computed peak is smaller by a larger factor than the magnitude required to interpret the experimental data. Experiments of this type should in fact be repeated with molecules in the gas phase.

Acknowledgements:

The continuing support of this research by the Natural Sciences and Engineering Research Council of Canada (NSERC) is gratefully acknowledged as are the contributions by P. Kaijser, J.-W. Liu and A.J. Thakkar to our research on X-ray and electron scattering from molecules.

References:

1. V.H. Smith, Jr., Phys. Script. 15, 147 (1977); Isr. J. Chem. 16, 187 (1977).
2. J.W. Liu and V.H. Smith, Jr., Chem. Phys. Lett. 45, 59 (1977); J. Phys. B 6, L275 (1973); Chem. Phys. Lett. 30, 63 (1975).
3. J.J. Bentley and R.F. Stewart, J. Chem. Phys. 62, 875 (1975); J. Comp. Phys. 11, 127 (1973).
4. D.A. Kohl and R.A. Bonham, J. Chem. Phys. 47, 1634 (1967).
5. A. Jaegle and A. Duguet, J. Chem. Phys. 10, 1554 (1972).
6. R.C. Ulsh, R.F. Wellenstein and R.A. Bonham, J. Chem. Phys. 60, 103 (1974).

7. S.N. Ketkar and M. Fink, Phys. Rev. Lett. 45, 1551 (1980); ibid, 47, 1968 (1981).
8. W. Kolos, H.J. Monkhorst and K. Szalewicz, J. Chem. Phys. 77, 1323 (1982). In their discussion of electron scattering, these authors refer to the usual elastic scattering cross-section as "quasi-elastic" and designate the fully elastic ones as "elastic."
9. K. Szalewicz, W. Kolos, H. Monkhorst and C. Jackson, J. Chem. Phys. 80, 1435 (1984).
10. I. Waller and D.R. Hartree, Proc. R. Soc. London A124, 119 (1929).
11. R. Benesch and V.H. Smith, Jr., in Wave Mechanics: The First Fifty Years, W.C. Price, S.S. Chissick and T. Ravensdale, eds. (Butterworths, London, 1973), pp. 357-3 7; Acta Cryst. A26, 579 (1970).
12. R.A. Bonham and M. Fink, High Energy Electron Scattering (Van Nostrand, New York, 1974).
13. J.J. Thomson and G.P. Thomson, Conduction of Electricity Through Gases, 3rd Ed. (Cambridge University Press, Cambridge, 1933). Vol. II, pp. 257-260.
14. D.G. Whiffen, Pure Appl. Chem. 50, 75 (1978).
15. A.J. Coleman, Int. J. Quantum Chem. Symp. 1, 457 (1967).
16. A.J. Thakkar and V.H. Smith, Jr., Chem. Phys. Lett. 42, 476 (1976).
17. A.J. Thakkar, A.N. Tripathi and V.H. Smith, Jr., Phys. Rev. A29, 1108 (1984).
18. P. Pulay, R. Mawhorter, D.A. Kohl and M. Fink, J. Chem. Phys. 79, 185 (1983).
19. D.A. Kohl, P. Pulay and M. Fink, Theochem. 17, 149 (1984).
20. W. Kolos and L. Wolniewicz, J. Chem. Phys. 49, 404 (1968).
21. J.-W. Liu, J. Chem. Phys. 59, 1988 (1973).
22. J. Epstein and R.F. Stewart, J. Chem. Phys. 66, 4057 (1977); ibid 67, 4238 (1977).
23. M. Fink, D. Gregory and P.G. Moore, Phys. Rev. Lett. 37, 15 (1976).
24. M. Breitenstein, A. Endesfelder, H. Meyer, A. Schweig and W. Zittalau, Chem. Phys. Lett. 97, 403 (1983); M. Breitenstein, A. Endesfelder, H. Meyer and A. Schweig, Chem. Phys. Lett. 108, 430 (1984).
25. F. Hirota, H. Terada and S. Shibata; Bulletin of the Faculty of Education, Shizuoka University, Natural Science Series, Vol. 31, p. 49 (1980).
26. M. Fink and C. Schmidekamp, J. Chem. Phys. 71, 5243 (1979); ibid, 71, 1243 (1979); ibid, 71, 5238 (1979).
27. Y. Sasaki, S. Konaka, T. Iijima and M. Kimura, Int. J. Quantum Chem. 21, 475 (1982).
28. A. Haberl and J. Hasse, Z. Naturforsch 29a, 1033 (1974).
29. A. Szabo and N.S. Ostlund, J. Chem. Phys. 60, 946 (1974).
30. W.R. Harshbarger, A. Skerbele and E.N. Lassettre, J. Chem. Phys. 56, 3784 (1971).
31. W.J. Janis, P. Kaijser, V.H. Smith, Jr. and M.H. Whangbo, Mol. Phys. 35, 1237 (1978).

32. S. Shibata, F. Hirota, N. Kakuta and T. Muramatsu, Int. J. Quantum Chem. 18, 281 (1980).
33. B.S. Sharma and A.N. Tripathi, J. Phys. B. At. Mole. Phys. 16, 1827 (1983).
34. L.C. Snyder and H. Basch 1972, Molecular Wavefunctions and Properties (New York: Wiley).
35. Shang-de Xie, M. Fink and D.A. Kohl, J. Chem. Phys. 81, 1940 (1984).
36. A.J. Thakkar, J. Chem. Phys. 81, 1943 (1984).
37. F. Hirota, N. Kakuta and S. Shibata, J. Phys. B 14, 3299 (1981).
38. P. Kaijser, V.H. Smith, Jr., A.N. Tripathi, W.P. Kraemer and G.H.F. Diercksen (to be published). A.N. Tripathi, V.H. Smith, Jr. and A.J. Thakkar (to be published).
39. A. Lahmann-Bennani, A. Duguet, H.F. Wellenstein and N. Roualt, J. Chem. Phys. 72, 6398 (1980), and A. Lahmann-Bennani, A. Duguet and H.F. Wellenstein, J. Phys. B. At. Mol. Phys. 12, 461 (1979).
40. J. Geiger and D. Moron-Leon, Phys. Rev. Lett. 42, 1336 (1979).
41. S. Geltman and R.K. Nesbet, Phys. Rev. A. 30, 1636 (1984).

QUANTUM CHEMISTRY BEYOND THE ALGEBRAIC APPROXIMATION WITH GAUSSIAN GEMINALS

Hendrik J. Monkhorst
Quantum Theory Project
Departments of Physics and Chemistry
University of Florida
Gainesvilles, Florida 32611

ABSTRACT. It is very hard to achieve chemical accuracies of 1 mhartree (or about 1 kcal/mole) for molecular energies. The electron correlation methods using expansions in virtual orbitals as computed from some zeroth-order problem usually converge very slowly. This is mainly due to the absence of explicit correlation in these expansions. A valuable remedy is presented in the form of Gaussian type geminal (GTG) expansions for pair functions in perturbation and coupled pair calculations. The combined use of Gaussian type orbitals (GTO) for Hartree-Fock orbitals and GTG's for pair functions enables the analytic evaluation of all many-electron integrals, regardless of the molecule or the number of electrons. Nonlinear parameters in the GTG's have been obtained by minimizing a new second-order energy functional which is computationally more efficient than a previous one. Saturated second- and third-order energies for He, Be, H_2 and LiH have been obtained, essentially matching previous accurate atomic results, and superseding decisively diatomic results. Coupled-pair results with the GTG basis obtained at the second-order level match other accurate Be and LiH results, and are far superior for He and H_2. Drastic, yet numerically reliable, approximate treatments for the strong orthogonality were introduced which dramatically reduce the computational efforts. Future developments are discussed.

R. J. Bartlett (ed.), Comparison of Ab Initio Quantum Chemistry with Experiment for Small Molecules, 463–488.

1. INTRODUCTION

The electron correlation problem continues to constitute the main effort of quantum chemistry. This activity is now moving from configuration interaction (CI) calculations to the implementation of more sophisticated methods, such as many-body perturbation theory (MBPT) and the coupled cluster (CC) method. Driven by computer technological advances, ever-larger basis sets of Gaussian-type orbitals (GTO's) are used and ever-larger molecules are tackled. An increasing emphasis on the accurate calculation of properties, including excited states, can be discerned; the present Conference Proceedings are clear evidence of this trend. Apart from a recognition that basis set effects on many properties can be substantial, it is also found that correlation corrections must be included for their accurate evaluation.

Essentially all recent perturbative and CC calculations are performed with the so-called "algebraic approximation" [1], i.e., with expansions for the correlated terms using virtual orbitals obtained from finite-basis set Hartree-Fock (HF) calculations. With a Møller-Plesset [2] partitioning of the many-electron Hamiltonian, it is now possible to obtain approximately 90% of molecular correlation energies [3,4]. Capturing the remaining ten percent requires a tremendous increase of computational effort, because the number of virtual orbitals to be included increases sharply. The fact that this is not only due to inadequacies in the molecular basis sets can be seen from a similarly slow convergence in atomic calculations using high accuracy HF orbitals. For example, in a partial wave expansion for the pair functions involved in the second-order correlation energy for the Ne atom [5], it was found that only 93% of that energy is obtained if spherical harmonics with $\ell \geq 4$ are neglected. (This amounts to an error of 27.6 mhartree, or 17.3 kcal/mole.) Reduction of the error to 1 mhartree requires the inclusion of partial waves with $\ell \leq 8$. This shows that in an "algebraic" MBPT calculation, orbitals through k symmetry would have been needed to achieve the "chemical accuracy" of 1 mhartree. Similar results hold for lighter atoms.

see references in Ref. [5]). This typical increase in virtual orbitals is computationally unmanageable at the present time, and it portends trouble for reaching chemical accuracy with the algebraic approximation, at least in its original form. Modifications for diatomics are under study [6]. We will return to this at the end of this article.

One way to avoid the intrinsically slow convergence of the algebraic approximation is to employ basis functions with explicit dependence on the interelectronic distance r_{12}. Such functions were used to obtain the best correlation energies for two-electron atoms [7] and diatomic molecules [8]. However, extension to many-electron atoms and polyatomic molecules is impractical, both in variational and perturbative calculations, because the resulting many-electron, many-center integrals are very difficult to evaluate. The single exception is the basis of explicitly correlated Gaussian-type geminals (GTG's), which are products of two GTO's (for electrons 1 and 2) and a correlation factor $\exp(-\gamma r_{12}^2)$. All many-electron atomic and polyatomic integrals can be evaluated in closed form, just like GTO integrals. The use of GTG's in quantum chemistry was already proposed in 1960, simultaneous to that of GTO's, by Boys and Singer [9]. Similar to the inefficiency of GTO's in accurate LCAO-SCF calculations (relative to Slater-type functions) the GTG's are less efficient for correlation than functions containing odd powers of r_{12}. Consequently, a careful optimization of the nonlinear parameters specifying the GTG's is required if practically short, yet accurate, GTG expansions are desired. This represents the major drawback to the use of GTG's and remains the single outstanding bottleneck to their routine application at the present time.

In 1970, Pan and King [10] were the first ones to apply the GTG's to the calculation of second-order correlation energies. Using a variational second-order energy functional due to Sinanoglu [11], and short geminal expansions, they obtained second-order correlation energies for He, Be, B^+ and Ne, amounting to 99.1%, 98.6%, 98.4% and 88.4% of the accurate values, respectively. Extension to molecules was made by Adamowicz and Sadlej [12] and to the He-He interaction potential by Szalewicz et al [13]. All these investigations testify to the need of a careful nonlinear optimization for accurate results. The strong orthogonality (SO) requirement on the geminal expansion (imposed by perturbation theory) generates a large number of four-electron integrals due to the Coulomb and exchange operators; their number scales like M^4K^2, where M and K are the number of GTO's (for expanding the HF orbitals) and GTG's (for expanding the pair functions), respectively. It is clear that only small orbital and geminal expansions can be used, lest the evaluation time of a single Sinanoglu functional become excessive.

A major breakthrough in the applicability of GTG's occurred when my coworkers and I proposed, and implemented, a new functional for the variational calculation of second-order correlation energies [14]. This functional has the same minimum as the Sinanoglu functional but the trial pair function need not be strongly orthogonal to the occupied HF orbitals. For sufficiently complete GTG basis sets the SO will be automatically achieved. No more than three-electron integrals occur; their number arising from the Coulomb and exchange operators scales like M^2K^2 with a proportionality constant much smaller than that in the Sinanoglu functional. As a result, much larger geminal basis sets can be used in practice, yielding very accurate second-order correlation energies for He, Be, H_2 and LiH. The results for He and Be even slightly surpass previous "basis set independent" results from partial-wave expansion and extrapolation techniques [15], whereas those for H_2 and LiH are unmatched.

In a subsequent series of publications we fully developed the theory and application of GTG's to the calculation of correlation energies for closed shell atoms and molecules, based on the Møller-Plesset partitioning. The systems He, Be, H_2 and LiH were considered throughout as examples. In the first paper [16] (hereafter referred to as I) we reported the theoretical and computational details connected with second-order energies computed from our new functional with extensive nonlinear optimization. An extensive study was made of convergence patterns resulting from various GTO and GTG expansion lengths. Using these optimal GTG expansions for the first-order pair functions, third-order correlation energies were also considered [17]. Hereafter, we refer to this work as II. The next paper in the series [18] (referred to as III) contains a detailed account of the calculation of coupled-pair-based correlation energies, using the GTG expansions obtained in I and applied in II. Having obtained these accurate, saturated (or basis-set independent) results, we could perform numerical experiments in search of simplifications in treating the strong orthogonality in MBPT and coupled-pair calculations (Ref. [19], referred to as IV). We found a combination of an approximate SO projection and the outright neglect of SO projectors in coupled-pair equations which leads to a dramatic reduction in computational effort, without any loss of accuracy. This finding is so significant that we now consider it possible for GTG expansions to become the approach of choice for routine, high-accuracy correlation energy calculations in quantum chemistry. At present, the only barrier to such a development is the problem of the systematic selection of nonlinear parameters specifying the GTG basis sets. At the end of this article, I will return to this question.

In this article I wish to highlight the characteristic features of the theory and computational methodology for the second-, third- and all-order coupled pair correlation energies with GTG expansions for pair functions. Most derivational and numerical details can be found in Parts I through IV; specific references to these will be made in the text. For example, Eqn. (30) in Part III will be indicated by Eqn. (III.30).

The next section contains an outline of the most important theoretical and computational aspects, along with a description of the most successful, approximate treatments of strong orthogonality.

This is followed by a results section, which amounts to an excerpt of the best results of I - IV. I close with an evaluation and assessment of future developments.

2. THEORY

2.1 General

We adopt the Møller-Plesset partitioning of the N-electron Hamiltonian,

$$H = H_o + V, \tag{1}$$

where the unperturbed Hamiltonian is the sum of one-electron Fock operators,

$$H_o = \sum_{i=1}^{N} f(i). \tag{2}$$

The exact, singlet wavefunction of a closed shell state can be expressed by

$$\Psi = \exp(T)\Phi \tag{3}$$

Here we choose to consider the ground state of an N-electron atom or molecule, for which the HF reference state Φ is an eigenfunction of H_o, built up from HF eigenfunctions $\phi_\alpha(\alpha = 1...N/2)$, and with eigen-energy E_o. In Part III, we have proven for closed shell states Φ that the cluster operator T of Eqn. (3) is spin free, i.e., no spin coordinates or operators need to appear in its representation. This means that the entire spin dependence of Ψ resides in Φ. This remarkable result dramatically simplifies the many-electron theory of closed-shell systems.

Rather than using second quantization, we will work in the first-quantized representation. The relationship between these representations was carefully analyzed in Part III. At the coupled-pair approximation level,

$$T \approx T_2, \tag{4}$$

and T_2 can be expressed as the symmetric sum

$$T_2 = \sum_{i<j}^{N} t(ij), \tag{5}$$

of strongly orthogonal, two-electron, spin-free cluster operators

$$t = 1/4 \sum_{\alpha,\beta=1}^{N/2} (|\tau^1_{\alpha\beta}><\phi^1_{\alpha\beta}| + |\tau^3_{\alpha\beta}><\phi^3_{\alpha\beta}|). \tag{6}$$

The integral operators $|\chi><\phi|$ have kernels $\chi(12)\ \phi^*(1'2')$ and the functions $\phi^s_{\alpha\beta}(12)$, s=1,3, are symmetrized or antisymmetrized products of ϕ_α and ϕ_β,

$$\phi^s_{\alpha\beta}(12) = \phi_\alpha(1)\phi_\beta(2) + (2-s)\phi_\alpha(2)\phi_\beta(1) \tag{7}$$

The spinless pair functions $\tau^s_{\alpha\beta}$, s=1,3 are obtained by solving the coupled pair equations. Their derivation has been presented in Part I, and Sections II-IV of Part III. Here we only provide the key steps. Starting from the Schrödinger equation

$$H\Psi = E\Psi, \tag{8}$$

and using Eqns. (3) and (4) for Ψ, premultiplication with $\exp(-T_2)$ gives an explicitly connected equation equivalent to Eqn. (8),

$$(H + [H,T_2] + 1/2\ [[H,T_2],\ T_2] - E)\Phi = 0. \tag{9}$$

In order to obtain the coupled pair equations, we project this equation against the operator U_2, whose action on any antisymmetric function $\Omega(1...N)$ is given by

$$U_2\Omega = \sum_{i<j}^{N} u(ij), \tag{10}$$

where

$$u(12) = q(12)\Gamma_{\Phi\Omega}(12). \tag{11}$$

The strong orthogonality operator q is defined by $q(12) = q(1)q(2)$, where

$$q(i) = 1 - p(i) \tag{12}$$

$$p = \sum_{\alpha}^{N/2} |\phi_\alpha\rangle\langle\phi_\alpha|, \tag{13}$$

and the two-particle transition density matrix $\Gamma_{\Phi\Omega}$ is defined by

$$\Gamma_{\Phi\Omega}(1'2'|12) = \binom{N}{2}\int\Phi^*(1'2'3..N)\Omega(123..N)d3..dN \tag{14}$$

Note that in Eqn. (11) q(12) operates only on coordinates (12) of $\Gamma_{\Phi\Omega}$. The result of the above U_2 projection is equivalent to

$$u(12) = 0, \tag{15}$$

with Ω now given by

$$\Omega = (H + [H,T_2] + 1/2\ [[H,T_2],T_2])\Phi. \tag{16}$$

The counterpart to this projection in second quantization is the projection against double excitations. To eliminate spin we integrate Eqn. (15) over spin coordinates. The result is

$$P_{\Phi\Omega}(1'2'|12) = 0, \tag{17}$$

where

$$P_{\Phi\Omega}(1'2'|12) = q(12)\int\Gamma_{\Phi\Omega}(1'\sigma_1,2'\sigma_2|1\sigma_1,2\sigma_2)d\sigma_1 d\sigma_2. \tag{18}$$

Eqn. (17) must be satisfied for all four points (1'2'; 12). From Eqn. (14) it is obvious that the points 1' and 2' appear as arguments of ϕ_α's only, which are linearly independent functions. Therefore, in order to obtain integrodifferential equations for $\tau^s_{\alpha\beta}$ of Eqn. (6) we can impose the following set of "two-point" equations:

$$P_{\Phi\Omega}\phi^{s}_{\alpha\beta}(12) = \int P_{\Phi\Omega}(1'2'|12)\phi^{s}_{\alpha\beta}(1'2')d1'd2'$$

$$= 0, \alpha, \beta = 1,2,..N/2, \quad (19)$$

or briefly

$$P_{\Phi\Omega}\phi^{s}_{\alpha\beta} = 0, \alpha, \beta = 1,2,...N/2. \quad (20)$$

In Section IV of Part III a detailed derivation of the integrodifferential equations resulting from Eqn. (20) is given. If the Møller-Plesset partitioning of Eqns. (1) and (2) is used, with the orbitals ϕ_α exactly satisfying

$$f\phi_\alpha = e_\alpha\phi_\alpha, \alpha = 1...N/2, \quad (21)$$

and if the strong orthogonality

$$q(12)\tau^{s}_{\alpha\beta}(12) = \tau^{s}_{\alpha\beta}(12) \quad (22)$$

holds for the pair functions of Eqn. (6), the equations for $\tau^{s}_{\alpha\beta}$ can be cast in the form

$$[f(1)+f(2) - e_\alpha - e_\beta]\tau^{s}_{\alpha\beta}(12) = R^{s}_{\alpha\beta}(12). \quad (23)$$

The right-hand side of Eqn. (23) can be logically written as

$$R^{s}_{\alpha\beta}(12) = -q(12)r^{-1}_{12}\phi^{s}_{\alpha\beta}(12)+L^{s}_{\alpha\beta}(12)+F^{s}_{\alpha\beta}(12)+Q^{s}_{\alpha\beta}(12). \quad (24)$$

The last three terms of this equation are lengthy, and will not be given here; they are defined in Eqns. (III.96), (III.97) and (III.93). $L^{s}_{\alpha\beta}$ is linear in the pair functions $\tau^{s}_{\alpha\beta}$ (arising from the second term in Ω of Eqn. (16)), whereas $F^{s}_{\alpha\beta}$ and $Q^{s}_{\alpha\beta}$ are quadratic in these functions.

Once coupled-pair functions have been obtained from Eqns. (22) and (23), the coupled-pair correlation energy can be calculated from

$$E_{XCP} = \sum_{\alpha=1}^{N/2} \varepsilon^1_{\alpha\alpha} + \sum_{\alpha<\beta}^{N/2} (\varepsilon^1_{\alpha\beta} + \varepsilon^3_{\alpha\beta}), \tag{25}$$

where

$$\varepsilon^s_{\alpha\beta} = s(1+\delta_{\alpha\beta})^{-1} \langle \phi_\alpha \phi_\beta | r_{12}^{-1} \tau^s_{\alpha\beta} \rangle \tag{26}$$

The subscript "XCP" in Eqn. (25) refers to pair energies from various approximate pair functions, obtained by dropping terms from $R^s_{\alpha\beta}$. If no terms are dropped, we speak of complete coupled pair (CCP) energies, hence X = C. If the hard-to-compute terms of $Q^s_{\alpha\beta}$ are dropped (they generate many connected four-electron integrals when Eqn. (23) is solved variationally; see Eqn. (III.93)), we speak of factorizable coupled pair (FCP) results, hence X = F. This very important and accurate approximation is given this name since the terms of $F^s_{\alpha\beta}$ lead to fully factorizable integrals over two electrons only, and short sums over occupied orbital labels. For a recent discussion of the FCP version in the second-quantized representation, refer to Paldus et al [20]. This version of coupled-pair theory is barely more involved computationally than the linear coupled pair (LCP) approximation (X=L), in which all quadratic terms are dropped from $R^s_{\alpha\beta}$. However, as shown in III, E_{LCP} seriously over-estimates the correlation energy. All three versions of coupled pair calculations have the significant property of being invariant to unitary transformations of the HF orbitals. For further discussion and a diagrammatic representation of Eqn. (24), refer to Section IV of Part III.

2.2 Second-order Energies and Nonlinear Optimization

When all terms in $R^s{}_\beta$ dependent on $\tau^s_{\alpha\beta}$ are dropped, Eqn. (23) reduces to

$$[f(1)+f(2)-e_\alpha-e_\beta]\tau^{[1]s}_{\alpha\beta}(12)=-q(12)r_{12}^{-1}\phi^s_{\alpha\beta}(12), \tag{27}$$

and we obtain the equation for first-order pair functions. Evaluation of Eqns. (25) and (26) provides the second-order correlation energy. These N2/4 uncoupled, linear integrodifferential equations play a special role in our approach, since they are used to obtain optimal GTG expansions for the pair functions. Adopting such an expansion with s-type GTG's,

$$\tau^{s}_{\alpha\beta}(12) \approx \sum_{k=1}^{k} c_k g_k(12), \tag{28}$$

where

$$g_k(12) = [1+(2-s)P_{12}]\exp[-\alpha_k(\vec{r}_1-\vec{A}_k)^2 - \beta_k(\vec{r}_2-\vec{B}_k)^2-\gamma_k r_{12}^2], \tag{29}$$

we are confronted with the task of determining the linear and nonlinear parameters of Eqns. (28) and (29), respectively. As indicated above, and explained in Ref. (14) and Part I, we successfully introduced the variational functional

$$J^{s}_{\alpha\beta}[\tilde{\tau}] = \langle\tilde{\tau}|h(1)+h(2)-e_\alpha-e_\beta|\tilde{\tau}\rangle+2\mathrm{Re}\langle\tilde{\tau}|q(12)r_{12}^{-1}\phi^{s}_{\alpha\beta}\rangle, \tag{30}$$

where

$$h = f + \Delta_{\alpha\beta}p, \tag{31}$$

$$\Delta_{\alpha\beta} = 1/2(e_\alpha+e_\beta)-e_1 + \eta. \tag{32}$$

η is an arbitrary positive parameter, with dimension of energy, and p is the projector of Eqn. (13). In Part I we proved that for all $\eta>0$, and Eqn. (21) exactly satisfied,

$$J^{s}_{\alpha\beta}[\tilde{\tau}] \geq \varepsilon^{[2]s}_{\alpha\beta}. \tag{33}$$

If $\eta=0$, the equality sign holds for $\tilde{\tau}=\tau^{[1]s}_{\alpha\beta} + \lambda\phi_1(1)\phi_1(2)$, with λ arbitrary. Therefore, $\tilde{\tau}$ is not guaranteed to be strongly orthogonal. However, when $\eta>0$, the equality sign holds for $\tilde{\tau} \equiv \tau^{[1]s}_{\alpha\beta}$, i.e., it satisfies Eqn. (27) and the SO condition of Eqn. (22) simultaneously. We see that the η parameter aids to enforce strong orthogonality; minimization of $J^s_{\alpha\beta}$ with respect to all parameters, and with ever larger GTG expansions, will cause Eqns. (22) and (27) to be gradually satisfied simultaneously. For this reason, we have called $J^s_{\alpha\beta}$ the weak orthogonality (WO) functional. Since GTO expansions were used for solving the HF equations (21), all integrals in Eqn. (30) can be evaluated in closed form. The time for evaluating a single $J^s_{\alpha\beta}$ with a given set of nonlinear parameters is approximately equal to $(aM^2K^2 + bM^4K)$, where $a \approx 5b$, and M and K are the number of GTO's and GTG's, respectively. This time is almost entirely spent on evaluating integrals; finding c_k of Eqn. (28) requires the solution of linear, inhomogeneous equations, which is relatively fast.

The approximate second-order energy $\tilde{E}^{(2)}$, computed as

$$\tilde{E}^{(2)} = \sum_{\alpha=1}^{N/2} J^1_{\alpha\alpha} + \sum_{\alpha<\beta}^{N/2} (J^1_{\alpha\beta} + J^3_{\alpha\beta}) =$$

$$= \tilde{E}^{(2)} \; [M,\eta,\vec{\xi}(K,M^{opt},\eta^{opt})], \tag{34}$$

has the indicated dependence on various parameters. All nonlinear parameters, totalling 3K for atoms and 5K for linear molecules for K GTG's per pair function, are collectively denoted by $\vec{\xi}$. For fixed K, a fixed, optimized HF orbital expansion of M^{opt} GTO's taken from the literature or obtained by us, and a fixed orthogonality forcing parameter η^{opt}, the $\vec{\xi}$ vector was extensively optimized. Powell's conjugate direction search technique [21] was used as implemented in a program provided by Dr. Harry F. King. A minimization threshold of 1μ hartree for $J^s_{\alpha\beta}$ was used throughout. The optimization is repeated for all $J^s_{\alpha\beta}$ separately. It was Pan and King's [10] and our [14,16] experience that a meaningful optimization does not require a very accurate HF energy, i.e., M^{opt} need not be large. This is fortunate since the time for calculating $J^{s}_{\alpha\beta}$ is proportional to M^4 for large M.

In Part I we gave an exhaustive account of the results of minimizations with the WO functional for second-order energies of He, Be, H_2, and LiH. Rather than quoting numbers here, let me summarize the results verbally, with emphasis on those that seem of general significance:

(i) A value of η^{opt} between 0.1 and 1 hartree seems optimal to balance the requirement of lowest $\tilde{E}^{(2)}$ value and the ability of the GTG's to satisfy the SO condition.

(ii) Short, relatively inaccurate GTO expansions can be used in the minimization; small values of M^{opt} can be chosen for which errors in the HF energy are at the mhartree level.

(iii) Results better than one mhartree are obtained with $K \geqslant 40$ per pair function; convergence patterns suggest errors of a few tenths of mhartrees at worst.

(iv) A remarkable insensitivity to M values is observed, as well as to variations of η values between zero and 100 hartrees. This is particularly true for the large K values. However, the measures for strong orthogonality, defined by

$$S = \langle \tilde{\tau} | p | \tilde{\tau} \rangle / \langle \tilde{\tau} | \tilde{\tau} \rangle, \tag{35}$$

albeit small, can vary many orders of magnitude, the largest value occurring for $\eta=0$.

(v) Comparison with published "algebraic" $E^{(2)}$ values shows our values to be superior by several mhartrees. The atomic results surpass those from partial wave expansions in the fourth significant figure (or at the 0.01 mhartree level).

2.3 Third-Order Energies

With the first-order pair functions obtained above it is possible to immediately compute the third-order correlation energy $E^{(3)}$, as defined by

$$E^{(3)} = \langle\Phi^{(1)}|V - \langle V\rangle|\Phi^{(1)}\rangle, \quad (36)$$

where $\langle V\rangle = \langle\Phi|V|\Phi\rangle$, and

$$\Phi^{(1)} = T_2\Phi, \quad (37)$$

with T_2 given by Eqns. (5-7) and satisfying Eqns. (22), (23) and (27). Using the relationship

$$\langle T_2^+T_2\rangle\langle V\rangle = \langle T_2^+T_2V\rangle, \quad (38)$$

$E^{(3)}$ can be given the explicitly connected form

$$E^{(3)} = \langle[T_2^+,[V_1T_2]]\rangle. \quad (39)$$

Derivation of working expressions is presented in Part II; refer there for details and numerical results. Here I summarize the main aspects of the calculations:

(i) Extremely accurate values for $E^{(3)}$ were obtained, provided $K \geqslant 20$ per electron pair function and a very large η^{opt} is used (100 hartrees or larger). The latter condition is needed to satisfy the SO condition (22) very accurately, resulting in only a weak dependence of $E^{(3)}$ on η.

(ii) An even weaker dependence on M, the size of the GTO basis, for $E^{(3)}$ than previously found for $E^{(2)}$ was observed.

(iii) The ratios $E^{(2)}/E^{(3)}$ found by us are significantly larger than the best "algebraic" results obtained earlier. This suggests that basis set calculations can tell us very little about MBPT convergence, and that the convergence of saturated MBPT expansions can be much faster than suggested by these calculations.

(iv) $E^{(2)}+E^{(3)}$ provides between 90 and 98 percent of the correlation energy. This seems to be a trend discernable in most MBPT calculations [3].

(v) Since calculation of $E^{(3)}$ requires the same integrals as $E^{(2)}$, its computation time is not much more than that for $E^{(2)}$.

.4 Coupled-pair Energies

To obtain coupled-pair energies we have to solve Eqn. (23), ow including terms in $R^s_{\alpha\beta}$ beyond the inhomogeneous one. These $^2/4$ equations are coupled, and nonlinear when $F^s_{\alpha\beta}$ and $Q^s_{\alpha\beta}$ are ept. We therefore have to resort to some iterative procedure to olve for $\tau^s_{\alpha\beta}$. One approach is the straightforward iteration proedure (SIP), generating the sequence of approximate pair functions $^{n]s}_{\alpha\beta}$, n = 0,1,2,... according to

$$[f(1)+f(2)-e_\alpha-e_\beta]\tau^{[n]s}_{\alpha\beta} = R^{[n-1]s}_{\alpha\beta}\{\vec{\tau}^{[n-1]}\} \qquad (40)$$

where the "vector" $\vec{\tau}^{[n]}$ collects all $\tau^{[n]s}_{\gamma\delta}$, and $R^s_{\alpha\beta}\{\vec{\tau}\}$ denotes the ependence of $R^s_{\alpha\beta}$ on $\vec{\tau}$. It should be remembered that $\vec{\tau}^{[n]}$ have to be trongly orthogonal. With $\vec{\tau}^{[o]} = \vec{0}$, $\vec{\tau}^{[1]}$ becomes the first-order air function. (Eqn. (40) reduces to Eqn. (27).) Further iterations give results closely related to MBPT. For example, the second iterate E_{CCP} (using $\vec{\tau}^{[2]}$ in Eqns. (25) and (26)) is equal to $E^{(2)} + E^{(3)} + E^{(4)}_Q$, where the last term is of fourth order with quadruply-excited intermediate states. Similarly, first, second and third iterate E_{LCP} gives $E^{(2)}$, $E^{(2)} + E^{(3)}$ and $E^{(2)} + E^{(3)} + E^{(4)}_D$, respectively, with $E^{(4)}_D$ denoting a fourth-order diagram with doubly excited states. For other order-by-order components, as well as a modified iteration procedure, refer to Section V.A of Part III.

We adopted the strategy of using the second-order energy optimized geminal expansion of Eqns. (28) and (29) for solving Eqn. (23), and their iterative counterpart Eqn. (40). Having fixed the nonlinear parameters, we are left with the task of determining the linear parameters c_k. In Eqns. (III.101) to (III.106) it is shown that this is possible with the unconstrained minimization of the WO functional

$$J[\tilde{\tau}] = \langle\tilde{\tau}|h(1)+h(2)-e_\alpha-e_\beta|\tilde{\tau}\rangle - 2\mathrm{Re}\langle\tilde{\tau}|R^{[n-1]s}_{\alpha\beta}\rangle, \qquad (41)$$

where h is defined by Eqns. (31) and (32). The associated Euler equation, derived from $\delta J = 0$, has a solution τ^{min} that satisfies Eqn. (40) and the strong orthogonality condition simultaneously, provided

$$\text{(i)} \quad [f,p] = o \tag{42}$$

$$\text{(ii)} \quad G^{(12)}_{\alpha\beta} = h(1) + h(2) - e_\alpha - e_\beta \text{ is positive definite.}$$

$$\text{(iii)} \quad q(12)\, R^{[n-1]s}_{\alpha\beta}(12) = R^{[n-1]s}_{\alpha\beta}(12) \tag{43}$$

(iv) the basis $\{g_k\}$ is complete.

J is a generalization of $J^s_{\alpha\beta}$ of Eqn. (30), except that its use is restricted here to the determination of τ^{min}.

In practice, in order to satisfy conditions (i) and (ii) above, we have used quite accurate representations for the HF orbitals and we imposed $\eta = 1$ hartree. Although the weaker condition $\eta>0$ is theoretically sufficient, practical difficulties might arise if η is too small. This is caused by $G_{\alpha\beta}$ for $\eta = 0$ having a null space spanned by $\phi_1(1)\phi_1(2)$ which leads to a possibly non-strongly orthogonal τ^{min} of the form

$$\tau^{min}(12) = q(12)\,\tau^{min}(12) + \lambda\phi_1(1)\phi_1(2). \tag{44}$$

This would cause the iterative procedure represented by Eqns. (40) and (41) to collapse discontinuously as $\eta\to 0$, provided $\{g_k\}$ is complete. In practice, the geminal basis is <u>not</u> complete, and the continuity of our algebraic equations requires the contamination by $\phi_1(1)\phi_1(2)$ to persist for small, positive values of η. As a result we observed a gradual deterioration of E_{XCP} as η approaches zero, with the most precipitous behavior occurring for the largest geminal sets. If $\eta >> 1$ hartree, the functional in Eqn. (41) overemphasizes the SO requirement at the expense of fulfilling Eqn. (40), and again poor results can occur. The value $\eta = 1$ hartree was chosen as a compromise between the SO requirement (leading to Eqn. (42) being well satisfied and S of Eqn. (35) being small) and a stable behavior of E_{XCP} calculations. For numerical details in support of this choice, refer to Part III, Sections VI and VII.

2.5 A Simplified Treatment of Strong Orthogonality

At this point our treatment of the SO projector q had two disadvantages. The first was the ambiguity about the proper choice of η. Although we found a rather weak dependence of our results on η, provided the geminal sets were not too small, we have no clear criterion for the "best" value. This situation makes it hard to decide upon results saturated to more than two or three significant figures.

The second drawback relates to the presence of q(1) and q(12) in $R^S_{\alpha\beta}$ (of Eqns. (III.93) to (III.97)), which cannot be dropped without violating the condition given by Eqn. (43). In $E^{(2)}$ calculations, q(12) generates a number of three-electron integrals proportional to M^4K, and in LCP or FCP calculations, this operator generates four-electron integrals, the number of which is proportional to M^4K^2. For accurate results, the M values must be quite large (so as to satisfy Eqn. (43) quite well), therefore causing substantial computing times for integrals as well as considerable I/O times in executing the iterations.

We were able to remedy both problems in a dramatic fashion. The η dependence was essentially eliminated with the introduction of an approximate SO projection whose time of evaluation is negligible when compared with other parts of MBPT or coupled-pair calculations. The time-consuming q(1) and q(12) in $R^S_{\alpha\beta}$ can be practically eliminated with a drastic, yet numerically safe, modification of the WO functional of Eqn. (41). These powerful solutions are described in Part IV. Here I wish to highlight their essentials.

The computational effort connected with q(12) arises from the intrinsic incompleteness of the geminal basis set; its action leads to functions outside the space spanned by that set. We therefore constructed an approximation to q(12) that projects within the "geminal basis" set space B, and has the property of coinciding with q when the geminal basis becomes complete.

If P_B is the orthogonal projection onto the subspace B spanned by the raw geminals g_k, k = 1...K, given by the explicit expression

$$P_B = \sum_{k,\ell=1}^{K} (\underset{\sim}{\$}^{-1})_{k\ell} |g_k\rangle\langle g_\ell| , \tag{45}$$

where $\underset{\sim}{\$}$ is the overlap matrix

$$(\underset{\sim}{\$})_{k\ell} = \langle g_k | g_\ell \rangle , \tag{46}$$

then the approximate SO projection can be expressed as

$$q_B = P_B q . \tag{47}$$

By its very definition, $q_B\tau$ always belongs to B, and therefore it is much easier to handle than $q\tau$. In practice, the linear coefficients in the expansion

$$q_B\tau = \sum_{k=1}^{K} c_k g_k \tag{48}$$

can be obtained by solving the system of linear equations

$$\sum_{\ell=1}^{K} (\$)_{k\ell} c_\ell = \langle g_k | q\tau \rangle . \tag{49}$$

These equations are obtained by simply equating $q\tau$ with $q_B\tau$, and projecting Eqn. (48) against all $\langle g_k|$, k=1..K. Eqn. (49) can also be derived from the minimization condition of the quadratic functional

$$D[\tilde{\tau}] = \|\tilde{\tau}-q\tau\|^2$$

$$= \langle\tilde{\tau}|\tilde{\tau}\rangle-2\mathrm{Re}\langle\tilde{\tau}|q\tau\rangle + \langle\tau|q\tau\rangle. \qquad (50)$$

in the subspace B.

q_B is neither idempotent nor hermitian. However, we proved that the action of q_B always improves the SO of τ, i.e., $q_B\tau$ is always closer to the subspace of strongly orthogonal functions than τ. I also point out that q_B is a poor approximation to q when the geminal basis is too small. But that situation is irrelevant in practice, since large geminal basis sets are needed for accurate results.

We applied q_B to the WO-functional-minimizing pair function τ^{min} before evaluating $E^{(3)}$ or the next-iterate $R^s_{\alpha\beta}$. Consequently, we refer to this approach as the WOP method (weak orthogonality plus projection). We observed a startlingly weak dependence on η. In fact, we now recommend the WOP method, with $\eta=0$, as the most accurate and reliable way of performing MBPT and coupled-pair calculations.

Next, to achieve an actual reduction of computational effort we must reduce the most time-consuming steps. For example, in the WO functional for first-order pair functions (Eqn. (30)), the linear term

$$2\mathrm{Re}\langle q(12)\tilde{\tau}(12)|r_{12}^{-1}\phi^s_{\alpha\beta}(12)\rangle \qquad (51)$$

is the most time-expensive one for large M, as it gives rise to $\sim M^4K$ three-electron integrals, whereas the quadratic term produces $\sim M^2K^2$ such integrals. If we express q in the form

$$q(12) = 1 - p(1)p(2) - p(1)q(2) - q(1)p(2), \qquad (52)$$

we observe that only the last two terms give rise to the three-electron integrals. We could therefore drop these terms, and consider the functional

$$G[\tilde{\tau}] = \langle\tilde{\tau}|h(1)+h(2)-e_\alpha-e_\beta|\tilde{\tau}\rangle+2\mathrm{Re}\langle\tilde{\tau}|1-p(1)p(2)|r_{12}^{-1}\phi_{\alpha\beta}^{s}\rangle \quad (53)$$

We refer to this functional as the super weak orthogonality (SWO) functional. If τ_{SWO} is the pair function for which G attains its minimum, the first-order pair function $\tau_{\alpha\beta}^{[1]s}$ can be obtained by SO projection:

$$\tau_{\alpha\beta}^{[1]s} = q\tau_{SWO}. \quad (54)$$

This procedure, which we call the SWOQ (super weak orthogonality plus q projection) approach, is proven and discussed in Section II.D of Part IV. The evaluation of the second term in Eqn. (53) requires only $\sim M^2K$ two-electron integrals. However, evaluating the second-order energy with Eqn. (25) again requires $\sim M^4K$ three-electron integrals, thus somewhat defeating the advantage of working with the SWO functional. It would be therefore adantageous if it is possible to replace q in Eqn. (54) by q_B. This we found to be the case; the resulting SWOP procedure (superweak orthogonality plus approximate projection) now requires only $\sim M^2K$ two-electron integrals to evaluate Eqn. (25), and it again largely eliminates the η dependence (see Tables 1 - 4 in Part IV).

The SWOP method can be easily generalized to include higher-order and coupled-pair calculations. Dropping q(1) and q(12) in the linear term $L_{\alpha\beta}^{s}$ of $R_{\alpha\beta}^{s}$ eliminates all four-electron integrals in the evaluation of $\langle\tilde{\tau}|L_{\alpha\beta}^{s}\rangle$. Only two- and three-electron integrals occur, their numbers scaling like K^2 and M^2K^2, respectively. As a result, the SWOP method affords the elimination of all four-electron integrals in LCP or FCP calculations. These integrals are still needed in CCP calculations because of the nonfactorizable $Q_{\alpha\beta}^{s}$ term. Their number scales like M^2K^3, and their evaluation is now the most time-consuming step. However, in Part III we have shown that the contribution of $\langle\tilde{\tau}|Q^s_{\alpha\beta}\rangle$ is very small and can be neglected for rather small systems.

The SWOP method represents the ultimate simplification of dealing with strong orthogonality, without significantly affecting the achievable accuracy given good orbital and geminal basis sets. It also has eliminated the need for the SO-enforcing parameter η when calculating the linear parameters c_k of Eqn. (28). The overall time consumption is extremely favorable. In fact, FCP calculations at the SWOP level take only slightly longer than third-order calculations, and are feasible also for larger systems than the ones considered thus far.

I refer to Part IV for an in-depth comparison of the WO, WOP, SWO, SWOP and SWOQ methods as implemented on Be and LiH for $E^{(2)}$, $E^{(3)}$ and E_{CCP} calculations.

SUMMARY OF RESULTS

In Table 1, I have collected our best results from in Refs. [16] through [19] and their comparison with the best results published by others, as well as with "experimental" correlation energies E_{corr}.

It is evident that our optimized $E^{(2)}$ results are superior to any results obtained previously. It is gratifying to note that, even for He and Be, our results are slightly better, although the methods in Ref. [26] are more suited to atomic correlation energy calculations. The $E^{(2)*}$ results for H_2 and LiH, taken from Adamowicz <u>et al</u>. [6], are significantly closer to our $E^{(2)}$ results than those published earlier also using the algebraic approximation. This is remarkable, since the evaluation method used (numerical HF orbitals, and Bethe-Goldstone level, optimized correlation orbitals) was not dedicated to optimal $E^{(2)}$ calculations. The same holds for $E^{(3)}$ values. I will return to these results.

The various E_{XCP} values show the usual trend of overestimation at the LCP level, and the near-equality of FCP and CCP results for

Table 1. Summary of the best of our and other published results, the latter ones being indicated by an asterisk (*). The internuclear distances for H_2 and LiH were 1.4 and 3.015 bohr, respectively. All energies are in mhartrees. E_{corr} are the "experimental" correlation energies, and E^*_{corr} are the best correlation energies from other methods.

	He	Be	H_2	LiH
$-E_{HF}$	2,861.67995[a]	14,573.02313[b]	1,133.629[c]	7,987.323[d]
$-E^*_{HF}$	2,861.68000[e]	14,573.02318[f]	1,133.630[g]	7,987.352[h]
$-E^{(2)}$	37.372[i]	76.35[j]	34.11[k]	72.18[l]
$-E^{(2)*}$	37.359[m]	76.29[m]	33.34[n]	70.25[n]
$-E^{(3)}$	3.621[o]	8.87[p]	4.38[q]	6.89[r]
$-E^{(3)*}$	3.631[s]	8.96[s]	5.07[n]	8.66[n]
$-E_{LCP}$	42.34[t]	98.34[t]	41.12[t]	82.74[t]
$-E_{FCP}$	42.00[t]	92.89[t]	40.41[t]	81.49[t]
$-E_{CCP}$	42.00[t]	92.87[u]	40.41[t]	81.51[v]
$-E^*_{CCP}$	38.2[w]	92.96[x]	40.37[n]	81.52[n]
$-E_{corr}$	42.044[y]	94.31±0.03[y]	40.85[y]	83.2±0.1[y]
$-E^*_{corr}$	42.044[y]	93.88[y]	40.80[y]	81.83[n]
$100E_{CCP}/E_{corr}$				
	99.90	98.5	98.92	97.9

[a] Schmidt and Ruedenberg (Ref. 22), M=20.
[b] Schmidt and Ruedenberg (Ref. 22), M=28.
[c] Optimized in this work, M=31 (Ref. 16).
[d] Optimized in this work, M=30 (Ref. 16).
[e] Gazquez and Silverstone, Szalewicz and Monkhorst (Ref. 23).
[f] Schmidt and Ruedenberg (Ref. 22), extrapolated.
[g] Kolos and Roothaan (Ref. 24); elliptic orbitals.
[h] Adamowicz and McCullough (Ref. 25); numerical molecular orbitals.
[i] Ref. 16; K=40, M=20.
[j] Ref. 19 (See Table I); K=60, M=13.

k Ref. 16 (See Table IV); K=20, M=31.
l Ref. 19 (See Table II); K=40, M=20.
m Malinowski, Polasik and Jankowski (Ref. 26); partial wave expansion through $\ell = 9$ and extrapolation, variation-perturbation treatment.
n Adamowicz, Bartlett and McCullough (Ref. 6); For H_2 and LiH, 45 and 70 numerical orbitals were used, respectively, with σ, π and δ symmetries.
o Ref. 17 (See Table IV); K=40, M=20.
p Ref. 19 (See Table IV); K=60, M=13.
q Ref. 17 (See Table VI); K=40, M=31.
r Ref. 17 (See Table V); K=40, M=20.
s Jankowski, Rutkowska and Rutkowski (Ref. 27); partial wave expansion and extrapolation.
t Ref. 18 (See Table VI); η = 1 hartree.
u Ref. 19 (See Table VII); M=13, K=60.
v Ref. 19 (See Table VIII); M=20, K=40.
w Ref. 29; see also footnote b to Table VIII of Ref. 18.
x Lindgren and Salomonson (Ref. 15); Non-factorizable diagrams not iterated, partial waves through ℓ=19.
y Ref. 18 (See Table IX).

---

small, closed-shell systems. As indicated earlier, the latter observation seems to have quite general validity and seems to be related to a (partial) cancellation of the nonfactorizable, quadratic T_2 terms (see Eqn. (III.93)) and T_4 terms [20]. This situation is extremely fortunate, particularly if it holds up in other molecules or atoms, because the neglect of these quadratic terms causes an enormous reduction in computational effort.

Except for He, in which case E^*_{CCP} was obtained with a relatively limited set of virtual orbitals [29], agreement between E_{CCP} and E^*_{CCP} is better than 0.1 mhartree. The Be result by Lindgren and Salomonson [15], obtained with an extensive partial wave expansion ($\ell \leq 19$), is not strictly comparable with E_{CCP}, since the nonfactorizable diagrams have not been iterated. This might explain the overshoot of 0.09 mhartree.

The recent results of E^*_{CCP} for H_2 and LiH by Adamowicz et al. [6] are in striking agreement with our E^*_{CCP} values. Although the actual close agreements should be considered fortuitous, the fact that E_{CCP} were obtained without explicitly correlated basis functions is significant. Using McCullough's [30] partial wave method for diatomic numerical HF and numerical MCSCF calculations, Adamowicz et al. [6] solved Bethe-Goldstone (BG) equations for all electron pairs. This provided numerical correlation orbitals, the union of which constituted the excitation space in second-quantized correlation calculations. Only orbitals with σ, π and δ symmetries were considered. This approach was inspired by an earlier observation by Adamowicz [31] that for H_2, BG-type functional is better suited in a GTG optimization than a second-order one since it provided more accurate values for higher-order ladder diagrams. The above close agreements are in contrast with disagreements between $E^{(2)}$ and $E^{(2)*}$, as well as $E^{(3)}$ and $E^{(3)*}$ for H_2 and LiH. However, $-(E^{(2)}+E^{(3)})$ values are close again: 38.49 and 38.41 mhartrees for H_2, and 79.07 and 78.91 mhartrees for LiH.

We must conclude that our GTG optimization strategy (of minimizing the second-order WO functional with respect to GTG parameters), while superior for $E^{(2)}$ and $E^{(3)}$ calculations, is not optimal for FCP (or CCP) calculations. Somehow the benefits of the explicit correlation in our GTG basis is partly lost when proceeding from $E^{(2)}$ to E_{CCP} calculations. Since, evidently, the pair functions should contain some explicit correlation for high-accuracy satisfaction of the CC equations, we must also conclude that neither our nor Adamowicz's CCP energies have converged. Improvements are in progress.

4. PERSPECTIVES

The feasibility and efficacy of the use of GTG expansions for electron correlation at the coupled-pair level has now been well established. As I discussed in the previous section, improvements in obtaining the best set of GTG's are still needed, and are being pursued. Alternative functionals and restrictions to the number of independent nonlinear parameters will be sought. Moreover, in order to handle molecules with first-row atoms, our GTO and GTG integral package is being extended to accomodate s, p, and d-type

GTO's and GTG's. Initial results for Ne are very encouraging [32]. Corrections beyond CCP will also be considered, in particular those related to T_1 and T_3 functions.

We can now look forward to reaching true chemical accuracy for correlation energies in small to intermediate molecules (less than about five first-row atoms) with any geometry. Although the use of GTG's seems to be crucial, it is possible that a mixture of numerical and analytic (expansion) methods will emerge as optimal for routinely achieving such accuracies.

Acknowledgements

This work has been supported by grants from NSF (CHE79006129, and CHE8207220), PANMR.I.9, Deutsche Forschungsgemeinschaft and NATO 008.80.

References:

1. S. Wilson and D.M. Silver, Phys. Rev. A14, 1949 (1976).
2. C. Moller and M.S. Plesset, Phys. Rev. 46, 618 (1934).
3. R.J. Bartlett, Ann. Rev. Phys. Chem. 32, 359 (1981) and references therein.
4. S. Wilson, Spec. Per. Rep. Theor. Chem. 4, 1 (1981) and references therein.
5. K. Jankowski, D. Rutkowska and A. Rutkowski, Phys. Rev. A26, 2378 (1982).
6. L. Adamowicz, R.J. Bartlett and E.A. McCullough, Phys. Rev. Letters, 54, 426 (1985).
7. C.W. Scherr and R.E. Knight, Rev. Mod. Phys. 35, 436 (1963); C.L. Pekeris, Phys. Rev. 115, 1216 (1959).
8. W. Kolos and L. Wolniewicz, J. Chem. Phys. 43, 2429 (1965); D.M. Bishop and L.M. Cheung, Phys. Rev. A18, 1846 (1978).
9. S.F. Boys, Proc. R. Soc. London Ser. A258, 402 (1960); K. Singer, ibid. 258, 412 (1960).
10. K.C. Pan and H.F. King, J. Chem. Phys. 53, 4397 (1970); 56, 4667 (1972).
11. O. Sinanoglu, J. Chem. Phys. 36, 3198 (1962).
12. L. Adamowicz and A.J. Sadlej, J. Chem. Phys. 67, 4298 (1977); 69, 3992 (1978); Acta Phys. Polon. A54, 73 (1978); Int. J. Quantum Chem. 13, 265 (1978).
13. K. Szalewicz, Thesis University of Warsaw, 1977; G. Chalasinski, B. Jeziorski, J. Andzelm and K. Szalewicz, Mol. Phys. 33, 971 (1977); K. Szalewicz and B. Jeziorski, Mol. Phys. 38, 191 (1979).

14. K. Szalewicz, B. Jeziorski, H.J. Monkhorst and J.G. Zabolitzky, Chem. Phys. Letters 91, 169 (1982).
15. P. Malinowski, M. Polasik and K. Jankowski, J. Phys. B12, 2965 (1979); I. Lindgren and S. Salomonson, Phys. Scr. 21, 335 (1980).
16. K. Szalewicz, B. Jeziorski, H.J. Monkhorst and J.G. Zabolitzky, J. Chem. Phys. 78, 1420 (1983), referred to as Part I.
17. K. Szalewicz, B. Jeziorski, H.J. Monkhorst and J.G. ZabolitZky, J. Chem. Phys. 79, 5543 (1983), referred to as Part II.
18. B. Jeziorski, H.J. Monkhorst, K. Szalewicz and J.G. Zabolitzky, J. Chem. Phys. 81, 368 (1984), referred to as Part III.
19. K. Szalewicz, J.G. Zabolitzky, B. Jeziorski and H.J. Monkhorst, J. Chem. Phys. 81, 2723 (1984), referred to as Part IV.
20. J. Paldus, J. Cizek and M. Takahashi, Phys. Rev. A30, 2193 (1984); J. Paldus, M. Takahashi and R.W.H. Cho, Phys. Rev. B30, 4267 (1984).
21. H.J.D. Powell, Comput. J., 7, 155 (1964).
22. M.W. Schmidt and K. Ruedenberg, J. Chem. 71, 3951 (1979).
23. J.H. Gazquez and H.J. Silverstone, J. Chem. Phys. 67, 1887 (1977); K. Szalewicz and H.J. Monkhorst, J. Chem. Phys. 75, 5785 (1981).
24. W. Kolos and C.C.J. Roothaan, Rev. Mod. Phys. 32, 219 (1960).
25. L. Adamowicz and E.A. McCullough, Int. J. Quantum Chem. 24, 19 (1983).
26. P. Malinowski, M. Polasik and K. Jankowski, J. Phys. B12, 2965 (1979).
27. K. Jankowski, D. Rutkowska and A. Rutkowski, J. Phys. B15, 4063 (1982).
28. S. Wilson and D.M. Silver, J. Chem. Phys. 66, 5400 (1977).
29. R.A. Chiles and C.E. Dykstra, Chem. Phys. Letters 80, 69 (1981).
30. E.A. McCullough, J. Chem. Phys. 62, 3991 (1975).
31. L. Adamowicz, Int. J. Quantum Chem. 13, 265 (1978).
32. K. Wenzel, J.G. Zabolitzky, K. Szalewicz, H.J. Monkhorst and B. Jeziorski, unpublished.

APPENDIX

SCHEDULE OF PLENARY LECTURERS FOR SYMPOSIUM ON COMPARISON OF *AB INITIO* QUANTUM CHEMISTRY WITH EXPERIMENT FOR SMALL MOLECULES: STATE-OF-THE-ART

Organizer: Rodney J. Bartlett, University of Florida

Monday, August 27, 1984
SESSION I: ELECTRONIC EXCITED STATES AND LIFETIMES

Chairman, Rodney J. Bartlett
University of Florida

Roald Hoffmann Cornell University	A Molecular Tool Kit for Bonding in Solids
D.A. Ramsay Herzberg Institute of Astrophysics National Research Council, Canada	Theoretical Problems Arising from the Study of Molecular Spectra
S.D. Peyerimhoff University of Bonn West Germany	Molecular Spectroscopy by *Ab Initio* Methods
Yngve Öhrn University of Florida	Spectral Properties of Small Molecules as Obtained with Polarization Propagator Methods

Monday, August 27, 1984
SESSION II: POTENTIAL ENERGY SURFACES

Chairman, Steve Binkley
Sandia Laboratories

C. Bradley Moore University of California, Berkeley	The Resolution of Molecular Dynamics?
Thom. H. Dunning, Jr. Argonne National Laboratory	Theoretical Studies of Reactions of Acetylene
Yuan T. Lee University of California, Berkeley	Dynamics and Mechanism of Elementary Chemical Reactions
Bowen Liu IBM San Jose	Recent Advances in *Ab Initio* Computation of Potential Energy Surfaces of Chemical Reactions

Tuesday, August 28, 1984
SESSION III: VIBRATIONAL SPECTROSCOPY

Chairman, George F. Adams
Army Research Office

H.F. Schaefer, III University of California, Berkeley	Infrared Spectra of Polyatomic Molecular Ions - Theoretical Predictions
R.J. Saykally University of California	High Resolution Infrared Spectroscopy of Polyatomic Molecular Ions
Peter Pulay University of Arkansas	Harmonic and Anharmonic Force Fields and Vibrational Spectra
P.R. Bunker Herzberg Institute National Research Council, Canada	The Rotation-Vibration Spectra of Quasilinear and Quasiplanar Molecules

Tuesday, August 28, 1984
SESSION IV: ESR AND NEGATIVE ION SPECTROSCOPY

Chairman, Vedene Smith
Queen's University, Ontario

W. Weltner, Jr. University of Florida	Molecular Information from ESR
E.R. Davidson University of Indiana	*Ab Initio* Calculation of Spin Hamiltonians
W.C. Lineberger JILA, University of Colorado	Negative Ion Spectroscopy
I. Shavitt Ohio State University	Experience and Problems with MCSCF and Multi-Reference CI Calculations for Small Molecules

Wednesday, August 29, 1984
SESSION V: METAL CLUSTERS

Chairman, Charles Bauschlicher
NASA Ames Laboratory

R.E. Smalley Rice University	Supersonic Cluster Beams and the Emerging Molecular Surface Science
W.A. Goddard, III California Institute of Technology	Mechanistic Considerations for Transition Metal Systems
V.E. Bondybey Bell Laboratories	Bonding and Structure of Small Metal Clusters
C.E. Dykstra University of Illinois	Accomplishing a High Accuracy Interface Between Theory and Spectroscopy

Wednesday, August 29, 1984
SESSION VI: VAN DER WAALS MOLECULES

Chairman, Henk Monkhorst
University of Florida

William Klemperer Harvard University	Non-Rigidity in Van der Waals's Molecules
Geerd Diercksen Max Planck Institute for Astrophysics	<u>Ab Initio</u> Calculations of Potential Energy Surfaces for Van Der Waal's Systems

Wednesday, August 29, 1984
SESSION VII: ALKALI METAL DIATOMIC POTENTIAL CURVES

Chairman, Ken Jordan
University of Pittsburgh

W.C. Stwalley University of Iowa	Spectra and Potential Energy Curves of Alkali Metal Diatomic Molecules: Experimental Results Compared with Theory
D.D. Konowalow State University of New York Binghamton	Potential Energy Curves and Electronic Transition Dipole Moment Functions of Alkali Metal Diatomic Molecules and Their Molecular Ions: Theory Compared with Experiment
R.A. Bernheim Pennsylvania State University	Electronic States of Li_2 and Li_2^+

George F. Adams B.H. Lengsfield III M. Page P. Saxe	STRUCTURE DETERMINATION FOR PENTAVALENT PHOSPHORUS COMPOUNDS
Robert Bernheim	METHYLENE
Louis Biolsi P.M. Holland J.C. Rainwater	RIGOROUS CALCULATION OF THE THERMOPHYSICAL PROPERTIES OF GROUND STATE LITHIUM ATOMS
Richard E. Brown G.D. Mendenhall	AB INITIO STUDIES ON THE DECOMPOSITION OF THE ALKYL HYPONITRITES
Luke A. Burke	SYNCHRONISM IN THE CONCERTED DIELS-ALDER REACTION
P.D. Dao S. Morgan A.W. Castleman, Jr.	STUDIES OF SPECTROSCOPIC SHIFTS IN VAN DER WAALS MOLECULES USING ONE- AND TWO-COLOR RESONANCE ENHANCED MULTIPHOTON IONIZATION
Carol A. Deakyne Michael Moet-Ner Cynthia L. Campbell Michael G. Hughes Sean P. Murphy	THE PROTON IN MIXED SOLVENTS: ACETONITRILE-WATER CLUSTER IONS
Brett I. Dunlap Boyd A. Waite	THEORETICAL STUDIES OF DISSOCIATION FOLLOWING LINEAR TO BENT TRIATOMIC PHOTODISSOCIATION
Gernot Frenking Dieter Cremer Andrej Sawaryn Paul v.R. Schleyer	STRUCTURES AND STABILITIES OF CBe_2, C_2Be and C_2Be_2
M.S. Gordon K.K. Baldridge T.A. Holme D.R. Gano	AB INITIO STUDIES OF INSERTION AND ABSTRACTION REACTIONS
A.R. Gregory K.K. Sunil K.D. Jordan	THE $1^2A'(\tilde{X}^2\Pi)$ STATE OF N_2O^+: COMPARISON OF AB INITIO THEORY WITH EXPERIMENT
Joshua B. Halpern	SPECTROSCOPY OF CYANOGEN
B. Andes Hess, Jr. L.J. Schaad J. Pancir	THEORETICAL STUDIES OF [1,η]-SIGMATROPIC REARRANGEMENTS INVOLVING HYDROGEN TRANSFER IN SIMPLE METHYL SUBSTITUTED CONJUGATED POLYENES

M. Thomas Jones	COMPARISON OF AB INITIO CALCULATIONS OF g-TENSORS FOR AROMATIC ORGANIC FREE RADICALS WITH EXPERIMENT
M. Thomas Jones Susan Jansen James Roble	ESR STUDIES OF ONE-DIMENSIONAL CONDUCTORS BASED ON TETRACYANOQUINODIMETHAN (TCNQ)
Joyce J. Kaufman John M. Blaisdell W. Andrzej Sokalski P.C. Hariharan	AB-INITIO CRYSTAL ORBITALS AND POLYMER ORBITALS
Joyce J. Kaufman P.C. Hariharan Cary Chabalowski Szczepan Roszak A. Laforgue	AB-INITIO CI AND COUPLED CLUSTER CALCULATIONS ON RDX (HEXAHYDRO-1,3,5,-TRINITRO-1,3,5, -TRIAZINE)
R.G. Keesee D.H. Evans R. Passarella A.W. Castleman, Jr.	AMMONIA-HALIDE ION COMPLEXES IN THE GAS PHASE
Mark Keil Gregory A. Parker	EMPIRICAL He + CO_2 POTENTIAL BY MULTI-PROPERTY FITTING IN THE IOSA
C.X. Liao C.L. Liao C.Y. Ng	A STATE-SELECTED STUDY OF THE SYMMETRIC CHARGE TRANSFER REACTION $H_2^+ + H_2$
Lawrence L. Lohr	A THEORETICAL INVESTIGATION OF THE GASEOUS OXIDES PO AND PO_2, THEIR ANIONS, AND THEIR ROLE IN THE COMBUSTION OF PHOSPHORUS AND PHOSPHINE
R.L. Martin H.J. Werner	RELATIVITY AND ELECTRON CORRELATION IN Cu_2
R. Ricki Mohr William N. Lipscomb	CORRELATION/POLARIZATION ADDITIVITY EFFECTS ON ENERGY AND GEOMETRY: FIRST-ROW SYSTEM ILLUSTRATED BY B2H7$^-$
H.J. Monkhorst B. Jeziorski K. Szalewicz J.B. Zabolitzky	ELECTRON CORRELATION WITH EXPLICITLY CORRELATED GEMINAL EXPANSIONS
Krishnan Raghavachari	THEORETICAL STUDY OF THE ROTATIONAL POTENTIAL SURFACE FOR ALKANES: BASIS SET AND ELECTRO CORRELATION EFFECTS

Lyn B. Ratcliff Daniel D. Konowalow Walter J. Stevens	ELECTRONIC TRANSITION DIPOLE MOMENT FUNCTIONS OF ALKALI METAL DIATOMIC MOLECULES
Lynn T. Redmon Bruce C. Garrett Michael J. Redmon	THEORETICAL STUDIES OF COLLISION INDUCED DIPOLE TRANSITIONS IN O_2 and S_2
Michael J. Redmon Lynn T. Redmon Bruce C. Garrett	THEORETICAL STUDIES OF $O(^3P)$ COLLISIONS WITH HF, HCl, AND CO
R.M. Regan C. Duda D.D. Konowalow	DAMPED DISPERSION ENERGIES OF INTERACTION FOR RARE GAS ATOMS
James P. Ritchie	ELECTRO DENSITY DISTRIBUTION ANALYSIS OF CH_3NO_2, CH_2NO_2, and NH_2NO_2
Celeste M. Rohlfing P. Jeffrey Hay Richard L. Martin	EFFECTIVE CORE POTENTIAL CALCULATIONS OF Ni, Pd, and Pt COMPOUNDS
A.M. Sapse D.C. Jain	AB-INITIO STUDIES OF COMPLEXES BETWEEN SOME HYDROCARBONS AND HF or NH_3
A.M. Sapse D.C. Jain	AN AB-INITIO STUDY OF NLi_2H and NF_2Li DIMERIZATION
Steve Scheiner Z. Latajka	THEORETICAL STUDIES OF H-BONDS AND Li-BONDS INVOLVING NH_3 and PH_3 WITH HF, HCl, LiF, and LiCl
Vedene H. Smith, Jr.	AB INITIO STUDIES OF X-RAY AND ELECTRON-SCATTERING FROM MOLECULES
K.K. Sunil K.D. Jordan	THEORETICAL INVESTIGATION OF THE IONIZATION POTENTIAL AND EXCITATION ENERGIES OF Cu and Zn
Donald G. Truhlar Franklin B. Brown Rozeanne Steckler David Schwenke	DYNAMICS CALCULATIONS BASED ON AB INITIO POTENTIAL ENERGY SURFACES
G.J. Vazquez S.D. Peyerimhoff R.J. Buenker	MRD-Cl STUDY OF THE ELECTRONIC STATES OF HO_2, HO_2^+, and HO_2^-

<u>D.K. Veirs</u> C. Schwartz R.J. LeRoy G.M. Rosenblatt	ACCURATE ISOTOPIC EFFECTS IN GROUND STATE MOLECULAR HYDROGEN: THEORETICAL CALCULATIONS AND EXPERIMENTAL MEASUREMENTS
<u>R.L. Wadlinger</u>	THE CHEMICAL BOND -- A MISSING INGREDIENT
<u>John Weiner</u> John Keller	DIRECT MEASUREMENT OF THE POTENTIAL BARRIER HEIGHT IN THE $B^1\pi_\mu$ STATE OF SODIUM DIMER
<u>R.L. Whetten</u> S.G. Grubb C.E. Otis A.C. Albrecht E.R. Grant	TOWARD AN EXPERIMENTAL DETERMINATION OF THE COMPLETE ELECTRONIC TERM SPECTRUM OF A POLYATOMIC MOLECULE
<u>William D. Laidig</u> Rodney J. Bartlett	THE DESCRIPTION OF FIRST ROW DIATOMIC POTENTIAL ENERGY SURFACES USING MULTI-REFERENCE COUPLED CLUSTER THEORY
<u>George D. Purvis III</u> William D. Laidig Rodney J. Bartlett	CAN LOCALIZED ORBITALS AND COUPLED CLUSTER METHODS PREDICT MOLECULAR THERMODYNAMICAL PROPERTIES TO CHEMICAL ACCURACY?
<u>G.A. Peterson</u>	THE COMPLETE BASIS SET CORRELATION ENERGIES OF FIRST ROW HYDRIDES
<u>H.H. Michels</u> R.H. Hobbs L.A. Wright	DISSOCIATIVE ATTACHMENT OF e + Li_2
<u>Ludwik Adamowicz</u> Rodney J. Bartlett	COUPLED CLUSTER CALCULATIONS FOR DIATOMIC NEGATIVE ANIONS WITH FULLY NUMERICAL ORBITALS
<u>S.P. Walch</u> C.W. Bauschlicher, Jr. B.O. Roos C.J. Nelin R.L. Jaffe	THEORETICAL STUDIES OF TRANSITION METAL DIMERS
<u>Jack Gelfand</u> Richard B. Miles Herschel Rabitz Thomas G. Kreutz Eric A. Rohlfing	MECHANISMS AND RATE CONSTANTS FOR THE VIBRATIONAL RELAXATION OF HD (v = 4,5, and 6) IN COLLISIONS WITH HD, ^{4}HE, and D_2
R.S. Mulliken <u>W.C. Ermler</u> <u>J.P. Clark</u>	CONFIGURATION INTERACTION CALCULATIONS OF POTENTIAL ENERGY CURVES AND TRANSITION MOMENTS OF $^1\Sigma_g$ and $^1\Sigma_u$ STATES OF N_2

Karen G. Lubic Takayoshi Amano	INFRARED DIFFERENCE FREQUENCY SPECTROSCOPY OF TRANSIENT SPECIES
S.C. Foster A.R.W. McKellar	DIODE LASER SPECTROSCOPY OF MOLECULAR IONS
Gary D. Bent A. Rossi	MCSCF STUDIES OF THE PHOTOCHEMISTRY OF BICYCLOBUTANE
D.A. Dixon A. Komornicki	MOLECULAR PROTON AFFINITIES

INDEX

GPSR Compliance

The European Union's (EU) General Product Safety Regulation (GPSR) is a set of rules that requires consumer products to be safe and our obligations to ensure this.

If you have any concerns about our products, you can contact us on ProductSafety@springernature.com

In case Publisher is established outside the EU, the EU authorized representative is:

Springer Nature Customer Service Center GmbH
Europaplatz 3
69115 Heidelberg, Germany

Batch number: 10815974

Printed by Printforce, the Netherlands